Nationalizing Science

Transformations: Studies in the History of Science and Technology
Jed Buchwald, general editor

Myles Jackson, *Spectrum of Belief: Joseph von Fraunhofer and the Craft of Precision Optics*

Alan J. Rocke, *Nationalizing Science: Adolphe Wurtz and the Battle for French Chemistry*

Nationalizing Science
Adolphe Wurtz and the Battle for French Chemistry

Alan J. Rocke

The MIT Press
Cambridge, Massachusetts
London, England

Set in Sabon by The MIT Press.
Printed and bound in the United States of America.

Library of Congress Cataloging-in-Publication Data

Rocke, Alan J., 1948–
Nationalizing science : Adolphe Wurtz and the battle for French chemistry / Alan J. Rocke.
p. cm. — (Transformations)
Includes bibliographical references and index.
ISBN 0-262-18204-1 (hc. : alk. paper)
1. Wurtz, Charles Adolphe, 1817–1884. 2. Chemistry—France—History—19th century. 3. Chemists—France—Biography. I. Title. II. Transformations (MIT Press).
QD22.W87 R63 2001
540'.944'09034—dc21   00-041854

for Cristine

Contents

Acknowledgments

Every author of a book incurs many debts in the course of research and writing, and in my case they are large indeed. After completing some smaller studies relating to this project in the late 1980s and the early 1990s, I received a grant in 1995 from the National Science Foundation that supported a semester's research leave, much of it spent in Paris. Three years later the writing was completed during a sabbatical semester offered by Case Western Reserve University. The University and its History Department have been a congenial working environment; I am particularly grateful for the material and personal support provided by John Bassett, Dean of the College of Arts and Sciences. Many other colleagues here have given generously of advice and help in this work, especially James Edmonson, Elisabeth Köll, Kenneth Ledford, Miriam Levin, Carroll Pursell, and Jonathan Sadowsky. Marie-Pierre LeHir went considerably beyond the collegial call of duty in helping me to improve my abilities in the French language, and John Barberet and Charlotte Sanpere also assisted in important ways.

As for European scholars, I owe an immense debt to Bernadette Bensaude-Vincent for her support and help in many ways, throughout the entire course of this project. I am no less grateful to her former student Natalie Pigeard, who has generously shared with me her deep understanding of the subject of Adolphe Wurtz, often directing me to manuscript sources which I would otherwise have found only with difficulty, if at all. Both of these scholars read the entire manuscript in draft, and offered many important suggestions. Although I have never met Ana Carneiro, I have benefited greatly from corresponding with her and especially from studying her outstanding doctoral dissertation on Wurtz's research school, completed in 1992 under the direction of Maurice Crosland. I have also received important scholarly and material assistance (as well as kind hospitality) in Paris and Strasbourg from Georges Bram, Elisabeth Crawford, Georges Frechet, Catherine Kounelis, and from two of Wurtz's descendents, Hubert Autrand and Denis Oechsner de Coninck (his great- and great-great-grandson, respectively, through his eldest child, Marie).

I am grateful to Christoph Meinel for many stimulating conversations on the subjects covered in this book, and for his kindness in sharing some of his unpublished work with me. In the spring of 1999, a period as visiting scholar at the Max-Planck-Institut für Wissenschaftsgeschichte in Berlin proved to be an extraordinarily valuable experience; I am particularly grateful to Ursula Klein for inviting me there, for many helpful conversations, and for important suggestions on a draft of chapter 10. I also want to thank Eric Francoeur, Britta Görs, Celia von Lindern, Hans-Werner Schütt, and Elisabeth Vaupel for their assistance.

For the actual pursuit of research, a scholar's greatest debts are to librarians and archivists on the one hand, and past workers in the field on the other. The staff at the Archives Nationales was invariably helpful to me, and Mme. Claudine Pouret and the other archivists of the Académie des Sciences make working there a scholar's delight. The large majority of the passages in manuscript documents that are cited in this study were found in those two extraordinary collections. Other archives and libraries whose collections supported this work included the Bibliothèque Nationale in Paris; the Archives du Bas-Rhin, the Archives Municipales, and the Bibliothèque Nationale et Universitaire in Strasbourg; the Bayerische Staatsbibliothek and the Deutsches Museum in Munich; the Berlin-Brandenburgische Akademie der Wissenschaften, the Staatsbibliothek Preussischer Kulturbesitz, and the Max-Planck-Institut für Wissenschaftsgeschichte in Berlin; the Institut für Organische Chemie at the Technische Hochschule in Darmstadt; the Vieweg-Verlag archive in Wiesbaden; and University College and the Wellcome Institute in London. In this county, I made use especially of Kelvin Smith Library of CWRU and its interlibrary loan and special collections divisions, the OhioLINK consortium, and Widener Library at Harvard. To all of these institutions and their hard-working professional staffs I give my heartfelt thanks, and my appreciation for permission to print extracts from manuscript sources.

Among Anglo-American historians of French science, I have been particularly inspired by the work of Maurice Crosland, Robert Fox, Frederic L. Holmes, Seymour Mauskopf, and Mary Jo Nye. Mary Jo Nye, in particular, has been wonderfully supportive of this work, and I am most grateful for her perceptive comments after reading the entire manuscript. I thank also Sy Mauskopf and David Cahan for valuable suggestions on Chapter 12. Larry Holmes and Trevor Levere invited me to present an early form of the material in chapters 1 and 2 at a symposium at the Dibner Institute of MIT in April 1996, which is appearing in *Instruments and Experimentation in the History of Chemistry,* ed. F. Holmes and T. Levere (MIT Press, 2000); I thank The MIT Press for permission to publish this material in revised form here. I also want to thank Laurence Cohen, Paul Bethge, and the other professionals at The MIT Press for their outstanding work.

My greatest debts are, as always, to family and friends, and especially to my wife, Cristine Rom.

Conventions and Abbreviations

All references to file boxes in the Archives Nationales are indicated by the numbers F17 (Ministère de l'Instruction Publique) or AJ16 (Académie de Paris, séries modernes) followed by a slash and a file-box number; location at the Archives Nationales in Paris is understood.

MIP stands for Ministère de l'Instruction Publique or Ministre de l'Instruction Publique. *Comptes rendus* stands for *Comptes rendus hebdomadaires des séances de l'Académie des Sciences. Annales de chimie* stands for *Annales de chimie et de physique. Annalen* stands for Justus Liebig's *Annalen der* [*Chemie und*] *Pharmacie.* 'Sorbonne' is used here as essentially synonymous with 'Faculté des Sciences de Paris', even though fields other than science were and are taught at the Sorbonne. In nineteenth-century chemical parlance, 'alcohol' signifies ethanol, 'aldehyde' signifies acetaldehyde, 'potash' signifies potassium hydroxide, 'soda' signifies sodium hydroxide, and 'carbonic acid' often signifies carbon dioxide gas.

My capitalization follows Anglo-American conventions. Similarly, Russian names are given in conventional Anglo-American transliterations, which often differ from familiar French or German transliterations.

The original French and German texts of translated passages are provided in the notes for passages from manuscript documents (and not for those that are available in printed sources).

Except where otherwise noted, translations are my own.

Nationalizing Science

Introduction

This book has, among its other goals, a biographical intent. The biographical approach to the history of science, long spurned as inevitably inclined toward hagiography, overly focused on the merely personal, and neglecting the agency of community and milieu, has experienced a resurgence in recent years.[1] Many historians of science have now come to view biography as offering possibilities for the exploration of interactions between the individual and society in the development of scientific ideas, and for a means of accessing concrete locally sited examples of investigation, discovery, and verification. It is the opposition between the contingent social world of human life and the apparently enduring product of the activity of that life that provides a focus of great interest in the study of history of science. How better to study that tension than to investigate exemplars of it, in the form of individual lives in science?[2] Along these lines, there is much to be said for accounts that attempt a strong integration between four biographical levels: the "merely personal" aspects of human life (such as upbringing, family, and personality), the cognitive realm of the development of scientific thought, the prosaic world of quotidian scientific work, and the professional milieu in which the scientist operates (including his or her political and social interactions).

The study of nineteenth-century French science offers particular challenges to the biographer. There are fine studies of the institutions of science available, as well as excellent examples of the "private science" of

1. M. Shortland and R. Yeo, "Introduction," in *Telling Lives in Science*, ed. Shortland and Yeo (Cambridge University Press, 1996).

2. See T. Söderqvist, "Existential Projects and Existential Choice in Science," in *Telling Lives in Science*.

noteworthy individual figures.[3] However, substantial and well-integrated biographies of the kind I have described are rare.[4] Among the sciences, chemistry has seen particularly little action. There are recent works on Joseph Louis Gay-Lussac, Marcellin Berthelot, and Jean-Baptiste Dumas,[5] but no modern monographic studies (and in many cases no serious studies whatever) on such major figures as Michel Eugène Chevreul, Louis Jacques Thenard, Antoine Balard, Jules Pelouze, Victor Regnault, Auguste Cahours, Edmond Frémy, Auguste Laurent, Charles Gerhardt, Henri Sainte-Claire Deville, François Marie Raoult, and, until very recently, Adolphe Wurtz.[6] Few of the studies that do exist have taken full advantage of the enormous quantity of untapped manuscript materials at such locations as the Archives Nationales and the Académie des Sciences in Paris.

This lacuna has been noticed and lamented. Maurice Crosland took note some years ago of the barriers to a clear understanding of the subject. Chief among these, he argued, was the absence of "detailed studies of certain key figures in science, of the scientific community, and of institutions. When these have been published, we shall have a clearer idea of what is meant by French science."[7] The situation he described then is only a little better today.

3. Examples of the first type in the Anglo-American literature: *The Organization of Science and Technology in France, 1808–1914*, ed. R. Fox and G. Weisz (Cambridge University Press, 1980); H. Paul, *From Knowledge to Power* (Cambridge University Press, 1985); M. Crosland, *Science under Control* (Cambridge University Press, 1992). Examples of the second type: F. Holmes, *Claude Bernard and Animal Chemistry* (Harvard University Press, 1974); G. Geison, *The Private Science of Louis Pasteur* (Princeton University Press, 1995).

4. One fine example is M. Crosland, *Gay-Lussac* (Cambridge University Press, 1978).

5. Crosland, *Gay-Lussac*; J. Jacques, *Berthelot 1827–1907* (Belin, 1987); L. Klosterman, "A Research School of Chemistry in the Nineteenth Century," *Annals of Science* 42 (1985): 1–80; M. Chaigneau, *J. B. Dumas, chimiste et homme politique* (Guy le Prat, 1984). For all their virtues, these three examples do not match my biographical model very well. Jacques's work is powerfully directed toward deconstructing the "myth" of Berthelot; Klosterman's is a focused study of a research school; Chaigneau's pioneering effort is frustratingly sketchy at many crucial points. Serious historical work on Dumas has scarcely begun.

6. See, however, A. Costa, *Michel Eugène Chevreul* (State Historical Society of Wisconsin, 1962); G. Roque et al., eds., *Michel-Eugène Chevreul* (Muséum National d'Histoire Naturelle, 1997); M. Blondel-Mégrelis, *Dire les choses* (Vrin, 1996); M. Novitski, *Auguste Laurent and the Prehistory of Valence* (Harwood, 1992).

7. M. Crosland, "History of Science in a National Context," *British Journal for the History of Science* 10 (1977), p. 111.

Similarly, Robert Fox and George Weisz believe that "we need to integrate the study of institutional structures with some serious account of the lives and motivations of individual scientists and the informal networks in which they worked."[8]

The dearth of modern archivally based studies and the complexities of the centralized and cumulated Parisian academic environment make such endeavors challenging. The first prerequisite for understanding networks of this type is to arrive at a clear time line of positions held for the most important members of the cast of characters; however, to judge from the available secondary literature (including nineteenth-century obituaries), even such basic data is still uncertain today. Professorships were often held briefly and/or cumulated, and were frequently handed off temporarily to younger substitutes; there were also many kinds of connections to institutions. (Consequently, a reference that states that a certain figure "began teaching" or "was appointed" at a certain institution in a certain year tells us less than it seems, even if the year is accurate.) Obituarists and biographers are often vague or inexact because they have not been able to determine the details; they also sometimes unintentionally transmit false information across generations of scholarship.

If one studies all the published obituaries and biographies of Jules Pelouze, for instance, one will remain in doubt as to when he was appointed to what position, or just when and where he founded his important private laboratory. Either the sources are silent at crucial points or they contradict one another; for instance, not one mentions the year of his elevation to professor at the École Polytechnique, and few agree on the time of founding of his lab (as it turns out, none is accurate). There never existed a more assiduous collector of historical details from obscure sources than J. R. Partington, but even he was forced to comment plaintively, after naming the various institutions with which Antoine Balard was associated through his career, that "I have found no dates for these appointments."[9] I am interested in this level of detail because the work that follows seeks to explore, among other things, the sorts of connections, networks, and filiations that

8. R. Fox and G. Weisz, "Introduction," in *Organization of Science*, ed. Fox and Weisz, p. 26.

9. J. Partington, *A History of Chemistry* (Macmillan, 1964), vol. 4, p. 96. Probably the best single source for such data is the annual government publication *Almanach royal [national, impérial]*, later editions of which add the words *Annuaire de la République Française*.

can help to exemplify and clarify how French science actually operated during the nineteenth century. Unless one knows clearly at all times which players had what influence and power in exactly which institutions, this sort of game cannot be played.

Influence and power are, of course, intimately connected with the availability of material resources. For a working research chemist over the last two centuries, "material resources" translates conventionally to "laboratory facilities," and the history of the academic chemical laboratory[10] in nineteenth-century France (and also, for comparative purposes, Germany) is a continuing concern in the chapters that follow. It was in the nineteenth century that government-supported scientific laboratories first became common and expected features of a modern or modernizing state, yet this crucial development has scarcely begun to be studied historically. In the preface to his outstanding edited book on this subject, Frank James rightly draws attention to the *chemical* origins of the laboratory, stressing both the importance of the rise of the scientific laboratory and its neglect as a historical subject.[11] Yet the book has almost nothing connected with France or Germany and little on chemistry; furthermore, none of the essays deal with the critical period for the story, the middle decades of the nineteenth century.

It is a commonplace of the secondary literature that in France government-sponsored laboratories were few and meager, and that this situation hampered the development of French science. However, here again, historically satisfying details of the actual laboratories are hard to find, even in archival sources. I suggest here that the deficiency in facilities was even more severe than has been claimed, and that the situation was particularly damaging for the science of chemistry (especially organic chemistry).

The laboratory is a part of the scientist's daily life, and I think it important for the historian to gain at least partial access to that quotidian level.

---

10. In this book, I use "academic" to refer broadly to the life of higher education and research, not to academies of science. I intend the phrase "academic laboratory" to designate any laboratory used by a scientist employed in an institution of higher education and active in research (including within the university system, the grandes écoles, and such research institutions as the Collège de France or the Muséum d'Histoire Naturelle). Private laboratories, which operated on the fringes of this system, played a very important role and will also be considered here.

11. F. James, ed., *The Development of the Laboratory* (American Institute of Physics, 1989), pp. 1–3.

Quotidian details not only provide fresh historicity; they also often introduce important additional factors that are vital to understanding the other action on the stage. For example, the characteristic practice of multiple professorships known as "cumul" meant multiple duties at multiple locations in Paris for the leading scientific figures, and the outlines of the system are well understood. But how did scholars actually perform their duties? How were financial pressures, time constraints, and logistics of commutes all related to central professional activities of teaching and research, and to career decisions of major scientists? The historian or the reader must, in his or her mind, follow Dumas or Wurtz on his daily rounds (ca. 1836 or ca. 1849, respectively) from a residence on the Left Bank, to the École Polytechnique (or Faculté de Médecine) in the Latin Quarter, to the École Centrale des Arts et Manufactures in the Marais on the Right Bank (where both also taught), to a private laboratory, and then back home again to understand the daily difficulties encountered in the putatively "conveniently centralized" world of Parisian science. Such an approach requires adding an appreciation of place, geography, and time to the more conventional considerations.[12]

Another aspect has to do with the material culture of science and its physical tools, such as analytical apparatus. Here I take advantage of the recent profusion of stimulating work in what some are calling "laboratory studies."[13] In the first two chapters of this book I pay close attention to the details of the development of organic analysis during the period 1811–1833 (after which period the techniques stabilized). The result of this process was the acquisition of the basic methodological tools with which the secrets of

12. For an example, see the first section of chapter 2 below. A recent collection of stimulating articles centering on the concepts of space and place in the history of science is *Making Space for Science*, ed. C. Smith and J. Agar (Macmillan, 1998); particularly helpful are the introduction by the editors and G. Gooday's paper "The Premises of Premises."

13. For which see, e.g., A. Franklin, *The Neglect of Experiment* (Cambridge University Press, 1986); James, *The Development of the Laboratory*; D. Gooding et al., eds., *The Uses of Experiment* (Cambridge University Press, 1989); D. Gooding, *Experiment and the Making of Meaning* (Kluwer, 1990); A. Pickering, ed., *Science as Practice and Culture* (University of Chicago Press, 1992); P. Galison, *Image and Logic* (University of Chicago Press, 1997); H.-J. Rheinberger, *Toward a History of Epistemic Things* (Stanford University Press, 1997); M. Heidelberger and F. Steinle, eds., *Experimental Essays—Versuche zum Experiment* (Nomos, 1998).

organic chemistry were unlocked, and so these developments are centrally important to our story. Along the way, we will see that the scientist's coordination of his research tools, the character of his teaching, the organization of his daily life, and the cultural characteristics of his collegial and wider social order were all curiously interconnected, and that only by appreciating some of the material details can the larger picture be most accurately assembled. This eclectic mandate has much in common with the broadly inclusive concept of laboratory work that Hans-Jörg Rheinberger calls "experimental systems."[14]

In this book I use the chemist Adolphe Wurtz (1817–1884) as a convenient focal point of the socio-politico-scientific network I want to investigate. I will leave for the chapters that follow the task of justifying the choice of this scientist; I will simply suggest here that it would be difficult to name a finer or more instructive example from the mid-nineteenth-century French chemical community. (No less an authority than Pierre Duhem referred to Wurtz—after speaking of Lavoisier, Dumas, Laurent, and Gerhardt—as a man "in whom intuition, vivacity, and the ardor of French genius all attained their acme."[15]) In exploring Wurtz's life and his connections to his social and scientific milieu, I do not neglect the details of his science. Indeed, I take as a fundamental axiom—as outlined above—that the only way fully to understand the life of the mind is also to understand the rest of the person, and vice versa. The strands of a life are not easy to tease apart, and they must be understood as interconnected in complex ways.

Another reason behind my choice of Wurtz indicates a major theme of this book. Wurtz grew up in Strasbourg, just a few miles from the border with the Grand Duchy of Baden. His native language was Alsatian, a German dialect, and his high German was as perfect as his Parisian French. He received some of his education at the laboratory of the great German chemist Justus Liebig, and his scientific career was devoted to introducing

14. Rheinberger, *Toward a History of Epistemic Things*; M. Hagner and H.-J. Rheinberger, "Experimental Systems, Objects of Investigation, and Spaces of Representation," in *Experimental Essays*, ed. Heidelberger and Steinle. An "experimental system" is "a basic unit of experimental activity combining local, technical, instrumental, institutional, social, and epistemic aspects" (Rheinberger, *Toward a History of Epistemic Things*, p. 238).

15. P. Duhem, *La science allemande* (Hermann, 1915). Quoted from English edition, *German Science* (Open Court, 1991), p. 31.

certain German ideas into Parisian scientific institutions. Consequently, Wurtz is an unexcelled vehicle with which to study the German influence on French science (and, to a lesser extent, the reverse).

The present work is, thus, a comparative essay on national and international science in the nineteenth century. In that sense it serves as something of a companion volume to a work which I published a few years ago, the leading figure in which was the German chemist Hermann Kolbe (1818–1884).[16] Wurtz and Kolbe had much in common. Born less than a year apart, they both died in 1884 at the age of 66. Each spoke German (or a German dialect) in his birth home; each was the eldest son of a Lutheran pastor in a small town near a major university city. Both remained sincere Protestants and became known as outspoken nationalists in their later years. Both were powerfully influenced by Justus Liebig. But there were important contrasts, too. Kolbe eventually found himself outside the mainstream of German chemistry in his opposition to atomic and structural ideas; Wurtz found himself outside the mainstream of French chemistry in his commitment to the same ideas. Another contrast was in their personalities and characters, but I will leave that for the reader to judge.

In the previous work I strove for the same sort of rich contextuality and integration that I have described above; in this goal I was assisted by the fortunate survival of about 850 letters from Kolbe's pen, many of them long, personal, and revealing. Wurtz's Nachlass is not nearly so rich, in part because many of his papers were destroyed by bombardment during the First World War. However, I have sought out Wurtz letters—and other documents of all kinds—wherever I could find them, and here again only the reader can judge whether I have achieved my goals.

Wurtz's mentors, Jean-Baptiste Dumas and Justus Liebig, were the most powerful men in Continental European chemistry during the middle decades of the nineteenth century. Consequently, the first two chapters discuss the early careers of both men. Each achieved early renown, and together, as wary rivals, they created, modified, and then vastly improved the theoretical and experimental tools that were required to construct organic chemistry— atomic and molecular theories, and methods for the analysis of organic substances. The later course of the science would be impossible to understand without this background. But the story is also remarkable for their personal

16. See A. Rocke, *The Quiet Revolution* (University of California Press, 1993).

and professional relations, and the relations of those around them. Liebig (and German chemistry) can be seen to have been far more oriented toward France and the French than has hitherto been appreciated, and Dumas and French chemistry were far more fundamentally influenced by the Germans than scholars have yet understood. Despite the powerful nationalist climate in both countries, science operated as an international institution to a degree that may surprise some readers.

Chapters 3–5 portray Wurtz's upbringing and his career up to 1853 in the difficult context of the late July Monarchy, the brief Second Republic, and the early Second Empire. I am interested in a full portrayal of the development of his scientific research and also of his successful navigation of the rough political waters of Parisian science in this era. Chapter 6 treats the crucial turning point in Wurtz's intellectual life: his conversion to the "reformed chemistry" of Laurent, Gerhardt, and Williamson. Few others in his collegial circle made this turn with Wurtz, and chapter 7 deals with his early efforts to persuade his colleagues of the advantages of the new system. These efforts were social and political in addition to scientific. Again, it is necessary to understand both of these aspects of a life in science in order to come to a full understanding of its history.

Chapter 8 describes the career of Wurtz's chief contemporary scientific rival, the somewhat younger Marcellin Berthelot, up to about 1864. Berthelot was strongly committed to scientism and to a kind of positivism, and he rejected most of the ideas that were at the heart of Wurtz's scientific convictions. But Berthelot and Wurtz were allies in their fight for proper material and financial support for their science; chapter 9 details that struggle during the 1850s and the 1860s. Chapter 10 returns to the issue of atoms versus chemical "equivalents," a story with large implications and one that was not resolved until after Wurtz's death.

Chapter 11 covers Wurtz's later career, from 1870 until his death in 1884, and the wider context of Parisian science in these years. Here the story is far more compressed than it was in chapters 3–7. There are three reasons for my decision to quicken the tempo of the historical narrative. For one thing, the Parisian science of the 1870s and the 1880s has been much more fully investigated by historians than that of the period from about 1840 until 1870; the earlier period, despite its critical importance for the fate of French science, has been quite neglected by historians and is therefore in greater need of detailed study. Another reason has to do with the trajectory of

Wurtz's career and of the development of French science. Wurtz's early years were far more eventful than his later period. My principal historical interest is in studying how his ideas took initial shape, how they were transformed and developed, and how these events interacted reciprocally with the wider scientific, political, and social context. One highly remarkable aspect of Wurtz's later career is the further development of his extraordinary research school. That story, however, has been well treated in the recent writings of Ana Carneiro and Natalie Pigeard, and this circumstance constitutes the third reason for a somewhat compressed treatment of the life of that research school.[17]

In my final chapter, I reflect on a number of historiographic issues. Wurtz failed to win a victory for his ideas in the collegial community of his day, and I consider the reasons for this defeat. A parallel but larger issue is the much-debated question as to whether French science as a whole declined relative to German science. The Wurtz story will be found to be instructive in judging this matter. If French science did decline relative to German science, it is also necessary to consider why that happened. Finally, I return again to the question of laboratories, and once more I argue for an eclectic and heterogeneous approach to scientific biography.

In writing about the Chemical Revolution, Frederic L. Holmes has offered a critique of the "typical" article on this subject: the article form itself "encouraged the tendency to pick out certain of [the event's] features as central, while displacing others toward the periphery." In contrast, Holmes believes that the Chemical Revolution was "a complex, multidimensional episode" that can only be properly encompassed in a "narrative broad enough to show how the various thematic strands that historians have isolated as critical factors were interwoven."[18] It is as such an effort toward a narrative conceived "on a grander scale," as Holmes puts it, that my account of a less successful revolution in French chemistry is told.

17. A. Carneiro, The Research School of Chemistry of Adolphe Wurtz (Ph.D. dissertation, University of Kent/Canterbury, 1992); Carneiro, "Adolphe Wurtz and the Atomism Controversy," *Ambix* 40 (1993): 75–95; N. Pigeard, L'Oeuvre du chimiste Charles Adolphe Wurtz (1817–1884) (thèse de maîtrise, Université Paris X Nanterre, 1993), passim; Pigeard, "Un alsacien à Paris," *Bulletin de la Société Industrielle de Mulhouse* 833 (1994):39–43; Carneiro and Pigeard, "Chimistes alsaciens à Paris au 19ème siècle," *Annals of Science* 54 (1997): 533–546.

18. F. Holmes, *Eighteenth-Century Chemistry as an Investigative Enterprise* (University of California Press, 1989), p. 118.

1

The Rise of Justus Liebig

From soon after the time of their first meeting as young men in Paris in 1823 until one of them died 50 years later, Justus Liebig (1803–1873) and Jean-Baptiste Dumas (1800–1884)—Adolphe Wurtz's two mentors—dominated the world of chemistry. In the vigor of their youth, both in concert and as rivals, and always in counterpoint to the fundamental influence of the Swedish chemist Jacob Berzelius (1779–1848), they built organic chemistry, almost from the ground up. The first decade of their careers, to about 1833, was devoted largely to developing the methodological and theoretical tool kit that would become essential to the science for the future. There is no better way to understand the development of post-Lavoisien French chemistry, especially relative to German chemistry, than to examine the early work and the early rivalry of these two fascinating individuals.

Liebig in Paris

Justus Liebig was the son of a materials merchant in Darmstadt, the capital of the Grand Duchy of Hesse.[1] He attended the universities of Bonn and Erlangen, where he became attracted to the Naturphilosophie of Friedrich Schelling and to chemistry as taught by the respected Karl Kastner. Liebig's letters home indicate that he was well satisfied with Kastner; however, he knew even then that he wanted more than German institutions could provide. Applying for a government grant at the instigation of Kastner, he stated that a period of study in Paris under "the greatest chemists of our

1. Among the standard biographies are the following: J. Volhard, *Justus von Liebig* (Barth, 1909); F. Holmes, in *Dictionary of Scientific Biography* 8 (1973): 329–350; W. Brock, *Justus von Liebig* (Cambridge University Press, 1997).

day" was necessary for completing his education; Kastner agreed with this assessment.[2] Liebig had been studying French since his first semester at university, since the language was, as he then put it, "necessary for a scientifically educated man."[3]

Liebig's grant, which was to have been for a six-month period, was renewed several times, and he ended up staying for 17 months (from late October 1822 until late March 1824). He resided in the Rue Harlay with another German student of chemistry, Karl Oehler. Liebig's initial impressions of Paris, a crowded, squalid, medieval city of 700,000, were uniformly negative. He deplored the "rotten meat," the "miserable vegetables," and the "depravity of the people." Life, he said, was embittered by a thousand unpleasantries. The French people "disgusted" him, and he was particularly appalled by the "hundreds of prostitutes" swarming around the theater district. Other foreign visitors to Paris in this period used remarkably similar expressions to describe their reactions.[4]

But within 2 months of his arrival, this 19-year-old German student had made the acquaintance of L. N. Vauquelin, L. J. Thenard, and J. L. Gay-Lussac, and he soon applied to these famous chemists for laboratory space. Not all the details of Liebig's first personal meeting with the Parisian chemical elite are known, but it seems that Carl Sigismund Kunth (1788–1850), professor of botany at Berlin, resident of Paris, and long-time collaborator of Alexander von Humboldt, provided the introductions[5]; Humboldt him-

2. Liebig to Grand Duke Ludwig [10 April 1822]; Kastner to Grand Duke Ludwig, 12 April 1822, in *Briefe von Justus Liebig nach neuen Funden*, ed. E. Berl (Gesellschaft Liebig-Museum, 1928), pp. 34–35.

3. Liebig to his parents, 19 November 1820 and 27 October 1821, ibid., pp. 12 and 29. This was a common conviction among Germans at this time. Georges Cuvier (cited in K. Kanz, *Nationalismus und internationale Zusammenarbeit in den Naturwissenschaften* (Steiner, 1997), p. 53) opined in 1811 that there were "no educated Germans" who had not mastered French, English, and Italian.

4. Liebig to his parents, 8 and 15 November 1822 and 1 January 1823, and Liebig to Schleiermacher, 17 January 1823, ibid., pp. 37, 38, 43, 45; and Liebig to Platen, 24 April 1823, in Volhard, *Liebig*, vol. 1, p. 37. For Paris in the Restoration and visitors' impressions thereof, see D. Jordan, *Transforming Paris* (Free Press, 1995), pp. 94–95.

5. "Ich habe auch Herrn Prof. Knuth [*sic*] aufgesucht . . . durch ihn habe ich den berühmten . . . Blume der Chemiker Gay-Lussac [words missing due to tears in the letter]" (Liebig to his parents, 1 January 1823, *Briefe*, p. 43); "Durch die Güte des Herrn Professor Knuth [*sic*], Mitarbeiter des Herrn v. Humboldt, hatte ich Gelegenheit, einer Sitzung der königlichen Akademie beizuwohnen, ich hatte hier die Freude,

Figure 1.1
Justus Liebig as a student at Erlangen in 1821, the year before his departure for
Paris. From a drawing by a fellow student. Source: E. F. Smith Collection,
University of Pennsylvania Library.

die Bekanntschaft von Vauquelin, Gay-Lussac und anderen vortrefflichen Männern
zu machen . . ." (Liebig to Schleiermacher, 17 January 1823, ibid., p. 46). How
Liebig contrived to gain an introduction to Kunth in the first place is not known.

self, the most prominent German scientist residing in Paris at this time, was then traveling in Italy. Liebig was already then trying to pry more money from his family and his government in order to lengthen his stay.[6]

By this time Liebig was certain that, however dissipated Paris's towns-people were, her scientists were indeed the best in the world; they combined an "exceptional mathematical sense" with a habitual avoidance of all unnecessary hypotheses; they also gave superb lectures with dazzling illustrative experiments. As Liebig wrote at the time to a friend in Hesse, the lectures of Gay-Lussac, Thenard, and P. L. Dulong had introduced him to a different, more challenging, but altogether more attractive sort of science; Liebig, thunderstruck, now regarded himself as a mere tyro at this better sort of chemistry and spoke fervently of his "metamorphosis" into an aspiring *real* chemist.[7] "It is a real shame how greatly the reputation of Germans has declined in physics, chemistry, and the other natural sciences," Liebig wrote to his friend Platen from Paris; ". . . there is scarcely a shadow left, and they fight over this shadow like mad dogs. Contemporary German chemists presume to play at philosophy, and thereby lose all their effectiveness. . . . The French and English proceed in the exact opposite fashion: here science is simply a mechanical stonework, the quasi-mathematical style of treating it allows no play of the mind[8] whatever; but [this method] is at the moment very good, it has led recently to the most magnificent discoveries, and is particularly useful for [practical] life."[9]

Liebig's letters demonstrate, in short, that he came to France expecting to find the highest-quality science in the world, and that he found there even more than he expected.

The views on French science expressed in Liebig's correspondence of 1823 are consistent with the views expressed in more detail in his autobiography. The "indescribable charm" to Liebig of the lectures of Gay-Lussac and others was due, in part, to a realist and phenomenalist style of explanatory theory that Liebig had never before encountered; German scientists followed a

6. Liebig to his parents, 1 January 1823; Liebig to Schleiermacher, 17 January 1823; Liebig to his parents, n.d. [ca. 17 January 1823], ibid., pp. 43–47.

7. Liebig to Schleiermacher, 17 January 1823, ibid., p. 45; Liebig to August Walloth, 23 February 1823, ibid., p. 49.

8. The Frenchified German word Liebig used here is "Räsonnement." In French, "raisonnement" simply means reasoning or use of logical faculties.

9. Liebig to Platen, May 1823, in Volhard, *Liebig*, vol. 1, pp. 42–43.

more idealist method, less explicitly experiential and more oriented toward the use of analogies and images. The logical clarity, the emphasis on experiments and on their interconnectedness, the tight organization of scientific knowledge, the avoidance of speculation, the mathematical method, and the view of nature as ruled by a tight causal web of natural laws all appealed to the young Liebig in a way that his previous German education had not.[10]

Liebig was not the only foreigner studying the sciences in Paris, and he was not alone in his admiration for his teachers. I have already mentioned two older Prussian scientists who were residing in Paris (Humboldt and Kunth). Several younger scientists were there too. Heinrich and Gustav Rose, brothers from Berlin, had studied in Paris in 1815–16; Gustav returned in 1823–24 during Liebig's and Oehler's residency, by which time two other Berlin chemists, F. F. Runge and Eilhard Mitscherlich, were there too. Rose and Mitscherlich, who were studying mineralogy and physics, roomed together that year; they seem to have had little contact with Liebig, and Liebig and Mitscherlich in particular did not get along. After Liebig's departure for home, other Germans came, including C. F. Schönbein in 1827–28, Gustav Magnus in 1828–29, and Robert Bunsen for most of 1832. Liebig returned to Paris with his friend and colleague Heinrich Buff in 1828; Magnus returned with his friend Friedrich Wöhler for 3 weeks in the late summer of 1833. What is striking about this list is the concentration of French-educated German chemists who later resided especially in Berlin, and the number that also spent a period in the laboratory of Jacob Berzelius (this category includes Heinrich Rose, Gustav Rose, Mitscherlich, Magnus, and Wöhler). No doubt Berzelius himself had something to do with this pattern. He spent 1818–19 in Paris, and he reportedly said, after hearing a lecture by Thenard, "I have been a professor of chemistry for twenty years, but only now do I see how it should be taught."[11] It is also important to note that the 18-year period from which we have been sampling is uncharacteristic in a longer perspective: before 1815 and after 1833 it was rare for German chemists to take educational tours in France.[12]

10. Liebig, "Eigenhändige biographische Aufzeichnungen," in H. von Dechend, *Justus von Liebig*, second edition (Verlag Chemie, 1963), pp. 20–22.

11. J. Dumas, *Discours et éloges académiques* (Gauthier-Villars, 1885), vol. 2, p. 302.

12. Only four German chemists traveled to Paris in the period 1789–1814: F. Stromeyer, H. A. Vogel, F. G. Gmelin, and L. W. Zimmermann. At least 15 of them went to Paris in the 18 years after 1815 (Kanz, *Nationalismus*, pp. 107–118).

The Scotsman Robert Christison studied with Gay-Lussac and Thenard in 1820, and has left us striking verbal portraits:

Gay-Lussac was perhaps the most persuasive lecturer I ever heard. His figure was slender and handsome, his countenance comely, his expression winning, his voice gentle but firm and clear, his articulation perfect, his diction terse and choice, his manner most attractive; and his lecture was a superlative specimen of continuous unassailable experimental reasoning.

Thenard impressed Christison less, though the force of his delivery was astonishing to him:

[Thenard's] matter was excellent; and he laid it down with a slap from his tongue and a blow with his fist, which made it irresistible. But the incessant vigour, sans relache sans repos, made one long for a little of his friend's no less persuasive quiet occasionally.[13]

Thenard, in a letter to Cuvier dated 1819, describes how it seemed from the other side of the lectern:

The large number of auditors who turn up from all over, the obligation to raise one's voice to be heard, the necessity of making continual efforts to capture attention, make the task difficult, but they become much more painful because of the heat from the furnaces with which the professor is surrounded. These discomforts are felt particularly in summertime.[14]

There were indeed a large number of auditors: in the period 1818–19 anywhere from 300 to 700 attended each of Gay-Lussac's and Thenard's lectures,[15] and in the 1840s audiences of more than 1000 were not uncommon in Dumas's classes. Louis Pasteur described these classes as "exactly like the theatre": one had to come early to get a good seat, there was much applause, and so on.[16] Schönbein, too, thought the lectures of Gay-Lussac "the best I have ever heard, and how they contrast with the bombast and emptiness [Wortkramerei und Sachlosigkeit] of the Germans." But Schönbein was

13. *The Life of Sir Robert Christison* (Blackwood, 1885), vol. 1, pp. 207, 239ff., cited in M. Crosland, *The Society of Arcueil* (Harvard University Press, 1967), p. 439.

14. F17/1933, cited and translated in M. Crosland, *Gay-Lussac, Scientist and Bourgeois* (Cambridge University Press, 1978), p. 147.

15. Crosland, *Gay-Lussac*, p. 147.

16. Pasteur to his parents, 9 December 1842, in *Correspondance de Pasteur, 1840–1895* (Flammarion, 1940), vol. 1, p. 81. He commented that there were always 600 or 700 people there.

Figure 1.2
Joseph Louis Gay-Lussac in the 1820s. Source: E. F. Smith Collection, University of Pennsylvania Library.

inconsistent in his nationalist stereotypes, for he also expressed an unfavorable contrast between Thenard's pretentious mannerisms and "the English simplicity and German modesty of Gay-Lussac's manners."[17]

In June of 1823, with permission from Thenard, Liebig began work in Vauquelin's former laboratory at the École Polytechnique, then occupied by the répétiteur H. F. Gaultier de Claubry, who assisted the young German chemist with apparatus and materials. Four weeks later he had finished his project, and with Thenard's help he wrote up the results. On 28 July 1823, Gay-Lussac read the paper at a meeting of the Académie des Sciences, with Liebig present as a guest.[18]

The subject of Liebig's paper was the fulminates, a family of explosive substances that had fascinated him since his earliest introduction to chemistry as a young boy. In the first published paper of Liebig's career, which had appeared in Germany the previous year, he had declared that fulminates consisted of ammonia combined with oxalic acid, but these just-completed experiments indicated that this view was erroneous. In addition to his own work, Liebig cited about a dozen previous workers, including such French luminaries as Claude Louis Berthollet, Antoine de Fourcroy, and Liebig's patron Thenard. The essential novelty of his paper was the discernment of a new acid at the heart of the fulminates, a substance which he named fulminic acid; he prepared and characterized several novel salts of this acid. Liebig even attempted to perform an elemental analysis of the salts (by distilling the samples with magnesia and trapping the decomposition products, ammonia and carbonic acid, in hydrochloric acid and limewater respectively). However, he expressed no confidence in this analysis, since, he said, the magnesia itself probably trapped some of these products. Liebig was soon to find that his distrust was justified.

Liebig's paper, even with its admittedly flawed analytical results, attracted plenty of attention. It was printed in the *Annales de chimie* and commented

17. G. Kahlbaum and E. Schaer, *Christian Friedrich Schönbein, 1799–1868* (Barth, 1899), pp. 81–82; W. Prandtl, *Deutsche Chemiker in der ersten Hälfte des neunzehnten Jahrhunderts* (Verlag Chemie, 1956), p. 201.

18. Liebig to his parents, 12 June, 30 June, n.d., 17 July, and 6 August 1823, *Briefe*, pp. 58–64; Liebig, "Sur l'argent et le mercure fulminans," *Annales de chimie* [2] 24 (1823): 294–317. On the last page of this article Liebig expressed his gratitude to Thenard and Gaultier.

upon by Gay-Lussac and Dulong, who proffered high praise.[19] In his annual report for 1823, Berzelius also reported at length and very favorably on this article.[20] It also impressed another important scientist. As Liebig told the story to A. W. Hofmann, after the meeting of the Académie des Sciences at which the paper had been read, as he was "engaged in packing up [his] specimens . . . a gentleman left the ranks of the academicians and entered into conversation with [him]." "With the most winning affability," Liebig continued, "he asked me about my studies, occupations, and plans. We separated before my embarrassment and shyness had allowed me to ask who had taken so kind an interest in me. This conversation became the cornerstone of my future."[21] The gentleman was Alexander von Humboldt, recently returned to Paris from Italy. Humboldt invited Liebig to dinner, with the purpose of introducing Liebig to more intimate acquaintance with his close friend and former collaborator Gay-Lussac. Poor Liebig, having failed to determine the identity of the unknown gentleman, never appeared.[22] However, the confusion was soon cleared up, and, with Humboldt's help, Liebig had entrée to mentorship and even friendship with Gay-Lussac.[23]

Early in 1824, shortly before Liebig was to depart for home, Gay-Lussac offered to collaborate with Liebig in his laboratory in the Paris Arsenal (Rue

19. In *Annales de chimie* [2] 24 (1823), p. 421.

20. J. Berzelius, *Jahresbericht über die Fortschritte der physischen Wissenschaften* [for 1823] 4 (1825), pp. 110–117.

21. A. Hofmann, "Life-Work of Liebig," in *Zur Erinnerung an vorangegangene Freunde* (Vieweg, 1888), vol. 1, p. 233.

22. In his reminiscences, Liebig said that Humboldt had returned just days earlier, that for this reason most of his friends did not yet realize he was back, and that this was why he failed to determine the man's identity. In fact, Humboldt returned to Paris about 5 months before this date. See *Briefe Alexander's von Humboldt an seinen Bruder Wilhelm* (Cotta, 1880), p. 90.

23. Volhard, *Liebig*, vol. 1, pp. 46–48. Liebig's reminiscence is in a long book dedication, written 17 years after the event, later (apparently) supplemented orally to his student and biographer Volhard. There is but a single sentence about this in Liebig's surviving correspondence: "Herr von Humboldt, the famous traveler, even came up to me [at the Académie des Sciences], and conversed nearly an hour with me." (Letter of 6 August 1823, *Briefe*, p. 64.) On the relevance of Humboldt's known homosexuality and Liebig's possible inclinations in that direction, see P. Munday, "Social Climbing through Chemistry," *Ambix* 37 (1990), p. 7. Liebig's affair with Platen is described on pp. 609–611 of X. Mayne, *The Intersexes* (privately printed, ca. 1908; Arno reprint, 1975) and on pp. 21–26 of Brock, *Liebig*.

de Sully, near the Bastille); the subject was a proper analysis of Liebig's new fulminate series, to be described in the next section. At the end of this period of direct mentorship with the most famous chemist in France, Liebig regarded himself as the Wunderkind of German chemistry. The tone of his letters home changed; he now regarded himself as a full-fledged member of the professional club, and he began corresponding with German chemists as an equal.[24] Hofmann wrote: "It was in Gay-Lussac's laboratory that Liebig conceived the idea of founding in Germany a chemical school, where he hoped to be to his younger fellow-workers what Gay-Lussac had been to him."

Partly through the advocacy of Humboldt, in the spring of 1824 Liebig was offered an ausserordentlicher[25] professorship at Giessen. He was appointed there in May, 2 weeks after his 21st birthday, without having completed a habilitation or even an Abitur,[26] and with some annoying doubts about his doctoral dissertation.[27] In 1825 he became ordentlicher professor, and he began immediately to pursue the sort of chemistry he had learned in France. But that French chemistry also had a Swedish flavor, especially in the rising importance of chemical atomism and of new methods of organic analysis.

The "Collaboration" between Gay-Lussac and Berzelius

The first substantial step toward a generally applicable method for determining the percentages of carbon, hydrogen, and oxygen in an organic substance—the prerequisite for calculating its chemical formula—was achieved

24. Liebig to his parents, 6 October 1823 and 16 January 1824; Liebig to Walloth, 18 January 1824; Liebig to Schleiermacher, 3 and 13 February and 25 March 1824, *Briefe*, pp. 69–74.

25. An ausserordentlicher professorship was junior, sometimes even unsalaried; an ordentlicher professorship was a regular and senior position.

26. The normal route to a professorship in Germany was to attend a neohumanist secondary school known as the "Gymnasium," whose leaving certificate was called the "Abitur," then earn the doctorate, and finally satisfy a teaching certification procedure known as "habilitation." Among other steps, this required a second dissertation.

27. Munday, "Social Climbing through Chemistry," pp. 9–11.

by Gay-Lussac and Thenard in their mutual work of 1810–11.[28] They oxidized the sample with intermixed potassium chlorate in a vertical tube that was strongly heated at the bottom, and collected the resulting permanent gases—carbon dioxide, excess oxygen, and nitrogen, if present—over mercury. The percentages of carbon and nitrogen in the sample were determined directly by the volumes of carbon dioxide and nitrogen; hydrogen and oxygen were determined indirectly. Using this method, Gay-Lussac and Thenard provided elemental analyses of 19 organic substances. The severe intrinsic difficulties of the method (many details of which I have omitted) and the high quality of the results testify to the superb skills of these chemists (especially Gay-Lussac, who apparently designed the apparatus and directed the experiments).

Berzelius began developing his own methods of organic analysis independently of the French chemists, but his early experiments in 1807 and 1808 were not successful. When the Gay-Lussac–Thenard collaboration was published, Gay-Lussac sent Berzelius a copy. Berzelius wrote back, praising the book highly, and added:

> I have been working long and hard on analyses of organic products. A hundred times I have begun experiments, which I have always then abandoned, despairing of being able to come to exact results; and as soon as I attempted to combine calculations with experiment, my hopes were further diminished.[29]

Despite the obstacles, Berzelius had just reached publishable results on analyses of oxalic and tartaric acids, achieved by destructive distillation with the brown oxide of lead (i.e., the peroxide), followed by capture of product water vapor with calcium chloride and carbonic acid with limewater.

Berzelius was justifiably distrustful of his own analyses, partly because lead peroxide provided too rapid an oxidation and could not be fully purified. He therefore borrowed from Gay-Lussac the use of potassium chlorate

28. J. Gay-Lussac and L. Thenard, *Recherches physico-chimiques* (Paris, 1810–11), vol. 2, pp. 268–350. Although elemental analysis is important in its own right, for it is the principal criterion for chemical identity, analysis was pursued much more avidly after the introduction of the atomic theory into chemistry—for which see A. Rocke, *Chemical Atomism in the Nineteenth Century* (Ohio State University Press, 1984).

29. Berzelius to Gay-Lussac, 25 September 1811, in *Jac. Berzelius Bref* (Almqvist and Wiksell, 1919), 3:ii, pp. 114–117.

as oxidizer, and provided some additional improvements. In order to slow the combustion and ensure full oxidation of the sample, he diluted the sample and oxidizer with admixture of common salt, placed it in a horizontal tube, and heated the tube strongly one section at a time, working gradually toward the back of the tube. Berzelius also decided to determine the two true products (water and carbonic acid) by condensed-phase capture (the former using a calcium chloride tube, the latter by absorption in potassium hydroxide solution), and to measure directly by weight rather than indirectly using volume. This modification of Gay-Lussac's procedure provided higher accuracy, greater simplicity, and a much more direct measurement strategy. Using this method, Berzelius performed precise analyses of 13 organic compounds. He published the results in 1814.[30]

Gay-Lussac appreciated the merits of Berzelius's approach, and he promptly adopted the horizontal combustion procedure and the calcium chloride tube for capturing water vapor—though he retained the volumetric measurement strategy for carbonic acid and nitrogen. But Gay-Lussac also had further improvements to suggest. One was the use of cupric oxide rather than potassium chlorate; cupric oxide was far more physically stable and just as good an oxidizing agent. Berzelius followed suit. A second suggestion pertained to the analysis of nitrogenous organic compounds. Gay-Lussac found that for these substances he was getting inconsistent measurements of nitrogen gas as a combustion product, because the nitrogen tended to be partially oxidized. To solve this problem, he added fresh filings of metallic copper at the front end of the combustion tube. When this copper was heated along with the rest of the tube, it provided a final-stage reducing agent that ensured that elemental nitrogen gas would be collected. The hot copper had no effect on the other products, carbonic acid and water.[31]

This first general method for organic analysis was thus developed between 1811 and 1815 in a series of rapid interacting steps, in what

30. J. Berzelius, "Experiments to Determine the Definite Proportions in Which the Elements of Organic Nature Are Combined," *Annals of Philosophy* 4 (1814): 323–331, 401–409; 5 (1815): 93–101, 174–184, 260–275; method described in 4: 401–408.

31. J. Gay-Lussac, "Recherches sur l'acide prussique," *Annales de chimie* 95 (1815), pp. 181, 184–186; Gay-Lussac, "Observation sur l'acide urique," ibid. 96 (1815), pp. 53–54.

amounted to an unplanned international collaboration between the two finest chemists of their day, Gay-Lussac and Berzelius.[32] This event came on the heels of the rise of stoichiometry and chemical atomism, and was closely connected with that development.[33] By 1815 the method had stabilized to the point that the only substantive difference between Berzelius and Gay-Lussac was in their treatment of carbonic acid. Both collected this product as a gas; Gay-Lussac then measured its quantity volumetrically (by absorbing it in liquid potash solution and noting the reduction of gaseous volume), whereas Berzelius measured it gravimetrically (by using a small lye-filled flask floating on the mercury in the receiver, whose gain in weight over 24 hours was the direct measure of carbonic acid absorbed). These two men were almost exact contemporaries and were then in the prime of their productive years; their personal relationship may be best described as respectful rivalry. Both were working within an essentially French cultural and scientific context, for Berzelius was, as his principal biographer H. G. Söderbaum put it, "a child of the Gustavian period" of the Swedish Enlightenment, and was "educated under the dominance of French taste" and Lavoisien chemistry.[34]

For the next 15 years, despite much scrutiny, few significant improvements were achieved in organic analysis, and the essential elements of the

32. A good source for the method in its ultimate form is J. Berzelius, *Lehrbuch der Chemie*, second edition (Arnold, 1827), 3:1, pp. 157–174. This volume was Berzelius's first full treatment of organic chemistry in his famous textbook, and this German version, edited and translated from the Swedish manuscript by Friedrich Wöhler, was the editio princeps, for the Swedish "first edition" followed behind the German. A good French description is C. Despretz, *Élémens de chimie théorique et pratique* (Méquignon-Marvis, 1830), vol. 2, pp. 742–757. F. Holmes ("The Complementarity of Teaching and Research in Liebig's Laboratory," *Osiris* [2] 5 (1989), p. 135) was the first to point out that the Gay-Lussac–Berzelius interaction produced a "standard apparatus . . . developed in part by Gay-Lussac and in part by Berzelius."

33. See Rocke, *Chemical Atomism*, chapters 2–4.

34. H. Söderbaum, "Berzelius und Hwasser, ein Blatt aus der Geschichte der schwedischen Naturforschung," in *Studien zur Geschichte der Chemie*, ed. J. Ruska (Springer, 1927), p. 177; Berzelius, *Autobiographical Notes* (Williams & Wilkins, 1934), pp. 16–38, 123–128, 179–180; A. Lundgren, "The New Chemistry in Sweden," *Osiris* [2] 4 (1988): 146–168; E. Melhado and T. Frängsmyr, eds., *Enlightenment Science in the Romantic Era* (Cambridge University Press, 1992), passim. Lundgren makes the important point that factors internal to Sweden strongly conditioned the early and generally favorable response to French oxygenist chemistry.

Gay-Lussac–Berzelius procedure did not change. Michael Eugène Chevreul was using the method, on Gay-Lussac's oral recommendation, as early as 1813, in the earliest stages of his classic work on the chemistry of fats.[35] The German chemist Johann Döbereiner adopted it immediately.[36] In England, Andrew Ure worried about his inability sufficiently to dry the sample, and recommended instead simply correcting for hygroscopic water.[37] William Prout suggested that adopting a modification of Gay-Lussac's abandoned procedure for hydrogen—measuring the loss of oxygen in the oxidizer to infer the quantity of water formed—was superior to capturing the water vapor product; he also recommended separate combustions for determining the separate products. He even eventually abandoned cupric oxide in favor of oxygen gas, although he was alone in doing so.[38]

Liebig and Dumas

As has already been noted, Alexander von Humboldt returned home from visiting Italy in time to play a crucial role in Liebig's sojourn in Paris. Ten months earlier, on his outward journey, he performed a similar office for a young French chemist. In September of 1822, traveling south, Humboldt passed through Geneva. The first thing he did upon arrival was to call on and introduce himself to the 22-year-old Jean-Baptiste Dumas, known to Humboldt only as the co-author of a recent paper on the physiology of

35. Gay-Lussac, *Annales de chimie* [2] 96: 53. This information suggests that Gay-Lussac introduced cupric oxide as early as 1812, or 1813 at the very latest, thus following up very quickly on his and Thenard's method published in 1811.

36. J. Döbereiner, "Ueber die Anwendung des Kupferoxyds zur Zerlegung organischer Substanzen und über die Zusammensetzung und Sättigungs-Capacität der Weinsäure," *Journal für Chemie und Physik* 17 (1816): 369–375. Because Döbereiner mentioned neither Berzelius nor Gay-Lussac here, he is usually given credit for independent discovery of the use of cupric oxide. I think this unlikely, because Döbereiner's article appeared nearly a year after Gay-Lussac's contribution was published in the premier chemistry journal of the day, and there was at this time little delay either in the mails or in publishing queues.

37. A. Ure, "On the Ultimate Analysis of Vegetable and Animal Substances," *Philosophical Transaction of the Royal Society* 112 (1822): 457–482.

38. W. Prout, "On the Ultimate Composition of Simple Alimentary Substances," *Philosophical Transactions of the Royal Society* 117 (1827): 355–388. For a discussion of Prout's participation in the development of organic analysis, see pp. 13–21 of W. Brock, *From Protyle to Proton* (Hilger, 1985).

blood. During Humboldt's stay in Geneva he used Dumas continuously as a sort of tourist guide and passive audience for his non-stop conversation. The effect of this concentrated exposure to one of the stars of French science was to whet Dumas's appetite for the metropolis; 4 months later, he traveled to Paris to seek his scientific fortune.[39]

There was a second curious parallel to Liebig's experience. After one of Dumas's first papers was read at the Académie des Sciences, an unknown but distinguished-looking gentleman invited Dumas to dinner. Dumas, a provincial Frenchman recently arrived from Switzerland, was abashed and failed to ask the man his name, but he soon determined that he had been speaking with the Marquis de Laplace. (This was on 18 August 1823, just 3 weeks after Liebig's similar misadventure with Humboldt.) Four months later, on François Arago's recommendation, Dumas was appointed Thenard's répétiteur at the École Polytechnique, to fill the vacancy left by Gaultier de Claubry's appointment at the École de Pharmacie; it is said that Dumas did not even know he was a candidate until after the election.[40] Apparently, and not surprisingly, Liebig and Dumas became acquainted during their 14-month overlap in Paris, but there seems to be no information whatever on this period of their acquaintance. Shortly after Liebig's return to Germany, André Marie Ampère recommended Dumas for a position at the Paris Athénée, a private institution similar to the Royal Institution of London and dedicated to public education. In Crosland's pithy summary, "at the beginning of his career Dumas was helped successively by nearly all the leading figures of the Arcueil circle."[41]

In Geneva, Dumas's research had been physiological and pharmaceutical; in Paris, influenced by the Arcueil circle, he began to move into chemistry. His first research after his arrival, in collaboration with the pharmacist

39. A. Hofmann got the story orally from Dumas and repeated it, without a date, in his important biography "Zur Erinnerung an J. B. A. Dumas" (*Erinnerung*, vol. 2, pp. 234–238). Humboldt's letters indicate that he left Paris on 13 September 1822 and arrived in Milan on 23 September (*Briefe Alexander's von Humboldt*, p. 90) On Dumas, see L. Klosterman, "A Research School of Chemistry in the Nineteenth Century," *Annals of Science* 42 (1985): 1–80; M. Chaigneau, *J. B. Dumas, chimiste et homme politique* (Guy le Prat, 1984).

40. These stories were also first told by Hofmann (ibid., pp. 239–241), who got them orally from Dumas. See also Chaigneau, *Dumas*, pp. 49–57. Dumas's salary was 1500 francs, and he began work in January 1824.

41. Crosland, *Society of Arcueil*, pp. 441–144.

P. J. Pelletier, was combustion analyses of nine important alkaloids, using the Gay-Lussac procedure. This was Dumas's introduction to organic-chemical research. Their results show reasonable accuracy, though the large molecular weights resulted in some uncertainty in deducing formulas.[42]

A few months after the Dumas-Pelletier paper appeared, Gay-Lussac invited Liebig to join him in a definitive analysis of the fulminates. This was an unusual, perhaps even an unprecedented step for the French chemist. Liebig later commented that, as far he knew, he was Gay-Lussac's first research student, and by this time the 44-year-old chemist had been famous for many years. Indeed, the French system of higher education made no provision for advanced research students, and Gay-Lussac never founded a research school. Liebig was probably correct that he was Gay-Lussac's first research student; the only other students closely associated with the master were Jules Pelouze, Edmond Frémy, and Victor Regnault, all three of whom were younger than Liebig.[43] In fact, Regnault and Gay-Lussac's son Jules became students of Liebig, and Pelouze became Liebig's associate.

Although Liebig's collaboration with Gay-Lussac lasted but 6 weeks (in February and March of 1824), the two men formed a close bond during this period. Liebig later recollected scenes from the Arsenal laboratory:

He used to say to me, "You must occupy yourself solely with organic chemistry, M. Liebig; that is what we are in want of." . . . I shall never forget the hours spent in the laboratory of Gay-Lussac! When we had finished a good analysis (you know, without my telling you, that the method and apparatus described in our joint memoir were by him alone)—when we had finished an analysis, he used to say to me, "Now, M. Liebig, you must dance with me, as I used to dance with Thenard when we had found something good." And then we danced!![44]

Liebig always regarded this period as decisive for his development as a scientist, and always looked to Gay-Lussac as his principal mentor. But this research also marked a stage in the older man's development. For one thing,

42. J. Dumas and P. Pelletier, "Recherches sur la composition élémentaire et sur quelques propriétés caracteristiques des bases salifiables organiques," *Annales de chimie* [2] 24 (1823): 163–191.

43. Crosland, *Gay-Lussac*, pp. 249–253.

44. "Liebig's recollection of Gay-Lussac and Thenard," a toast at the International Banquet of Chemists in Paris, reported from translated shorthand in *The Laboratory* 1, no. 16 (1867): 285. I have modified the passage slightly in light of the French version, transcribed from privately held Gay-Lussac family papers (Crosland, *Gay-Lussac*, pp. 278–279).

it provided the opportunity for his first extended description of the analytical method he had been using for the previous 9 years. Although the method was by this time very well known ("très-connu"[45]), the collaborators now described it in detail, including such accessory operations as a method of ensuring a perfectly dry sample before the combustion is to begin and a convenient new way to collect and measure the gaseous products. A third innovation provided a method of performing the combustion in a vacuum so as to eliminate atmospheric nitrogen that would otherwise affect the measurement of nitrogen from the sample.

Upon completing the analysis, Liebig and Gay-Lussac calculated a formula by converting percentages of each element into numbers of atoms, using assumed atomic weights. This was only the second paper since Gay-Lussac had begun to use the word "atome"; before, he had preferred to specify composition solely in terms of elemental "volumes." (This was also only the second paper Gay-Lussac had published since the death of his anti-atomist patron and friend Berthollet.) Here Gay-Lussac betrayed the influence which the rising fortunes of the chemical atomic theory, and especially the work of Berzelius, had had upon him. Most of his contemporaries among the Parisian chemists, including Thenard, Ampère, Chevreul, and Dulong, had long since become chemical atomists.[46]

The fulminate analyses were atypical in one respect: Liebig and Gay-Lussac determined nitrogen by a second method, quantitative measurement of hydrogen cyanide released in acid hydrolysis of the compounds. For his general method of determining nitrogen, Gay-Lussac had since 1815 usually applied an accessory combustion, in which the relative amounts of carbonic acid and nitrogen given off were measured. Since this procedure resulted only in measurement of a ratio of volumes from an unweighed sample, it was thereafter routinely denominated, somewhat inappropriately, a *qualitative* determination of nitrogen. The percentage of carbon in the compound was already known from the

---

45. J. Liebig and J. Gay-Lussac, "Analyse du fulminate d'argent," *Annales de chimie* [2] 25 (1824), p. 294.

46. On pp. 109–115 of *Chemical Atomism* I suggested that Gay-Lussac may have been giving implicit expression here to ideas which he had held for many years but had suppressed in courtesy to Berthollet. It is certainly true, however, that Gay-Lussac was genuinely opposed to excessive speculation, so that there was also an internal inhibition against embracing atomism.

principal combustion, so that the nitrogen/carbonic acid volume ratio could then be used to calculate the percentage of nitrogen in the substance. Dumas and Pelletier, among others, had used this procedure; as they later realized, their results were not good, presumably because complex mixtures of nitrogen oxides were forming.

By the end of his stay, Liebig must have already considered himself a respected junior colleague in the Parisian chemical elite and Gay-Lussac's chief student, despite his youth and his status as a foreigner. Certainly in 1824 he was higher than Dumas in the pecking order; although 3 years younger, he had come to Paris 3 months earlier, he was known to the elite at the time of Dumas's arrival, and he was more clearly professionalized into chemistry. (Dumas continued to publish papers on physiology even after his transfer to Paris.) However, by the time of Liebig's departure both men had gotten a good taste of organic chemistry, and of organic analysis à la Berzelius and Gay-Lussac. They were to remain rivals in this field for the next 50 years.

In October of 1828, Liebig returned to France, on a commission from the Hessian government to investigate the production of sugar from beets. Although he was in Paris only 2 weeks, he found the opportunity to spend some time working in Gay-Lussac's laboratory. Dumas invited him to lunch, and in general he was treated with extraordinary kindness and generosity (much remarked on in his letters to home). This was when he first met Pelouze (who was then studying with Gay-Lussac), with whom he formed a close lifetime friendship.[47] During the first few years after his Paris sojourn, Liebig must have regarded himself almost as a non-resident member of the Parisian chemical community. He spoke fluent French, corresponded regularly with Gay-Lussac and other Parisians in their own language, took Pelouze, Regnault, and Jules Gay-Lussac as advanced students and collaborators in Giessen, and regarded the *Annales de chimie*,

---

47. In his éloge of Pelouze, Dumas wrote that Liebig knew Pelouze from Gay-Lussac's laboratory, and it is often stated that they met during Liebig's first period in Paris. See, e.g., J. Dumas, *Discours et éloges académiques* (Gauthier-Villars, 1885), vol. 1, p. 154. However, Pelouze arrived in Paris for the first time a year after Liebig left, and his letters to Liebig (Liebigiana IIB, Bayerische Staatsbibliothek) make more than one reference to their first meeting in 1828. On Liebig's 1828 trip to France, see Volhard, *Liebig*, vol. 1, pp. 122–128.

rather than any of the German journals, as his first outlet for publications.[48] He even maintained good relations with Dumas.

Liebig was certainly regarded as Frenchified by Berzelius and his prize student, Friedrich Wöhler. In Liebig's first paper after his return, Wöhler claimed to smell "a Parisian aroma" of "fast and sloppy" work; his picric acid paper was "entirely à la française, that is, he immediately deduces important theoretical conclusions from partial observations and incomplete analyses." A subsequent paper on the action of chlorine on organic compounds was interesting, but was "more a series of excerpts from a laboratory notebook than a completed project." By 1830, Wöhler had become more charitable, but still he criticized Liebig for "laboring somewhat under the error of the French school, failing to base his calculations on fully secure experiments, and in general being too hasty with calculations. . . . Liebig is quite industrious, and it is only too bad that he works in the French-German fashion like L. Gmelin, and in general offers a mass of small new facts, without providing absolutely secure reliable results or properly concluded investigations."[49] Such prejudicial stereotypes of the French style of scientific research were well represented in Germany and elsewhere.[50] In any case, Berzelius agreed with Wöhler that Liebig's work showed him to be too often "thoughtless" and "careless," and that he was "fast but incomplete" or "fast but sloppy" in his research.[51]

But this was nothing when compared to Wöhler's and Berzelius's negative early assessment of Dumas. "Dumas hunts after discoveries, as in

48. For instance, although Liebig published his papers on picric acid in Schweigger's *Journal für Chemie* and in Poggendorff's *Annalen der Physik*, the corresponding French-language articles in the *Annales de chimie* were earlier, presumably because Liebig had a habit of sending them directly to Gay-Lussac as enclosures in letters.

49. Wöhler to Berzelius, 11 December 1825, 17 May 1828, 18 May 1829, 14 February and 26 March 1830, in *Briefwechsel zwischen J. Berzelius und F. Wöhler*, ed. O. Wallach (Engelmann, 1901) (hereafter cited as "Wallach"), vol. 1, pp. 101, 218, 258, 287, 291.

50. The zoologist L. F. Froriep, powerfully impressed by French science (1803), nonetheless drew an unfavorable comparison between the German scientist, who published monographic studies after long deliberation, and the French scientist, who published far too freely and "gladly replaces the ideas of the morning by those that occur to him in the evening" (quoted in Kanz, *Nationalismus*, p. 136).

51. Berzelius to Wöhler, 13 January 1826, 9 April and 9 July 1830, in Wallach, pp. 106, 292, 304.

general all Frenchmen do," wrote Berzelius. "Science [for him] is a public hunting preserve, and therefore the game that is bagged is [his] legal property." Dumas was a schemer ("sehr intriguant") and a "Charlatan"; "instead of discussing difficult questions, this chemical dancing master tries to shine with new explanations."[52] Wöhler reported to Berzelius documentation of several outright instances of Dumas's plagiarism ("Plagiat," by which Wöhler meant that Dumas sometimes tried to claim priority for discoveries that were not new).[53] When Heinrich Rose visited Paris again in 1830, he was highly impressed with all he saw and all he met, including Dumas; but he subsequently learned that at the same time Dumas was flattering him he had a paper in press that was critical of Rose's work. After this Rose called Dumas a "Jesuit" (referring not to his Catholicism but to his duplicity), and among the German elite the epithet stuck.[54]

New Challenges for Organic Analysis in the 1820s

Celebratory Liebigiana has promoted the notion that there was no elemental organic analysis worthy of the name before Liebig's innovations. This is clearly not true, as the preceding discussion and other recent scholarship[55] demonstrate. In the 1820s authors of papers on the subject usually omitted all details, merely specifying combustions performed according to the "customary procedure" or the "usual method" (obviously assumed to be well known to their readers); this referred to the Gay-Lussac–Berzelius standard procedure or to some insubstantial modification thereof. Nor was the customary procedure reserved for the great masters of the day. Berzelius even asserted that "no particularly great skill of the operator is necessary in order to reach a reasonably reliable result" using his method. But Berzelius weakened his point by proceeding to lambaste the sloppy work

52. Berzelius to Wöhler, 18 July and 22 November 1826, 9 April 1827, 22 January 1831, and 24 January 1832, in Wallach, pp. 132–133, 155–156, 170–172, 335, 396.

53. Wöhler to Berzelius, 11 March 1827, in Wallach, p. 168.

54. Related in Wöhler to Berzelius, 7 November 1831 (Wallach, p. 319).

55. Holmes, "The Complementarity of Teaching and Research in Liebig's Laboratory."

56. Berzelius, *Lehrbuch*, pp. 173–174.

and hasty conclusions that characterized many recent investigations.[56] And in the same year the preceding words were published (1827), William Prout, one of the finest analysts of his or any other day, declared that attaining precision in elemental organic analysis "has always proved a most difficult problem."[57] William Henry agreed, stressing that "considerable skill" was needed to perform these analyses, even with the latest improvements.[58] Liebig certainly was not satisfied with the state of the art. His papers of the late 1820s repeatedly lament the uncertainties and difficulties of analyses, his own as well as others'.

Not only chemists but also pharmacists were intensely interested in analysis. As a boy, Liebig had served briefly as a pharmaceutical apprentice in Heppenheim, a few miles from his home in Darmstadt, and before his arrival in Paris it had been his intention eventually to direct a pharmaceutical institute, similar to several earlier successful German models. Dumas, too, was briefly an apprentice pharmacist in his home town of Alais (in southern France, about 30 miles from Nîmes), and his first teacher (and co-author) in Geneva had been the pharmacist Auguste Le Royer. Only recently has it been properly appreciated how fully the science of chemistry achieved professionalization, especially in Germany, from roots in pharmacy.[59]

The hot subject in Parisian pharmaceutical research in the half-dozen years preceding the arrival of Liebig and Dumas was a group of nitrogenous organic compounds known subsequently as alkaloids. In these years

57. Prout, "On the Ultimate Composition of Simple Alimentary Substances," p. 357.

58. W. Henry, *Elements of Experimental Chemistry*, supplement to second American edition, from ninth London edition of 1823 (Desilver, 1823), pp. 122–128. Henry added: ". . . some practice in [these operations] is necessary to enable a person, who is even conversant in the general processes of chemistry, to obtain accurate results. A single experiment should never be depended upon; but the analysis of each substance should be several times repeated, and a mean taken of those which do not present any very striking disagreement. . . ."

59. The locus classicus for this thesis, widely accepted today, is B. Gustin, The Emergence of the German Chemical Profession, 1790–1867 (Ph.D. dissertation, University of Chicago, 1975). E. Homburg ("The Rise of Analytical Chemistry and its Consequences for the Development of the German Chemical Profession (1780–1860)," *Ambix* 46 (1999): 1–32) has recently provided compelling arguments for the claim that advances in analytical chemistry sparked dramatic changes in German academic laboratories and educational practices in the 1820s.

it seemed that a new alkaloid was isolated almost every month, and all were pharmacologically interesting. It had been none other than Gay-Lussac who had created the excitement, even though he had not been a discoverer himself, for it was he who first clearly signaled that the new substances were the first examples of a new chemical category: organic bases derived from plants.[60] The first attempt to provide good elemental analyses of many of these compounds was the paper by Dumas and Pelletier discussed above. But this paper had flaws, which were soon recognized even by the authors.

One problem was the difficulty of determining nitrogen content accurately, even under the best conditions. Combustions could produce complex mixtures of nitrogen compounds, contrasting starkly with the simplicity of the other two products, carbonic acid and water. Alkaloids presented two further complications. Their nitrogen content was low, commonly around 5 percent, which meant that analyses required measuring and characterizing tiny amounts of gas; the results were easily compromised by inadvertent admixture of atmospheric nitrogen derived from the sample tube. Furthermore, the molecular weights of these compounds were high. For instance, Dumas and Pelletier calculated that a morphine molecule had 107 atoms in it; Liebig's recalculation a few years later lowered the presumed number to 78, but this was still a very large molecule indeed, and it was typical of the category.[61] The problem is that small uncertainties in measured percentages of carbon, hydrogen, nitrogen, and oxygen, which present no difficulties in determining smaller formulas, lead in larger formulas to significant uncertainties, since with high numbers of atoms the various percentage intervals corresponding to the various possible formulas are so closely spaced. The net effect of the discovery of alkaloids around 1820, then, was to pose a new challenge for the art of organic analysis.

60. J. Lesch, "Conceptual Change in an Empirical Science," *Historical Studies in the Physical Sciences* 11 (1981): 307–328.

61. Dumas and Pelletier, "Recherches sur la composition élémentaire et sur quelques propriétés caracteristiques des bases salifiables organiques"; Liebig, "Ueber einen neuen Apparat zur Analyse organischer Körper, und über die Zusammensetzung einiger organischen Substanzen," *Annalen der Physik* [2] 21 (1831), p. 18. The present formulation of the molecule, using a two-volume rather than Liebig's four-volume formula, is half about half as large, and contains 40 atoms.

Unlike Dumas, Liebig did not confront the problem of alkaloid analysis at the beginning of his organic-chemical career, but rather was led to it gradually by working with other nitrogenous organic compounds. His collaboration with Gay-Lussac on the fulminates has been discussed above. Wöhler, who in 1824 was studying with Berzelius in Stockholm, announced the isolation of a new compound, silver cyanate, in the same month as the Liebig–Gay-Lussac paper appeared; remarkably, the percentage composition was virtually identical to that reported for silver fulminate, although the salt had completely different properties. This coincidence prompted a two-year controversy between Liebig and Wöhler (hitherto unknown to one another) in which, in effect, each man accused the other of faulty analyses.

This dispute was played out with some drama. Liebig later told Wöhler that the Parisians regarded the two Germans as implacable enemies, which was never the case. "If I am interpreting it correctly," Wöhler wrote to Berzelius, "this putative enmity relates less to L. and to my negligible chemical personality than to our respective teachers and schools that have educated us, or, more precisely, have attempted to educate us [i.e., Gay-Lussac and Berzelius]."[62] The dispute was settled when all four men were finally convinced of the others' accuracy, and the identity of composition of these two different substances became one more example of the emergent phenomenon of "isomerism."[63] For present purposes, it is only necessary to underline the circumstance that much of the dispute concerned the fine details of chemical analysis. However much the Gay-Lussac–Berzelius method had achieved a certain standardization, there was still plenty of room for distrust over specific analyses, even those done by the masters.

In the winter of 1826–27, Liebig, having settled one issue, picked up an unfinished project that he had begun with Gay-Lussac before his departure from Paris. The action of nitric acid on a variety of organic materials produced an interesting new nitrogen-containing organic acid (which 6 years later Dumas named "picric acid"), whose elemental analysis appeared to require fractional atoms. Liebig published the proposed formula containing half-atoms, but noted that since such a formula contradicted the atomic

---

62. Wöhler to Berzelius, 14 February 1830, in Wallach, p. 287.

63. The dispute is treated in detail, with relevant documentation, on pp. 171–173 of Rocke, *Chemical Atomism.*

theory he preferred to assume that his analysis was not sufficiently precise. Indeed, 10 months later he published a more refined analysis that contained no fractions. He then commented that, although he did not regard the atomic theory as epistemologically certain, it was precise and satisfying, and it could not be replaced.[64]

Liebig's concern over the deficiencies in analytical methods, especially for nitrogenous compounds, began to get acute just when he was starting to form much closer bonds with chemists outside the Parisian orbit. His rapprochement with Wöhler, which followed their first meeting in April of 1826, led quickly to an extraordinarily close friendship (by 1830 they were on "du" terms), and his first meeting with Berzelius in September of 1830 had a similar effect in his relationship with the older Swedish chemist. One mark of this drift was Liebig's adoption that year of Berzelian atomic weights, used by Wöhler and other (mostly German) Berzelians, in place of the conventional equivalents preferred by most French and English chemists.

Liebig continued to collide (inadvertently) with Wöhler in his research on nitrogen-containing organic compounds. In an effort to avoid competing with his new friend, he chose a related but distinct compound to investigate, a nitrogenous substance he named hippuric acid. Although he eventually arrived at an analytical result he thought was precise, he was troubled by the fact that the small proportion of nitrogen made it difficult to achieve a "sharply defined result." This problem was similar to that presented by the alkaloids, and he pointed out that an analysis of morphine which he had just carried out differed from one reported by Dumas and Pelletier.[65]

Early in 1830, Liebig published a critique of Prout's new apparatus for elemental organic analysis. He argued that it was considerably more complicated and no more accurate than the standard approach; however, he also used the occasion to point out the deficiencies of the Berzelius–Gay-Lussac

64. J. Liebig, "Mémoire sur la substance amère produite par l'action de l'acide nitrique sur l'indigo, la soie, et l'aloès," *Annales de chimie* [2] 35 (February 1827): 72–87; "Sur la composition de l'acide carbazotique," *Annales de chimie* [2] 37 (1828): 286–291. Liebig's first formula was $C_{12\frac{1}{2}}N_{2\frac{1}{2}}O_{16}$, his second $C_{15}N_3O_{15}$; the modern formula, translated into Liebig's conventions, would be $C_{12}N_3O_{14}$.

65. J. Liebig, "Ueber die Säure welche in dem Harn der grassfressenden vierfüssigen Thiere enthalten ist," *Annalen der Physik* [2] 17 (1829): 389–399.

method. One difficulty had always been that only very small samples (on the order of a tenth of a gram) could be burned, because otherwise the resulting volumes of gases were too large to handle easily—and of course, the smaller the sample, the smaller is the maximum precision attainable. Gay-Lussac had introduced a modification which Liebig thought useful, namely carrying out a separate combustion solely for determining hydrogen (from product water vapor); a much larger sample could be used for this measurement, since the water was the one product collected in the condensed phase rather than volumetrically. Nitrogen was really the most problematic component, Liebig emphasized; Prout's new apparatus did nothing at all to ameliorate this, since it was designed to analyze only non-nitrogenous compounds. This was the context in which Liebig commented that what was needed was not a new apparatus (meaning the one introduced by Prout) but a better method for nitrogen.[66]

This comment should not be taken to indicate that Liebig thought all was well with organic analysis. On the contrary, the above discussion has shown that Liebig was much troubled by the inaccurate and cumbersome character of many aspects of the art, especially (but not solely) nitrogen determinations. In September of 1830, Liebig published an analysis of camphor, in which he once more lamented that only approximate numbers were so far achievable (owing to the relatively large molecular weight and the volatility of the substance).[67] He later wrote to Berzelius:

. . . I think it quite probable that my analyses [of indigo and picric acid] are faulty, as well as all analyses of nitrogenous compounds whose carbon content is calculated volumetrically, because in such cases everything depends on the relative [volume] ratio of nitrogen and carbonic acid, which for compounds that easily form nitric oxide in combustion is so difficult to determine.[68]

The relevant published literature of the 1820s contains many such incorrect, discordant, or insecure analyses, even from the hands of the elite. A comparison of these results with modern data reinforces the sense that elemental organic analysis continued to be, as Prout noted, a most difficult problem.

66. J. Liebig, "Ueber die Analyse organischer Substanzen," *Annalen der Physik* [2] 18 (1830): 357–367.

67. J. Liebig, "Ueber die Zusammensetzung der Camphersäure und des Camphers," *Annalen der Physik* [2] 20 (1830): 41–47, on 43, 45.

68. Liebig to Berzelius, 14 September 1833, in *Berzelius und Liebig, Ihre Briefe von 1831–1845*, ed. J. Carrière, second edition (Lehmann, 1898), p. 71.

Of course, comparing two analyses of a substance undertaken almost 200 years apart requires one to assume identity of composition and perfect purity of sample, and so such evidence must be used with caution. The fact remains, however, that many if not most pre-1830 formula determinations fail to match modern figures, whereas most post-1830 determinations are identical to those accepted today. The early-nineteenth-century techniques used to identify and purify substances were constant during this watershed period, but the analytical methods were not. A change in method ca. 1830 clearly made a difference.

In sum, in the 1820s highly accurate and secure results were achievable, but not always, and not with all kinds of compounds. This situation changed suddenly in the fall of 1830, when Liebig invented what only a few months earlier he had declared unnecessary: a new apparatus.

Introducing the Kaliapparat

On 19 December 1830, Wöhler wrote to Berzelius: "I am curious to learn [Liebig's] new method of organic analysis. He can apply it to very large quantities of sample."[69] Liebig had obviously developed the innovation during the autumn of 1830; his correspondence with Berzelius had not yet begun. But Berzelius did not have to remain curious very long, for 3 weeks later, in his first letter to the Swedish chemist, Liebig wrote:

I have felt compelled to invent a new apparatus for my analyses, which permits one to subject to combustion not just a few tenths of a gram of sample, as is now customary, but any arbitrarily large quantity. The carbonic acid is captured in a specially constructed vessel in which the absorption is complete, and in which the acid can be weighed directly and without the slightest loss.

Liebig proceeded to say that he had thoroughly tested the new method by using it on well-known compounds such as racemic acid and urea, and asserted that it made organic analysis just as simple and precise as inorganic. He continued:

You will tell me that a father does not easily scold his child, and so I have praised the apparatus more than it deserves. I am therefore very curious to hear your opinion, after you read the description of it, which would be much too tedious for a letter, in Poggendorff's journal.[70]

69. Wallach, p. 327.
70. Liebig to Berzelius, 8 January 1831, in Carrière, *Briefe*, pp. 4–5.

The article appeared in the January 1831 issue of the *Annalen der Physik*.[71] Liebig began by pointing out the general limitations of organic analysis: small samples burned in the combustion tube yield less accurate results, but large samples yield unwieldy amounts of gas; for small formulas this does not matter, but in large-weight compounds—such as alkaloids—the limitation becomes so severe as to make an accurate analysis impossible. Furthermore, the determination of nitrogen content is beset with special problems, including the impossibility of excluding all traces of atmospheric nitrogen and the difficulty of collecting pure nitrogen gas from the combustion.

Liebig's principal solution to these interconnected difficulties was to devise a means by which to capture in condensed phase and to measure gravimetrically both carbonic acid and water in a single operation, and then, if required, to carry out a second "qualitative" volumetric determination of nitrogen. (He suggested insubstantial improvements to this second procedure, but he regarded it as still difficult and inclined to give unreliable data.) By collecting condensed carbonic acid and water in "real time," Liebig could increase sample size dramatically, thus increasing precision by the same factor; moreover, the strategy was simpler and more direct than the previous versions. Of course, since 1811 the water vapor product had always been captured in condensed phase and by the same method Liebig used, namely a calcium chloride tube. However, all analysts had hitherto collected the carbon dioxide product as a gas, and no one had attempted to capture it directly and immediately in a condensed phase.

To do this Liebig developed his potash apparatus, a triangular piece of glass tubing connected to the end of the combustion train, in which five bulbs were blown. Three of the bulbs, arrayed in a horizontal line at the base of the triangle, held a potassium hydroxide solution that strongly attracted the carbon dioxide and condensed it in the form of potassium carbonate; the other two bulbs, flanking and situated above the base line, prevented overflow of the bubbling solution during the course of the combustion. The apparatus was carefully weighed before the analysis began. After the combustion was complete, the operator broke the upturned tip off the back of

71. J. Liebig, "Ueber einen neuen Apparat zur Analyse organischer Körper, und über die Zusammensetzung einiger organischen Substanzen," *Annalen der Physik* [2] 21 (1831): 1–47.

the combustion tube, and then from the front end of the Kaliapparat he sucked "for a short time a certain portion of air" through the train; this ensured that any residual product carbonic acid and water vapor would end up in the Kaliapparat and calcium chloride tube, respectively.[72] Then the potash apparatus was weighed again, the increase in weight being equal to the carbonic acid released by the sample. With this device Liebig could capture almost any quantity of carbonic acid, and so he could burn almost any quantity of sample. The increase in precision and ease of measurement was dramatic.

But Liebig was aware that his debts to previous analysts were deep; he modestly commented in introducing the discussion, "the only thing that is new about this apparatus is its simplicity, and the complete reliability it affords."[73] He knew that Berzelius had long been weighing carbonic acid, rather than measuring its volume as Gay-Lussac did, but *collecting* it volumetrically, as all workers had hitherto done, imposed serious restrictions. He wrote to Berzelius:

After I had begun to concern myself preferentially with organic analysis, I quickly came to the conclusion that only your method of determining carbon by the *weight* of carbonic acid promised entirely secure results in all circumstances, and since then all my efforts have been devoted to making this process more easily accessible; this was the way my apparatus came to be.

Liebig went on to point out that a reliable gravimetric determination of carbon can serve as a control on the "qualitative" determination of nitrogen (the separate measurement of the nitrogen/carbon dioxide volume ratio in order to calculate percent of nitrogen in the compound), for if calculations based on this ratio differ from the those based on the gravimetric measurement, then the purity of the evolved nitrogen gas cannot be trusted. "This control has never before been used, and for this reason it is possible that very many earlier analyses are faulty; here I do not exclude my own analysis of picric acid, in any case I plan to repeat it."[74]

72. Ibid., p. 6. Liebig left these directions vague, probably intentionally. Too little air sucked through the train would fail to deposit all remaining combustion products, but too much would lead to excess product, due to the inadvertent collection of naturally occurring atmospheric water vapor and carbon dioxide. The right amount of sucking could only be learned by experience or expert advice.

73. Ibid., pp. 4–5.

74. Liebig to Berzelius, 14 September 1833, in Carrière, *Briefe*, p. 71.

Figure 1.3
A modern reproduction of Liebig's "potash bulb" apparatus. Photograph taken by the author in 1984 at the Justus-Liebig-Museum in Giessen.

There was also a contingent instrumental reason for the sample-size limitations, hence precision limitation, of the Gay-Lussac–Berzelius procedure. Any process that required volumetric collection of evolved gases required a closed system ending in a pneumatic trough, and mercury was usually necessary as the fluid, since so many gases are at least partially soluble in water. The weight of the mercury generally places the entire combustion train under pressure; this pressure can be minimized by clever arrangements,[75] but it cannot be eliminated. It is also necessary to heat the combustion tube strongly, in order to produce complete combustion. Hot glass and fragile couplings under pressure are a recipe for inconvenience, or even disaster. One could, and did, wrap the combustion tube in sheet-steel to prevent blow-holes, but this procedure had its own disadvantages and in any case there was a limit to such palliatives.

The introduction of Liebig's Kaliapparat eliminated the need for a pneumatic trough, and enabled the entire process to take place at atmospheric

75. The Liebig–Gay-Lussac collection method, outlined in their collaborative article of 1824, was designed to do exactly this.

pressure. The tube could also be heated more strongly, insuring complete combustion, for it no longer mattered if the glass softened. Finally, Liebig's technique of sucking air through the tube at the end of the combustion to collect residual combustion products could only be performed in such an "open" system; until Liebig's elegant solution, the problem of residual gases had always posed a difficult dilemma for analytical accuracy. Thus, Liebig's move to an "open" atmospheric-pressure system resulted in several important advantages.

Liebig argued for the superiority of his new method partly by demonstrating the large number of accurate analyses that could be performed by semi-skilled hands. Immediately upon describing the apparatus and procedure, he stated that "Herr Hess, one of my students, has at my suggestion undertaken an analysis of racemic acid, as his first assignment of this kind." The percentage composition determined by this unpracticed student could be compared to that just published by none other than Berzelius himself, and Liebig judged it "exact."[76] He then proceeded to report analyses of two additional well-studied substances (urea and cyanuric acid), as well as 13 alkaloids, the ultimate test of organic analysis. The quality of this surprisingly large number of analyses was very high, measured both by contemporary historical and modern standards.[77]

Frederic L. Holmes, who has studied the role of Liebig's students in the immediate aftermath of the invention of the Kaliapparat, concludes that the new method "enabled [Liebig] almost at once to accelerate the pace of his personal research ... [and he] noticed that students could learn very quickly to produce reliable results with the new apparatus"; in general, the method "quickly changed research practices in the field at large." Nonetheless, Holmes portrayed the innovation as consisting of only "small modifications" and "a further refinement" of an existing method that Liebig found satisfactory; he doubts that Liebig anticipated in advance any profound effect of the apparatus on the field, or that the method soon became so routine that any student could produce publish-

76. Liebig, "Ueber einen neuen Apparat," p. 7.

77. A comparison of Liebig's, Dumas's, and modern results is revealing. For morphine, Dumas and Pelletier arrived at the formula $C_{15}H_{20}NO_{2\frac{1}{2}}$; Liebig's was $C_{17}H_{18}NO_3$; the modern formula is $C_{17}H_{19}NO_3$. For the sake of comparison, Dumas's and Liebig's formulas have been stated using modern atomic weights.

able results.[78] The above discussion suggests, however, that Liebig was well aware of the importance of the invention, and he argued that it placed good analyses within the grasp of unpracticed hands.

There is some real ambiguity in the story. Part of it is due to Liebig's introduction of his method with the phrase "There is nothing new about this apparatus except" simplicity and reliability. There is a sense in which this is literally accurate, but in a larger sense Liebig knew otherwise. German scientists in publications were often inclined to the rhetorical device known as litotes—an understatement calculated for effect, expressing exaggerated (but ultimately insincere) modesty. Similar rhetoric occurs in other landmark German chemical papers, such as Wöhler's synthesis of urea and August Kekulé's structural and benzene theories.[79] The more flamboyant French tended to follow different rhetorical strategies than the Germans, and in the nineteenth century this literary and stylistic difference was the source of innumerable misunderstandings.

I will return to this point at the end of the next chapter. There and elsewhere, I will argue that a consideration of differences in national cultures, styles, and institutions is essential for understanding the attitudes and behavior of our protagonists, and the trajectory of the historical action. And it is by such examples that the dynamic relationship between the ever-present nationalist and internationalist impulses in science can be seen in play.[80]

78. Holmes, "Liebig's Laboratory," pp. 139–142.

79. A. Rocke, *The Quiet Revolution* (University of California Press, 1993), pp. 171 and 239–241; Rocke, "Hypothesis and Experiment in the Early Development of Kekulé's Benzene Theory," *Annals of Science* 42 (1985), pp. 364–368.

80. For an excellent recent discussion of this dynamic relationship, see Kanz, *Nationalismus*.

2

The Rise of Jean-Baptiste Dumas

Dumas's Response to Liebig's Innovation

Liebig sent a French version of his Kaliapparat paper to Gay-Lussac for inclusion in the *Annales de chimie*. Since Liebig had found multiple occasion in it to criticize Dumas's and Pelletier's 1823 paper, Gay-Lussac gave the paper to his younger colleague for review. It was published in the fall of 1831, with Dumas's response immediately following. Dumas's response was mild but inconsistent. Analysis of non-nitrogenous organic compounds by the traditional method, Dumas averred, was "the simplest of operations," and no improvements were necessary. As for Liebig's arrangement for determining nitrogen, Dumas claimed that his own (very similar) method was distinctly superior, and he provided many details. Both men had taken the basic idea—combustion of the sample followed by in situ reduction of the nitrogen oxide mixture—from Gay-Lussac's 1815 paper. Virtually all Dumas's comments were related to the nitrogen method. However, despite his cavalier dismissal of the need for any innovations, Dumas declared in his very first paragraph that Liebig's method was "destined without any doubt to change the state of organic chemistry in the very near future." How the contradictory judgments contained in this response can be reconciled is difficult to say, and they provide an indication of how frustrating Berzelius and many Germans found it to read Dumas's papers.[1]

1. Liebig, "Sur un nouvel appareil pour l'analyse des substances organiques; et sur la composition de quelques-unes de ces substances," *Annales de chimie* [2] 47 (1831): 147–197; Dumas, "Lettre de M. Dumas à M. Gay-Lussac, sur les procédés de l'analyse organique," *Annales de chimie* [2] 47 (1831): 198–213.

Early in October of 1831, Dumas sent Liebig a copy, hot off the press, of the issue of the *Annales de chimie* that contained Liebig's article and his own response. Dumas told Liebig that he had discovered that he had erred in critiquing Liebig's method for determining nitrogen; he had looked over Liebig's paper too fast, and had not understood (or even properly read) what was there. He could now see, Dumas wrote apologetically, that their nitrogen methodology was essentially the same. And Dumas promised that he would correct his "blunder" by writing a note to be inserted in the next issue.

One reason for Dumas's propitiation of Liebig is revealed in this letter: he felt that he badly needed Liebig's influence to be elected to a new vacant position in the Académie des Sciences. To judge from Liebig's published comments, Dumas feared that his countrymen had gained the false impression that Liebig valued P. J. Robiquet's scientific work over his own, which might lead the Académie members to elect his rival for the seat. Dumas was convinced that Liebig's influence was determinative:

In the current state of chemistry in France, it is Germany that determines the opinion. . . . Of all the chemists I have met, it is you whose character and ideas inspire the most attachment in me. . . . From Germany, one cannot gain an idea of my position here. I am the only one in Paris who reads your papers. I cannot find anyone with whom to talk chemistry, for no one stays current with what is happening in the science.[2]

Liebig responded cordially.[3] In November Dumas published, as he had promised, a long footnote that corrected the errors and misleading impressions in his earlier response. He conceded some flaws in his alkaloid article of 1823; however, despite his conciliation of Liebig, he continued to maintain the superiority of his own procedures.[4]

Despite the artful dodging of this initial response, Dumas almost immediately adopted the Kaliapparat. When he wrote these letters he was still

2. Dumas to Liebig, n.d., postmarked 5 October 1831, Liebigiana IIB, Bayerische Staatsbibliothek, Munich: "Dans l'état où est la chimie en France, c'est l'allemagne qui forme l'opinion. . . . De tous les chimistes que j'ai rencontrés, vous êtes celui dont le caractère, les idées m'inspirent le plus d'attachement. . . . On ne se fait pas en allemagne une idée de ma position. Il n'y a que moi à Paris qui lise vos mémoires. Je ne puis trouver personne à qui parler chimie, car personne ne se tient au courant de ce qui se passe dans la science."

3. Liebig to Dumas, 23 October [1831], Archives de l'Académie des Sciences, Paris.

4. Dumas, *Annales de chimie* [2] 47 (1831): 324–325n.

Figure 2.1
Jean-Baptiste Dumas as a young man, ca. 1830. Source: E. F. Smith Collection,
University of Pennsylvania Library.

measuring carbonic acid by volume,[5] but in a paper published early in 1832 (within 4 months of the publication of his first reactions to Liebig's Kali-apparat method) he wrote that "for some time" he had been using Liebig's device "with full success."[6] In addition to the direct influence of Liebig, Dumas may also have been swayed by the fact that Jules Gay-Lussac, the son of Joseph Louis, studied with Liebig in the fall of 1831 and learned the new method from him. A Kaliapparat was in Paris as early as the beginning of December, brought there by Charles Oppermann, an Alsatian Liebigian, who taught Jules Pelouze the method.[7] Jules Gay-Lussac and Pelouze became the first Parisians to publish a paper (actually just a note, published at the end of 1831 or the beginning of 1832) in which the Kaliapparat was employed for analysis.[8] Dumas's laboratory bench at the École Polytechnique was next to Pelouze's, and he may have learned the method third-hand from his neighbor.

More details are revealed in a letter of January 1832 in which Dumas lamented his situation to Liebig.[9] Unlike Liebig, Dumas had no laboratory in his residence, and he had two workplaces, the École Polytechnique and the École Centrale des Arts et Manufactures, where he needed to be present every day. This created severe logistical difficulties. He lived in a house bordering the Jardin des Plantes, which was reasonably convenient to the École Polytechnique, also located on the Left Bank, near the Panthéon. But the École Centrale was located 3 kilometers to the north, near the Place de Thorigny, in the Marais on the Right Bank. At the time there were almost no good streets running perpendicular to the Seine; furthermore, nearly all

5. Dumas, "Recherches sur la liqueur des Hollandais," *Annales de chimie* [2] 48 (1831): 185–198.

6. "Depuis quelque temps . . . avec un plein succès . . ." (Dumas, "Sur l'esprit pyro-acétique," *Annales de chimie* [2] 49 (1832): 208–210).

7. Liebig to Pelouze, 27 November 1831, Dossier Pelouze, Archives de l'Académie des Sciences, Paris. "M. Oppermann [the bearer of this letter] a travaillé 2 ans dans mon laboratoire et vous pouvez de lui apprendre la méthode dont je me sers pour les analyses organiques plus parfaitement, que par la description que j'en ai fait. Je serais flatté si on reconnaîtrait son utilité à Paris."

8. Gay-Lussac and Pelouze, "Sur la composition de la salicine," *Annales de chimie* [2] 48 (1831): 111. The issue date was September 1831, but the actual date of appearance may not have been before close to the end of the year.

9. Dumas to Liebig, n.d. (postmarked 23 January 1832), Liebigiana IIB.

of the bridges crossing the river were in the hands of private entrepreneurs who charged hefty tolls.[10] The fact that Dumas's commutes to the École Centrale were on foot, and through dangerous neighborhoods, was noted by one of his students, and August Wilhelm Hofmann later mentioned the "extraordinary loss of time" experienced by Dumas's successor Wurtz in the latter man's similar daily commutes between the École Centrale and the Faculté de Médecine during the period 1845–1850.[11]

Commutes cost more than just time and money; they also slowed the pace of research. Any analysis in Dumas's lab at the École Polytechnique took a day for the combustion and another day for the accessory nitrogen determination. On the one hand, this meant that slow absorption of carbonic acid gas in potash solution (by the older method) was no disadvantage, for the apparatus had to sit overnight in mid-analysis anyway. However, he added, if an analysis could be done in 4 hours or less, this would provide a compelling reason for him to adopt it; moreover, working with gases inevitably caused "d'ennui et de fatigue" obviated by Liebig's gravimetric procedure. Consequently, Dumas indicated his intention to do analyses in Liebig's fashion and promised to give Liebig credit when he did so. Here again he did not omit the opportunity to lament what he characterized as the unenviable condition of chemistry in France, and the ignorance of cutting-edge foreign chemistry of "nos pauvres chimistes."[12]

It appears, then, that Dumas's strong attraction to the new method was due in part to his fractured professional life.[13] Historians have become accustomed to thinking of the French practice of cumul as a convenient and lucrative lifestyle for those who were lucky enough to cumulate positions,

10. City streets "crossing Paris were virtually impossible to use" (D. Jordan, *Transforming Paris* (Free Press, 1995), pp. 94, 112).

11. D. Colladon, *Souvenirs et mémoirs* (Aubert-Schuchardt, 1893), p. 191; A. Hofmann, "Erinnerungen an Adolph Wurtz," in Hofmann, *Zur Erinnerung an vorangegangene Freunde* (Vieweg, 1888), vol. 3, pp. 221–222.

12. Dumas to Liebig, n.d., postmark 23 January 1832, Liebigiana IIB.

13. Frank James has pointed to a parallel situation in the life of Michael Faraday. He has argued that Faraday's discovery of electromagnetic induction was only made possible by his release from consulting work at the firm of Pellatt and Green, some 3 miles distant from the Royal Institution, and his hire at Woolwich Arsenal, a much less time-consuming occupation. See James, "The Military Context of Chemistry," *Bulletin for the History of Chemistry* 11 (1991): 36–40.

but Dumas's testimony suggests otherwise. His life was so complicated that he needed the speed and simplicity of the Liebigian method.

In a long article published in March of 1834, Dumas discussed all the available methods for organic analyses, giving decided preference to those that weighed carbonic acid over those that measured its volume. The method "that merits preference in every respect is that which uses the ingenious absorption apparatus of M. Liebig." The potash-bulb device "simplifies organic analysis to such a degree, and gives results so precise, that it can be regarded as one of the most invaluable acquisitions which analytical chemistry has made for many years."[14] In a paper published in April, Dumas wrote:

The procedure to follow in all research on the composition of the organic bases has been very nicely traced in the remarkable paper in which M. Liebig described his invaluable condenser; here it will suffice to repeat some examples from this paper.[15]

As Dumas went, so went Parisian chemistry. The Gay-Lussac–Berzelius method of volumetric capture of carbonic acid died in the early 1830s.

Berzelius's Response

Berzelius read Liebig's paper in April of 1831, and studied it with care. He had time to insert a review of the article into his annual report for calendar year 1830, just then being printed, and he praised the new apparatus and the new analyses highly.[16] In his first letter to Liebig he focused entirely on the data, not mentioning the apparatus (which Berzelius presumably had not yet attempted to construct); he wrote that it was "quite inconceivable" to him how Liebig had been able to do so much in such a short time. But Berzelius was particularly interested in Liebig's critique of Dumas and Pelletier in the paper:

The unreliability of the French analyses, which is brought out so strikingly in your work, is a damned curious thing. For when one reads the paper of Dumas and

14. Dumas, "De l'analyse élémentaire des substances organiques," *Journal de pharmacie* [2] 20 (March 1834): 129–156. This was an advance publication of the text in *Traité de chimie appliquée aux arts* (Béchet, 1835), vol. 5, pp. 26–28.

15. Dumas, "Détermination du nombre d'atomes qu'une matière organique renferme," *Journal de pharmacie* [2] 20 (1834), p. 211.

16. Berzelius, untitled review, *Jahresbericht über die Fortschritte der physischen Wissenschaften* [for 1830] 11 (1832): 214–215.

Pelletier, considers the [nitrogen/carbonic acid] controls and the lovely agreement which they achieved in everything, it is clear that wherever Dumas is present, the results have been helped along with the pen, and when he couldn't figure out the correct answer by calculation he took a false one as his model.[17]

In a wide-ranging response to Berzelius's letter, Liebig touched on some other measurements recently published by Dumas, namely measurements of vapor densities:

As little as I trust Dumas's work, the calculations of this tightrope dancer seldom fail to meet the mark; I have become convinced . . . that Dumas is right; it always irks me that this fellow, in spite of his unclean, impossible and sloppy manner of working, nonetheless shakes masterpieces from his sleeve, for which, to be sure, his pen deserves the most credit.[18]

In 1832 Liebig reported to Berzelius that he had reduced the size of the Kaliapparat to the point that it could be weighed on any balance; he offered to send one to Berzelius. Berzelius gratefully accepted the gift, commenting: "You are at the moment certainly the greatest master in the art of carrying out precise organic analyses which we now have." Owing to various postal difficulties, it was more than 9 months before the gift arrived. Liebig had sent two Kaliapparate, of which only one survived the journey; Berzelius noted that this made little difference, since he could now blow as many as he liked using the one surviving specimen as a model.[19] About 18 months later, Berzelius wrote to Wöhler:

We are using Liebig's apparatus daily for these [investigations]. It is a magnificent instrument. By means of small insubstantial modifications we are now to the point that the result that one arrives at cannot be incorrect, and that one neither obtains water in excess nor loses any carbonic acid.[20]

By this time Wöhler had long since adopted the device. In November of 1831 he visited Liebig's lab in Giessen for 2 weeks of intense collaborative

17. Berzelius to Liebig, 11 February and 22 April 1831, in *Berzelius und Liebig*, ed. J. Carrière, second edition (Lehmann, 1898) (hereafter "Carrière"), pp. 6–7.

18. Liebig to Berzelius, 8 May 1831, Carrière, p. 11. J. Partington (*A History of Chemistry*, vol. 4 (Macmillan, 1964), p. 339) mistranslates an archaic form of the word "nevertheless" in this phrase (demongeachtet = demungeachtet) as "with the devil's help."

19. Liebig to Berzelius, 6 November and 22 December 1832, and 30 May 1833; Berzelius to Liebig, 27 November 1832, 15 January, 21 May, and 30 August 1833, in Carrière, pp. 43, 46, 49–51, 60, 66, and 68 (quotes on pp. 50–51, 68).

20. Berzelius to Wöhler, 20 March 1835, in Wallach, vol. 1, p. 609.

work. Liebig taught him the use of the new apparatus, which Wöhler reported to Berzelius as "superb" [ganz vortrefflich], and they carried out many analyses together. This was "a side of chemistry to which I was hitherto a complete stranger."[21] After his first experience of working directly with Liebig, Wöhler had even more respect for his friend:

He is, by the way, the best and most honest fellow in the world, and in chemistry has virtually unrivaled zeal. The days with him passed like hours, and I count them among the happiest of my life. His organic apparatus seems quite splendid to me; he is moreover a master of organic analysis and performs it with a pedantic exactitude. But as regards inorganic analysis, such as filtrations, use of the lamp, etc., one spots the imperfect French methods. He uses neither a filter stand, nor good filters, nor usually the lamp. He knew no better, but was immediately ready and happy to come over to the Swedish flag.[22]

The collaboration was so welcome to both men that Wöhler returned to Giessen 2 months later for another 2-week stint, and then again for 4 weeks in July and August of 1832.[23]

In April of 1831, Liebig became co-editor of Geiger's *Magazin der Pharmacie* (renamed *Annalen der Pharmacie* in January of 1832), which he then began to use as his personal publication organ. The first issue of the renamed journal contained no fewer than three contributions that mentioned analyses performed in the new manner.[24] Liebig's students had already published articles in other journals on studies in which the Kaliapparat had been used,[25] and the flow of such articles, by Liebigians

21. Wöhler to Berzelius, 24 November 1831, in Wallach, p. 381.

22. Wöhler to Berzelius, 1 December 1831, ibid., p. 387. Berzelius wrote Liebig (13 December 1831): "I am happy to learn that you have made Wöhler into a proselyte for organic analyses. He was always before somewhat disinclined toward this kind of work" (Carrière, p. 19).

23. Wöhler to Berzelius, 17 January and 19 August 1832, in Wallach, pp. 399 and 448.

24. C. Pfaff and J. Liebig, "Ueber die Zusammensetzung des Caffeins," *Annalen* 1 (1832): 17–20; Wöhler and Liebig, "Ueber die Zusammensetzung der Schwefelweinsäure," *Annalen* 1 (1832): 37–43; Pelouze and Gay-Lussac, "Ueber die Zusammensetzung des Salicins," *Annalen* 1 (1832): 43. All three papers emphasized the use of the Kaliapparat. On Liebig's takeover of the journal, see U. Thomas, "Philipp Lorenz Geiger and Justus Liebig," *Ambix* 35 (1988): 77–90.

25. "Herr Hess" was mentioned in Liebig's landmark article. See also C. Oppermann, "Einige vergleichende Versuche mit dem sogenannten Baumwachs und Bienenwachs," *Magazin für Pharmacie* 35 (1831): 57–64.

and others, continued in the new journal.[26] Robert Bunsen must have learned the method when he spent some days in Giessen just at the time (summer 1832) that Liebig was working with Wöhler on the oil of bitter almonds. Liebig was on holiday with his student Jules Gay-Lussac in Heidelberg in April of 1832; he met Mitscherlich and Magnus there, and also in Berlin 6 months later, and taught them the new method.[27] The community of Berzelians in Berlin—not just Mitscherlich and Magnus, but also Heinrich and Gustav Rose and other chemists—quickly adopted it. The relevant French chemists also converted, as we have seen. The English, not yet oriented toward organic chemistry in general or toward organic analysis in particular, were slow to move. Perhaps the last study in which the older method is known to have been used was one published by Pelletier in 1833.[28]

The transmissions in the year after the publication of Liebig's method were of two different types. One was bodily communication from the hand of the master (or from a surrogate master), such as Liebig directly to Opperman (whence to Pelouze, Dumas, and Gay-Lussac), or directly to Jules Gay-Lussac, to Wöhler, to Bunsen, to Mitscherlich, or to Magnus, among many others. The second (less frequent) was long-distance transmission, as happened with Berzelius. It is interesting to note that even the supreme master Berzelius waited until he had an authentic Kaliapparat blown by Liebig before he proceeded to reproduce the method; of course, at that time he had only the sketchy report in the 1831 article at hand, for no detailed description had yet appeared. Once Berzelius had the glassware, it did not take him much time with the device before he was able consistently to produce results that "cannot be incorrect," without ever having learned the technique directly from someone who had mastered it.

26. C. Ettling, "Beiträge zur Kenntniss des Bienenwachses," *Annalen* 2 (1832): 253–267; Wöhler and Liebig, "Untersuchung über das Radikal der Benzoesäure," *Annalen* 3 (1832): 249–282; R. Blanchet and E. Sell, "Ueber die Zusammensetzung einiger organischer Substanzen," *Annalen* 6 (1833): 259–313; A. Boutron-Charlard and T. Pelouze, "Mémoire sur l'asparamide et sur l'acide asparamique," *Annales de chimie* [2] 52 (1833): 90–105.

27. See Carrière, pp. 41–43; Wallach, p. 431.

28. J. Pelletier, "Untersuchung über die elementare Zusammensetzung mehrerer näheren Bestandtheile der Vegetabilien," *Annalen* 6 (1833): 21–34.

The same has been true of modern historically sensitive reproductions of the Kaliapparat method.[29] The fact that one can learn quickly and relatively painlessly how to produce astonishingly good results from a mere printed description suggests much about the intrinsic characteristics of the method itself. It seems that not too much intangible, tacit, and bodily "gestural knowledge" of the kind so well investigated by Otto Sibum and by Simon Schaffer was needed to work the device.[30] If one could blow the complex shape of the glass tube, and if one had access to a sufficiently complete description of the process (as everyone did after 1837), one did not need personal instruction. And all of this is implicit testimony to the ingenious simplicity of the device, and to the method that was built around it.

Other than his reduction of the size of the Kaliapparat to facilitate weighing, Liebig retained the original apparatus and method virtually unchanged throughout the 1830s; this is clear not only from the essential identity of the two-page description of it in January of 1831 with the detailed monograph that appeared 6 years later, but also from his explicit statements and those of his friends.[31] By March of 1833 Liebig was prepared to expand his claims about the accuracy achievable with his appa-

29. At the University of Western Ontario, Melvyn Usselman and Christina Reinhart have recently performed Kaliapparat combustions, following Liebig's 1837 instructions with care. Their results are even more precise than Liebig himself claimed for the method. This research will be published soon. Similar modern experiments with similar results are related in W. Conrad, Justus von Liebig und sein Einfluss auf die Entwicklung des Chemiestudiums und des Chemieunterrichts an Hochschulen und Schulen (Ph.D. dissertation, Technische Hochschule Darmstadt, 1985), p. 17.

30. O. Sibum, "Reworking the Mechanical Value of Heat," *Studies in the History and Philosophy of Science* 26 (1995): 73–106; Sibum, "Working Experiments," *Cambridge Review* (May 1995): 25–37; S. Schaffer, "Experimenters' Techniques, Dyers' Hands, and the Electric Planetarium," *Isis* 88 (1997): 456–483. These circumstances also appear to work against H. Collins's extended critique of experimental replication in *Changing Order* (Sage, 1985).

31. In a letter to Wöhler of 9 February 1837, Liebig stated that the only change made from the original apparatus of 1831 was the use during the final aspiration of a diagonal glass tube inserted over the broken end of the combustion tube, to prevent CO_2 contamination from the charcoal fire (A. Hofmann, ed., *Aus Justus Liebig's und Friedrich Wöhler's Briefwechsel in den Jahren 1829–1873* (Vieweg, 1888), vol. 1, p. 99). Wöhler wrote Berzelius (12 February 1837) that it was his understanding that Liebig still used the apparatus "in its original configuration and simplicity" (Wallach, p. 674).

ratus.[32] In fact, he made the "very unfortunate discovery" that most of the analyses performed before the Kaliapparat was introduced were inaccurate, his own as well as others', for he found that the volumetric procedure, for mysterious reasons, tended to give a slightly larger value for carbonic acid than the Kaliapparat did. Among other consequences, he was now forced to admit that his pre-1830 analyses of turpentine, uric acid, indigo, picric acid, and camphor were wrong, and Dumas's were right.[33] In exculpation of their teacher on this occasion, Liebig's students Blanchet and Sell commented that only now does science possess an analytical method—Liebig's—which always gives the same results, no matter who undertakes the analysis.[34] Liebig himself often emphasized this point, as well, only taking care to remark that one must always operate on fully purified and homogeneous samples; and that analysis is merely a tool of the scientific investigator, and no number of mere analyses strung together constitute a proper research program.[35]

Dumas's Career Breakthrough

We have seen that from the beginning of 1832 Dumas enthusiastically adopted the use of the Kaliapparat. However, Dumas—like Liebig—always emphasized the urgent need for an accurate method for nitrogen, which had so confounded the best analysts of his day. By early 1833 he had devised a modification of the existing procedure that constituted a notable

32. F. Holmes ("The Complementarity of Teaching and Research in Liebig's Laboratory" (*Osiris* [2] 5 (1989), p. 143) states that Liebig made no claims regarding higher accuracy until this time. I had no more success than Holmes in finding an explicit statement of this kind from Liebig earlier than 1833. Nonetheless, Liebig must have been conscious that his procedure was intrinsically more precise than the volumetric approach.

33. Liebig to Berzelius, 15 March, 30 May, and 14 September 1833, Carrière, pp. 52, 62, and 71; Berzelius to Liebig, 30 August 1833, Carrière, p. 67. Berzelius commented: "It's good to have a confirmation of Dumas's analyses, for it is impossible to rely on D.'s results, since he so often lets himself be misled by theoretical predictions, and never publicly confesses his errors."

34. Blanchet and Sell, "Zusammensetzung," pp. 304–305.

35. Liebig, ed., *Annalen* 22 (1837): 50–52; Liebig, "Ueber die vorstehende Notiz des Hrn. Akademikers Hess in Petersburg," *Annalen* 30 (1839): 313–319. Holmes has rightly emphasized this point in his discussion in "Liebig's Laboratory."

advance.[36] Dumas placed lead carbonate in the back of the combustion tube, with the sample and the oxidizer arranged as usual. Application of a vacuum followed by strong heating of the carbonate resulted in a surge of carbonic acid gas through the tube which flushed out all vestiges of residual air. The tube was again placed under vacuum, and the combustion was carried out as usual. The nitrogenous gases produced in the combustion were reduced by passing over hot activated metallic copper. Finally, another surge of carbonic acid from heated carbonate ensured that any residual sample nitrogen was flushed out of the tube into the collector. The collector was a graduated bell jar in a pneumatic trough, whose mercury was topped by a layer of potassium hydroxide solution. The potash absorbed all the carbonic acid from the carbonate, as well as all the carbonic acid and water vapor from the sample, leaving (in principle, at least) the nitrogen alone to be measured volumetrically.

By this procedure Dumas was confident of eliminating even the smallest traces of contaminant nitrogen from residual air in the sample, and of quantitative collection of pure nitrogen from the sample. He emphasized that this was the first nitrogen method that yields results "with a rapidity and certainty at least equal to that now obtainable" for carbon and hydrogen (by which he meant with the Kaliapparat). Previous methods have been "radically inexact," Dumas asserted—rather contradicting his confidence of two years earlier.[37] Dumas's nitrogen method could be, and almost immediately generally was, coupled with the Kaliapparat procedure to provide an excellent two-step analytical method for elemental analysis of nitrogenous organic compounds. But the Dumas nitrogen method in its original form was not particularly simple to carry out, and despite the schematic clarity, contamination of sample with air and product with oxides was hard to eliminate. For these reasons, the method did not displace its competitors until it was gradually simplified. The systematic errors of the method were

36. Dumas, "Recherches de chimie organique," *Annales de chimie* [2] 53 (1833): 164–181. The paper was read on 5 August 1833, but Dumas commented that he had earlier taught the new nitrogen procedure to Pelouze and Boutron-Charlard, who used it in their paper on aspartic acid (read 11 March 1833). This was the first publication of Dumas's innovative method for nitrogen; most citations in the secondary literature incorrectly refer to Dumas's letter to Gay-Lussac published in the *Annales de chimie* in 1831, which discusses only the older method.

37. Ibid., pp. 171–172.

still troublesome as late as the beginning of the twentieth century, even though by then it was long since acknowledged as the best general method of its type.[38]

Dumas's fundamental work on organic analysis coincided with a period in which he rapidly rose through the hierarchy of the Parisian elite. In 1826 he had published a landmark paper on vapor densities, drawing important implications for the atomic theory, and from that time onward he was a principal promoter of organic chemical theory. Hard-working, prolific, eloquent, persuasive, and ingenious, as well as a superb experimentalist, Dumas quickly became famous by a profusion of chemical studies of all kinds. He also became ever more socially well-connected, through the good offices of his patron Louis Jacques Thenard. His financial situation was eased on 18 February 1826, when he married Herminie Brongniart, daughter of the wealthy naturalist and factory owner Alexandre Brongniart; the couple immediately moved into Brongniart's roomy apartments at the Muséum.[39] About this time Dumas began writing what would be an outstanding and voluminous *Traité de chimie appliquée aux arts* (eight volumes, 1828–1846). It is said that the early volumes of this work were based on his lectures in the Athénée; the theoretical portions are of extraordinary interest, and to this day have not been sufficiently appreciated. In 1829 he added an additional duty (as mentioned earlier), namely professor at the École Centrale des Arts et Manufactures, a private school of applied sciences which he co-founded that year with Théodore Olivier, Alphonse Lavallée, and Eugène Péclet.[40] In 1830, much overworked, he resigned his position at the Athénée.

38. V. Meyer and P. Jacobson, *Lehrbuch der organischen Chemie*, second edition (Veit, 1907), vol. 1, pp. 21–26.

39. On Dumas and the Brongniarts, see L. de Launay, *Une grande famille de savants, les Brongniart* (Rapilly, 1940), pp. 140–150. Launay states that Dumas suffered heavy debts for many years, which would help to explain his powerful ambition for multiple positions.

40. C. de Camberousse, *Histoire de l'École Centrale des Arts et Manufactures* (Gauthier-Villars, 1879), pp. 11–36. The founders were concerned about the supremacy of English technology and trades; their goal was to provide training for civil engineers, parallel with the military engineering tradition of the École Polytechnique. Capital was provided by Lavallée, along with a luxurious seventeenth-century mansion, the Hôtel de Juigné on the Rue des Coutures-Saint-Gervais. An excellent modern monographic treatment of the École Centrale is J. Weiss, *The Making of Technological Man* (MIT Press, 1982).

Figure 2.2
The École Centrale des Arts et Manufactures was housed in the Hôtel Juigné,
a.k.a. Hôtel Salé (1626), in the Marais, from its founding in 1829 until 1884.
Today this building is the home of the Musée Picasso. Photograph taken in March
1999 by the author.

Neither of Dumas's two positions at the beginning of the 1830s, répétiteur at the École Polytechnique and professor at the École Centrale, was
very remunerative, and his professional position was not yet stable. He
was much distressed by what he saw as the inattention of French chemists
to modern chemistry, particularly organic, and by the monopolization of
remunerative positions by a few powerful figures. He wrote to Liebig:

My existence is difficult in Paris, by the nature of my character. I never leave my
laboratory or my study. I do not ask anything from anyone at all. Such habits of life
result in my current situation of having a position [as répétiteur at the École
Polytechnique] worth 2000 francs for my entire income. I am absolutely resolved
never to kowtow to anyone to obtain a better position. The result is that I will long
be without one, and that intrigue has an upper hand in supplanting me. So do not
be surprised at anything, if in the future you hear about me in matters that you may
regard as examples of favoritism. I have made up my mind. Cumul is killing science in France, and to correct the abuses new chairs are being created in favor of
those who have nothing. I will never open my mouth against those who cumulate

positions, nor will I ever consent to the creation of a position in favor of my worthless self. The poor people have too many taxes already.[41]

Dumas's professional situation was on the brink of dramatic alteration even as he wrote these words. During a period of only 6 months in late 1831 and early 1832, he satisfied all the requirements for a medical doctorate, and the same year he qualified for a science doctorate. In addition to his two doctorates he was also elected to the Académie des Sciences—despite his fears expressed to Liebig—and was a candidate for a professorship at the Muséum d'Histoire Naturelle. As it happened, he lost out to Gay-Lussac in this position, but Gay-Lussac felt obliged to resign one of his two other professorships in order to accept the position at the Muséum. The position Gay-Lussac vacated was professor of physics at the Sorbonne, which then went to Dulong. Dulong had been agrégé, and his promotion made room for another junior scholar at the Faculté des Sciences. Baron Thenard, as Dean of the Faculté and a member of the Conseil Académique, had the clout to ensure his protégé's nomination. Thus was Dumas appointed agrégé (1832), then professeur adjoint (1836), and finally professeur de chimie (1841).

Even the amazing Dumas was not able to win all these important degrees and candidacies in so short a time without a good deal of help, and there is no doubt that he received it, both officially and unofficially. His most important supporters were his father-in-law Brongniart, and Baron Thenard. When Thenard retired from the École Polytechnique in 1836, Dumas was promoted from répétiteur to professeur, adding yet another

---

41. Dumas to Liebig, n.d., ca. November or December 1831, Liebigiana IIB: "Mon existence est difficile à Paris, par la nature de mon caractère. Je ne sors pas de mon laboratoire ou de mon cabinet. Je ne demande jamais rien à qui que ce soit au monde. Cette habitude de vie fait que je suis là avec une place de 2000 fr. pour tout potage. Je suis parfaitement résolu à ne jamais plier le dos devant personne, pour obtenir une place meilleure. Il en résulte que je serai longtemps sans en avoir et que l'intrigue a beau jeu pour me supplanter. Ainsi, ne vous étonnez de rien, si l'avenir vous rend témoin en ce qui me concerne de choses que vous regarderez peut-être comme des passe-droits. J'en ai pris mon parti. Le cumul tue les sciences en France, pour en corriger l'abus, on crée de nouvelles chaires en faveur de ceux qui n'ont rien. Jamais je n'ouvrirai la bouche contre ceux qui cumulent, jamais je ne consentirai à la création d'une place en faveur de mon chétif individu. Le pauvre peuple a bien assez d'impôts sans cela."

centrally important position to his portfolio. From 1834 Thenard had invited Dumas to replace him periodically at the Collège de France, where Dumas's brilliant historical and theoretical *Leçons de philosophie chimique*, printed in 1836, added to his luster.[42] In short, in the early to mid 1830s Dumas was converted quickly from an up-and-coming young man to a leading member of the Parisian academic chemical elite.

Liebig therefore was not alone in closely following the rise of his sometime friend and sometime rival. No one was more proud and more confident of analytical skill and reliability than Liebig, with good reason; and yet Liebig was astonished and abashed to admit on several occasions in the early 1830s that the "fast and sloppy" Dumas had published better analyses than he. When Liebig realized once more that Dumas's analysis of chloral was superior to the best that he could produce, Liebig discovered one source of Dumas's legerdemain. Dumas, unlike Liebig, was always careful to supplement gravimetric analysis with vapor density determinations, which he had pioneered, and which provided a valuable control. "I have come to the conclusion," Liebig wrote to Berzelius, "that a chemist who neglects [vapor densities] from laziness (as I did) gives the advantage to those who do not. . . . I am now wiser; I am convinced that Dumas's conclusions, although not analytical [i.e., gravimetric], cannot reasonably be improved upon, and I will not hesitate to admit this publicly."[43] He reiterated to Pelouze that vapor densities were a "perfect control, extremely simple and expedient," and he now recognized them as a necessary adjunct to gravimetry.[44]

In an unplanned international collaboration, Gay-Lussac and Berzelius had invented the first practical method of combustion analysis of organic compounds in the period 1811–1815. In a curiously parallel haphazard international collaboration during the years 1826–1833, Dumas and Liebig had vastly improved it: Dumas by methods for determining vapor densities and nitrogen content, and Liebig by the invention of his Kaliapparat for

42. M. Chaigneau, *J. B. Dumas* (Guy le Prat, 1984), pp. 106–114; Crosland, *Gay-Lussac* (Cambridge University Press, 1978), pp. 156–158; Crosland, *The Society of Arcueil* (Harvard University Press, 1967), pp. 441–444.

43. Liebig to Berzelius, 22 July 1834, Carrière, p. 93.

44. Liebig to Pelouze, 14 June 1834, Dossier Dumas, Académie des Sciences, cited in Holmes, "The Complementarity of Teaching and Research in Liebig's Laboratory," p. 65.

carbon content. Both learned from the other, and adopted many of the other's techniques. So did everyone else in European chemistry. The future would soon demonstrate the extraordinary fecundity of these analytical innovations.

The Standardization of Organic Analysis

Until 1837, Liebig's Kaliapparat method had been communicated largely by direct means, one chemist showing another the procedures in chains of transmission that led back to Liebig himself. Despite the inefficiency of such transmission, by the mid 1830s his new device was widely employed by European analysts, especially in Germany and France. Berzelius provided a fine summary of the "state of the art" in his classic textbook of chemistry. In a section on organic analysis written around 1835 and published in 1837, he characterized Liebig's procedure as leaving "nothing to be desired," and stated that it had brought elemental analysis to such a state of simplicity that the procedure was "one of the easiest operations," the previous methods being no longer of any but historical interest. In fact, the simplicity and low skill demands of the Kaliapparat led Berzelius to predict a swell of very bad analyses—bad not because of the analytical procedure itself, which was unproblematic even in unpracticed hands, but because the preliminary steps of identification and purification were not properly carried out. In other words, Berzelius was suggesting that the challenging and problematic character of the earlier methods had had the beneficial concomitant that only sophisticated and conscientious chemists performed analyses at all; this internal protection now no longer applied. In contrast to Berzelius's extravagant praise of Liebig, he simply listed Dumas's nitrogen method among several alternatives, including Liebig's own version.[45]

Despite Berzelius's praise of Liebig, publication of this section of his textbook caused consternation in Giessen. Since Liebig had published nothing in detail on the method—his 1831 paper was sketchy—Berzelius had been

45. Berzelius, *Lehrbuch der Chemie*, third edition (Arnold, 1837), vol. 6, pp. 49, 55, 60–63. That this section was written in 1835 can be verified by reading the relevant passages in the Berzelius-Wöhler correspondence, for Wöhler was still serving Berzelius as translator and editor. As was the case for the second edition, the German third was the editio princeps of this important work.

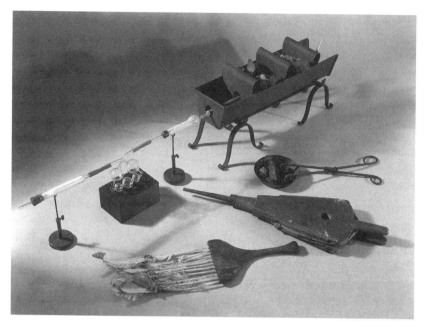

Figure 2.3
Liebig's combustion train, ca. 1840, with accessory implements. The first calcium chloride tube (center) was designed to capture all the water produced in the combustion. The second calcium chloride tube (left) was optional; weighed together with the potash-bulb device, it ensured no loss of water from the potash solution. The upturned end of the combustion tube (upper right) was to be snipped off at the end of the combustion. Source: Photograph Collection, Deutsches Museum, Munich.

forced to turn to a detailed discussion that had just appeared in Eilhard Mitscherlich's textbook.[46] But Mitscherlich and Liebig were not friendly with each other, and Berzelius's innocent adoption of a number of Mitscherlich's modifications gave Liebig a bad case of heartburn. (Berzelius had been Mitscherlich's mentor; he was aware that Liebig had taught Mitscherlich the Kaliapparat method directly, and assumed that Mitscherlich's modifications had been devised by Liebig; he also underestimated the depth of enmity between the two men.) Liebig's secretiveness may well have been intentional, to retain a degree of monopoly on the method, but he could now see that his tactic (if such it was) had become self-defeating.

46. E. Mitscherlich, *Lehrbuch der Chemie*, second edition (Mittler, 1834), vol. 1, p. 205–210. According to the preface, this volume was printed in November 1832.

These events led Liebig to decide that it was time for a full account of the method from his own pen, and he was given the appropriate occasion when he had the opportunity to write an article on "Analysis, organic" for a chemical handbook he was helping to edit. The article appeared in the third fascicle of the first volume of the handbook, published in late summer 1837, but an offprint was available as early as April of that year. Liebig assured the publisher that the fascicle would sell very well, since he was, as he put it, giving away all his secrets in it.[47] Liebig's article was really a small monograph, a classic of analytical chemistry. It is filled with descriptions of ingenious auxiliary devices and procedures, all designed to increase simplicity, reliability, and accuracy.

After the publication of this monograph, anyone with minimal chemical expertise could fashion the devices and perform the operations from the detailed instructions readily available in bookstores. The essential superiority of the method and its now openly accessible details ensured that it would be employed anywhere chemistry was done. The method can be found in German as well as non-German textbooks throughout most of the rest of the century, little altered from Liebig's description of 1837.[48]

The method was initially slow to spread to Great Britain, largely because British chemists generally failed to participate in the ferment in chemistry (especially organic) of the 1820s and the 1830s. Prout, who published his last analysis in 1827, was still dismissive about the Kaliapparat 10 years later.[49] The respected London chemist J. F. Daniell provided an additional unintentional demonstration of how out of touch the English had become

47. Liebig, "Analyse, org.," in *Handwörterbuch der reinen und angewandten Chemie* (Vieweg, 1836–1842), vol. 1, pp. 357–400 (third fascicle, 1837); Liebig, *Anleitung zur Analyse organischer Körper* (Vieweg, 1837); Liebig, *Instructions for the Chemical Analyses of Organic Bodies* (R. Griffin, 1839). For precise publication information, see M. Schneider and W. Schneider, "Das 'Handwörterbuch' in Liebigs Biographie," in *Orbis pictus*, ed. W. Dressendörfer and W.-D. Müller-Jahnke (Govi-Verlag, 1985). The Liebig citation is from his letter to Eduard Vieweg of 11 January 1837 (ibid., pp. 251–252). The *Handwörterbuch*, a landmark scientific publication, was sold piecemeal in fascicles over the course of many years; this has led some scholars to misdate the various volumes and the articles within them.

48. For example, C. Fresenius, *Anleitung zur quantitativen chemischen Analyse*, third edition (Vieweg, 1853), pp. 342–391; G. Fownes, *A Manual of Elementary Chemistry, Theoretical and Practical*, tenth edition (Lea, 1874), pp. 448–457.

49. W. Brock, *From Protyle to Proton* (Hilger, 1985), pp. 18–20.

by describing the Gay-Lussac–Berzelius method as standard, with no mention of Liebig or Dumas, in his 1839 textbook.[50] That same year William Gregory declared that "for a good many years previous to 1836, no organic analyses were published by any British experimenter"; in 1840, two English chemists working in Berlin averred more generally: ". . . but little of what is done abroad, especially in Germany, seems to find its way into England, or, at least, until after the lapse of some years."[51] These sentiments were shared by many foreign observers. On the approaching death of Humphry Davy, Berzelius commented: "England could not bear to lose Davy and [William] Wollaston simultaneously; her science is on a much lower level than many other countries. Faraday and Turner, despite all their virtues, are far from being able to replace the deceased." On Edward Turner's new textbook, Berzelius wrote 2 months later:

What abominable rubbish it is; no knowledge of anything that was not done in England by Englishmen, and with the imperious tone that accompanies true ignorance. . . . It is strange to see how far the French and English chemical literature falls short of the German. I would never have believed it.[52]

The deficiencies began to be addressed in the late 1830s, largely owing to Liebig's impetus. Several British chemists studied in Giessen starting around 1835, including Thomas Thomson, Jr., W. C. Henry, R. Kane, William Gregory, Lyon Playfair, and John Stenhouse. Upon returning to their homeland, their professional efforts (writing of books and papers, editing of journals, teaching students, and the founding of the Chemical Society and the Royal College of Chemistry), resulted in a quickening of English chemistry, especially in the organic realm. Liebig's influence in Britain was also much fostered by his visit to England, Ireland, and Scotland, for 2 months in the summer of 1837.[53]

50. J. Daniell, *An Introduction to the Study of Chemical Philosophy* (Parker, 1839), pp. 316–317. There is a fuller discussion in Daniell's second edition (1843, pp. 606–610), in which Liebig's method was introduced as "the more usual process." But Prout's method is also described, with no clear preference given.

51. Gregory, "Translator's Preface" to *Instructions*, p. iii; W. Francis and H. Croft, letter of 13 December 1840 from Berlin, *Philosophical Magazine* [3] 18 (1841): 202.

52. Berzelius to Wöhler, 1 May and 10 July 1829, in Wallach, pp. 253 and 267.

53. See J. Volhard, *Justus von Liebig* (Barth, 1909), pp. 131–149; W. Brock and S. Stark, "Liebig, Gregory, and the British Association," *Ambix* 37 (1990): 134–147.

Thomas Thomson described Liebig's method in his *Chemistry of Organic Bodies* of 1838. He wrote: "Professor Liebig has published this article as a separate pamphlet. Were any person to favour us with an English translation of it, he would contribute essentially to promote the prosecution of vegetable chemistry in Great Britain, and would be conferring an important favour on the British chemical public."[54] Liebig's student Gregory supplied this desideratum (prefaced with an apology for its appearance 2 years after the original). This English version was widely read. It was echoed shortly thereafter by Thomas Graham's treatment of the Liebigian method in his widely influential textbook.[55] The same year this last account appeared, Gregory published a polemic about the sad state of British chemical education, pointing out that British students were generally going to study with Liebig or Wöhler in order to find there what they could not obtain at home.[56] But Gregory and several of his German-educated chemical friends had started a process that soon raised the level of British chemistry markedly.

A French translation of Liebig's monograph also shortly appeared; this was published together with a critical essay by the peripheral and iconoclastic French organic chemist F. V. Raspail, but Raspail's opprobrium was directed not to the analytical method, which in fact he praised highly, rather to Liebig's theoretical inclinations.[57] Thereafter, as in England, the Kaliapparat method entered the mainstream French chemical literature. We have seen that Dumas himself recommended it wholeheartedly in two articles published in 1834; in the 1840s all of the principal French chemical textbooks strongly urged its use. The fact that two of Liebig's research students, Regnault and Pelouze, were authors of perhaps the two most prominent French textbooks of the day, did not hurt.[58]

54. T. Thomson, *Chemistry of Organic Bodies* (Bailliere, 1838), p. vi.

55. T. Graham, *Elements of Chemistry* (Bailliere, 1842), pp. 698–708.

56. W. Gregory, *Letter to the Right Honorable George, Earl of Aberdeen . . . on the State of the Schools of Chemistry in the United Kingdom* (Taylor & Walton, 1842), esp. pp. 18–22 and 28–29. See also R. Bud and G. Roberts, *Science versus Practice* (Manchester University Press, 1984).

57. Liebig, *Manuel pour l'analyse des substances organiques* (Baillière, 1838).

58. Pelouze and Frémy, *Cours de chimie générale* (Masson, 1848–1850): "The simplest of all analytical methods, and one which has replaced those we have just discussed, is due to M. Liebig. . . . [It] is as simple as it is exact; it has much contributed to the very great progress made by organic chemistry in recent years" (vol. 3, pp.

The Art of Analysis in International Perspective

In Germany there was never any doubt about the immediate dominance of the Liebigian method, which acquired virtually mythic proportions among chemists soon after its introduction. Carl Löwig wrote in 1844 of the "new epoch" thereby introduced into chemistry, and such rhetoric was typical of the German textbook writer of this period.[59] Some of Liebig's students were in the habit of sporting five-bulbed lapel pins as a badge of honor.[60] Many chemists in the nineteenth century cherished their "potash bulbs," sometimes employing a single apparatus for 20 years or more.[61] Forty-four years after the event, Hofmann called it the greatest of Liebig's discoveries, and the "main source" of the rise of chemistry, "one of the chief glories of our age."[62] Depictions of the Kaliapparat figure prominently in the logo of the American Chemical Society and on the jacket of the journal of the Fachgruppe Geschichte der Chemie of the Gesellschaft Deutscher Chemiker. Although organic analyses today are done by specialized commercial laboratories rather than by individual research chemists, the method is essentially the same as Liebig's.

Liebig's Kaliapparat method represented a complete transition to gravimetry and thus to an essentially *chemical* methodology. Gay-Lussac was a chemist in the fullest sense, but he made his initial reputation in the physics of gases and of heat; he taught physics in the Paris Faculty of Sciences from its origin in 1809 until 1832, and was in the physics section of the Académie des Sciences. Eudiometry, a method of measurement associated as much with

---

49–60); Regnault, *Cours élémentaire de chimie*, second edition (Masson, 1849–50), vol. 4, pp. 7–17. Regnault's work was translated into German and edited by Adolf Strecker; it was perhaps the most popular organic chemical textbook in Germany in the 1850s.

59. Löwig, *Chemie der organischen Verbindungen*, second edition (Vieweg, 1846), vol. 1, p. 141 (preface dated September 1844).

60. O. Krätz and C. Priesner, *Liebigs Experimentalvorlesung* (Verlag Chemie, 1983), p. 11.

61. One of Adolphe Wurtz's students wrote: "I myself used one given me by Wurtz for twenty years [ca. 1851–1871], and it almost broke my heart when it finally cracked in my hands." (Scheurer-Kestner to Hofmann, in Hofmann, "Wurtz," pp. 224–225.)

62. Hofmann, "The Life-Work of Liebig," in *Erinnerung*, vol. 1, p. 229.

physics as with chemistry, was more natural to him than the gravimetric procedure, which was canonically chemical; Crosland speculates that we may see here "a late strand of French Cartesianism," in which "extension" is instinctively regarded as the proper measure of matter.[63]

Berzelius was quickly persuaded of the merits of Gay-Lussac's volumetric approach. Nonetheless, we have seen that a fully chemicalized procedure—carbonic acid captured in condensed phase by a chemical reaction, the formation of carbonate, rather than physically as a gas—had essential advantages. It is also very curious to note that Liebig's chemical procedure had such advantages over the physical method as regards accuracy and precision, since in this century it was precisely a turn back to physics that provided a quantum leap in analytical accuracy.

Liebig's method fit its context well in other respects. From his student days, Liebig planned a career in which he would instruct *groups* of students in pharmaceutical and chemical practica. The Kaliapparat method was superbly suited to this pedagogical strategy, which interacted strongly with his research program, and developed into a characteristic German style of group research.[64] As we will see, French chemists did not develop research groups of the German type until the late nineteenth century, and so were not able to exploit the method in the way that Liebig and other German chemists did. Furthermore, Dumas's method for nitrogen, which has some parallels to Liebig's for carbon, had intrinsic limitations and never achieved the kind of mass-production simplicity that use of the Kaliapparat afforded.

There is a purely theoretical "fit" as well, that likewise worked better in Germany than in France. The Kaliapparat method developed hand in hand with chemical atomism.[65] For one thing, it was the inability to arrive at

63. Crosland, *Gay-Lussac*, pp. 92–114, 117–118 (quote on p. 93). On the Cartesian tradition in French chemistry, see H. Metzger, *Newton, Stahl, Boerhaave et la doctrine chimique* (Alcan, 1930), pp. 19–26.

64. Holmes, "Liebig's Laboratory."

65. Ursula Klein has argued convincingly that it was precisely in these years (1827–1834) that chemical formulas were first used by Liebig and Dumas as "paper tools" in a fashion that immediately caught on in organic chemistry, and that they caught on precisely because of their great heuristic value. Consequently, Klein sees a true revolution in this development. Klein's compelling viewpoint is complementary to the argument made here. See also Klein, Experimente, Modelle, Paper-Tools, Habilitationsschrift, Universität Konstanz, 1999; "Paving a Way through the Jungle of Organic Chemistry," in *Experimental Essays*, ed. M. Heidelberger and F. Steinle

secure atomistic formulas for the alkaloids that was a major reason for Liebig's unhappiness with earlier analytical methods, and the proximate incentive for developing a new one. For another, the emergence of the phenomenon of isomerism just at this time posed another sort of challenge for elemental analysis; atomic theory provided a route out of the conundrum. Chemical atomism of the Berzelian type influenced German chemists quickly and decisively. Already by Liebig's Paris years it was becoming influential, and by the time of Liebig's own conversion to Berzelian atomic weights—in the very year he invented the apparatus—it was dominant. These events were undoubtedly connected.[66] The French response to Berzelian atomism was more muted and ambivalent. Although there was a prevailing school of chemical atomists in Paris by the 1820s, these ideas were vitiated by continuing rhetoric of "volumes" as a putative synonym for "atoms" and by a strong anti-metaphysical tendency that made atomistic pronouncements seem suspect in tone and epistemological character. It is reasonable, therefore, to believe that there may have been a greater sense of urgency in Liebig's mind than in Dumas's.

The introduction of the Kaliapparat serves as a convenient benchmark for the gradual passing of chemical leadership from France to Germany. Within a year or so of this invention, Liebig met Berzelius for the first time and began corresponding with him, converted from equivalents to atomic weights, became personally and professionally close to the German Berzelian school, and acquired editorship of a journal that permitted him to relinquish the *Annales de chimie* as his preferred publication organ. All these factors loosened his ties to Paris, at the same time that he and his compatriots began to flex their own scientific muscles. Indeed, he and the German Berzelians (Wöhler, Magnus, Heinrich Rose, Gustav Rose, and C. G. Gmelin) were all publishing prolifically in the early 1830s.

Some leading active French chemists were simultaneously becoming worried about the state of health of the science in their country. In his earliest

---

(Nomos, 1998); "Paper-Tools and Techniques of Modelling in Nineteenth-Century Chemistry," in *Models as Mediators,* ed. M. Morgan and M. Morrison (Cambridge University Press, 2000).

66. It is easy to see why Gaston Bachelard could refer to instruments as "materialized theories." There is now a substantial literature on the "co-generation" of experimental and theoretical entities.

extant letter to Liebig, dated 1832, Pelouze fretted about his ignorance of German, and vowed to study the language daily—a resolution that quickly faded. "Nothing new is being done in Paris, at least as far as I know," he reported. "It seems as though discoveries and truly useful facts are personally reserved for you. We are all astonished at your prodigious activity and at the experiments with which you never cease to enrich chemistry." In 1833 he asked Liebig for scientific news, "in which Germany is so rich today," and in another letter wrote, "In the chemical world, we no longer speak of anything in Paris but your experiments." A year after that he described the universal admiration for Liebig's "prodigious fecundity" shared by all the chemists "of our poor country." [67]

Later in 1834, Pelouze lamented: "In terms of chemistry, we have absolutely nothing new in France that you don't know about. All light now comes from Germany, and to my great misfortune I don't understand your language." That Pelouze was not simply insincerely currying favor with Liebig is indicated by the fact that he wrote in nearly identical terms to Wöhler, with whom he was not close nor from whom he could expect favors.[68] Pelouze may have imbibed his respect for the Germans from his mentor. In a letter dated September 1826, Gay-Lussac informed Liebig that he had nearly accepted the latter's offer to welcome his son to Giessen, for the purpose of learning German; he added that he had insisted that *all* his children, even his daughters, learn the language.[69] Gay-Lussac was one of

67. Pelouze to Liebig, 19 June 1832, 8 March 1833, 14 June 1833, and 11 April 1834, Liebigiana IIB, Bayerische Staatsbibliothek: "On ne fait rien de nouveau à Paris, au moins que je sache. Il parait que les faits vraiment utiles et les découvertes vous sont personnellement réservés. On est étonné partout de votre prodigieuse activité et des expériences dont vous ne cessez d'enricher la chimie" (19 June 1832). "On ne parle plus à Paris, dans le monde chimique, que de vos expériences" (8 March 1833).
68. Pelouze to Liebig, 20 November 1834, ibid.: "En fait de chimie nous n'avons absolument rien de nouveau en France que vous ne connaissiez. Toute lumière vient de l'allemagne maintenant et pour mon très grand malheur je ne comprends pas votre langue." Cf. his words to Wöhler: "Il n'y a à Paris absolument rien de nouveau en chimie; tout le monde d'ailleurs profite des vacances pour voyager. C'est, comme par le passé, en Allemagne que l'on voit paraître tout ce qu'il y a de neuf." Pelouze to Wöhler, 6 August 1836, Dossier Pelouze, Archives de l'Académie des Sciences, Paris.
69. Gay-Lussac to Liebig, 10 September 1826, in Gay-Lussac file, Sammlung Darmstaedter, Staatsbibliothek zu Berlin Preussischer Kulturbesitz. On Gay-Lussac's interests in foreign languages and travel, see also Crosland, *Gay-Lussac*, pp. 36–38, 271, 275.

Figure 2.4
Jules Pelouze, ca. 1836. Source: E. F. Smith Collection, University of Pennsylvania
Library.

the few French chemists in his generation to have mastered German, and perhaps the only one to have traveled to Germany for study.[70] That pattern began to change around 1830, thanks to Liebig.

As was noted above, Dumas was also writing to Liebig at the same time (1831 and 1832) and in very similar terms as Pelouze, decrying the ignorance of German developments, and indeed of important discoveries anywhere outside of France, by "our poor chemists." He related that Antoine Bussy, professor at the School of Pharmacy, had declared to him the other day that organic analyses were useless. "Voilà comme à Paris," Dumas wrote sarcastically, "on comprend la chimie."[71] Like Pelouze, Dumas knew little German, but he recognized the vital importance of scientific developments across the Rhine, and he managed to remain au courant with the Liebig-Berzelius school. To be sure, such professions of respect for Liebig and German chemistry and such denigrations of Parisian science on the part of up-and-coming French scientists cannot be regarded as neutral descriptions, unproblematic judgments, or even sincere avowals. However, the frequency of such judgments in this correspondence suggests that some leading French chemists were indeed concerned about the health of their national community as early as 1830. Liebig even reported to Berzelius that Gay-Lussac himself recognized France's insularity and consequent decline in status on the competitive world stage:

A certain mental indolence inhibits the French, to their shame, from becoming acquainted with the work of foreigners; Gay L. shares it and considers that this way of thinking must gradually lead to the decline and extinction of all scientific sensibility in France; all his letters to me are filled with reproaches on this point and especially against himself, but that is no character flaw, and must be excused considering his other qualities.[72]

Like Pelouze, Dumas, and Gay-Lussac, who recognized the high quality of foreign science while struggling with feelings of patriotic loyalty, Liebig and some of his compatriots had deeply divided feelings. Liebig never forgot his transforming experience as a young man in Paris, and his loyalty to his mentor Gay-Lussac never waned. The Napoleonic period engendered not only

70. K. Kanz, *Nationalismus und internationale Zusammenarbeit in den Naturwissenschaften* (Steiner, 1997), pp. 142–143.

71. Dumas to Liebig, n.d., postmarked 23 January 1832, Liebigiana IIB.

72. Liebig to Berzelius, 4 August 1831, in Carrière, p. 15.

animosity in Germany but also a surprising degree of respect and emulation. Many if not most German scientists of this period were Francophones (Alexander von Humboldt is only the best known of the type), whereas a vanishingly small number of French scientists were able to move easily in German language or culture.

But Liebig, along with many of his countrymen, increasingly discovered his own powers, and began to see the French scientific establishment in an altered light. Humboldt's permanent return to Berlin in 1827 can stand as the symbol of the beginning of a shift of the center gravity of European science eastward. I argued above that Liebig's analytical innovations were connected with his drift away from the Parisian orbit in the late 1820s. The change was particularly evident by 1832, when he published a brutal denunciation of the French chemical establishment, lambasting them for arrogance, chauvinism, provincialism, rhetorical bombast, and scientific thievery. [73] Only Gay-Lussac, Dulong, Arago, and Chevreul escaped censure. The harshness of these judgments was slightly ameliorated by Liebig's argument that much of the behavior he described was an inevitable product of the structure of the French scientific establishment. Since university lectures were free, there was no opportunity for young French scientists to earn a livelihood early in their careers, as German Privatdozenten could; furthermore, the monopolistic power of the Académie des Sciences, "the source of all remunerative positions," led to an unseemly scramble for success. This was why French scientific papers all seemed so arrogant and self-promoting, Liebig thought. In contrast, it might be said that German scientists tended to wear their sobriety and modesty on their sleeves, vaunting it and at times almost overplaying the card. To a German scientist used to German rhetoric, the literature from Paris made the French sound considerably more brash and presumptuous than they were. When German scientists visited Paris and spoke directly with their French counterparts, their impressions were invariably quite different, and much more favorable. [74]

73. Liebig, "Bemerkungen zu vorhergehenden Abhandlung," *Annalen* 2 (1832): 19–30, esp. pp. 19–22.

74. See Wöhler's fascinating comments on his trip to "Babylon" in his letters to Berzelius of 13 and 27 October 1833, in Wallach, pp. 526–538. Since Liebig had attacked Thenard first and foremost, and since Thenard and Gay-Lussac were no longer friends, some contemporaries thought that the attack by Gay-Lussac's protégé was at least partially directed toward Thenard's protégé Dumas. Wöhler speculated

Liebig's attack surely must have created anger and resentment in Paris, but it did not destroy his standing there as one of the leaders of European chemistry. In 1836 Pelouze wrote once more to Giessen:

You speak to me, my dear friend, of the spirit of nationalism in France and of its presence even in science; well, you must be French, for I tell you—and I would be the last person in the world to flatter you—that no indigenous chemist, no matter who they are, has as many admirers as you do in Paris.[75]

In this period, the "spirit of nationalism" in science, although real enough, was often overbalanced by powerful cosmopolitan and internationalist instincts.[76] To some degree, at least, Liebig still managed to retain his status as an honorary Frenchman.

---

to Berzelius (pp. 529–530) that Thenard had been told by "friends" that Gay-Lussac himself may have instigated the attack.

75. Pelouze to Liebig, 27 December 1836, Liebigiana IIB: "Vous me parlez, mon cher ami, de l'esprit de nationalité en France et de son existance jusque dans les sciences, alors vous devez être français, car je vous dis, sans vouloir le moins du monde vous flatter, que pas un indigène, quel qu'il soit, n'a autant d'admirateurs que vous à Paris."

76. In this I agree with Kanz's position in *Nationalismus*, pp. 11–18, 231.

3
The Education of an Alsatian Chemist

One of the most interesting French chemists of the nineteenth century was born in Alsace, a German-speaking French province that borders the Rhine. As a student of both Liebig and Dumas, Adolphe Wurtz (1817–1884) played a central role in the Parisian chemistry of the second half of the century. In order to properly assess his importance, we must understand the milieu from which he arose.

The Alsatian Environment

Adolphe Wurtz spent his boyhood in Wolfisheim, a village 3 miles west of Strasbourg, where his father, Johann Jacob Würtz (Jean Jacques Wurtz), was a pastor.[1] The elder Wurtz (1787–1845)[2] had been raised as the only son of Lutheran parents in a modest Strasbourg home. After attending the Protestant seminary[3] in Strasbourg and then traveling around Switzerland and northern Italy, he returned home to begin his career. His first congregation was in Bergzabern, in the Palatinate, about 60 miles north of Strasbourg, but the transfer of this region from France to Bavaria in 1816 led to his reappointment in Wolfisheim. J. J. Wurtz is described by his son's biographers as quiet, cultured, and serious, perhaps even somewhat severe. He developed his intellectual abilities by "tireless exercise"; his profound sense of duty to his flock left little time for his family. His sermons, elevated in tone and filled with classical allusions, were probably insufficiently

---

1. The following biographical details are taken from the two principal biographies of Adolphe Wurtz: C. Friedel, "Notice sur la vie et les travaux de Charles-Adolphe Wurtz," *Bulletin de la Société Chimique* [2] 43 (1885): i–lxxx; A. Hofmann, "Erinnerungen an Adolph Wurtz," in *Zur Erinnerung an vorangegangene Freunde* (Vieweg, 1888), vol. 3.

appreciated by his congregation of mostly simple farmers. Regarding the tone of Wurtz's religious beliefs we have testimony from the two principal biographers of his son; they are not entirely consistent. According to Charles Friedel, "he was not satisfied with a traditional faith; he sought the truth with passion and perseverance."[4] A. W. Hofmann notes that he maintained a strict interpretation of the Augsburg Confession, the foundation of Lutheranism, and that he successfully inculcated this creed in his children.[5] Other testimony suggests that his beliefs included elements of liberal rationalism.[6]

J. J. Wurtz's wife, Sophie (1794–1878), came from a large and remarkable Strasbourg family. Her father, Johann Jacob (Jean Jacques) Kreiss (b. 1764), was senior pastor at the Jung St. Peterskirche (Église Saint-Pierre-le-Jeune), a Lutheran church in the heart of the old city. One brother, Theodor, became an eminent professor of Greek and Latin at the theological faculty in Strasbourg; another, Adolph, became a pastor at Breuschwickersheim, about 2 miles west of Wolfisheim. Both exerted positive influences on their sister's children; Theodor became like a second father to Adolphe after the untimely death of J. J. Wurtz in 1845. Sophie Wurtz possessed a lively intelligence and an "unshakable serenity," and her optimistic energy unfailingly brightened the household.[7] Charles Adolphe Wurtz (Karl Adolph Würtz), the eldest child, was born on 26 November 1817 in the Strasbourg parsonage (9 Grande Rue de l'Église, near the

2. The birth year of Jean Jacques Wurtz is taken from Adolphe Wurtz's birth certificate, a transcription of which is in the Wurtz Dossier at the Académie des Sciences, Paris. The document also informs us that Wurtz's grandfather, born in 1761, was also named Jean Jacques.

3. Hofmann reported that the elder Wurtz attended the University of Strasbourg ("Wurtz," p. 175). However, the University had been disbanded in 1793, and its legal successor was the Protestant seminary. A Protestant faculty of theology was not reestablished until 1818. See J. Craig, *Scholarship and Nation Building* (University of Chicago Press, 1984), pp. 9–11; P. Leuilliot, *L'Alsace au début du XIXe siècle* (S.E.V.P.E.N., 1959–60), vol. 2, pp. 179–186.

4. Friedel, "Wurtz," p. ii.

5. Hofmann, "Wurtz," p. 211.

6. H. Strohl, *Le protestantisme en Alsace* (Éditions Oberlin, 1950), pp. 406–407.

7. Hofmann, "Wurtz," p. 177. Hofmann, a close friend of Adolphe Wurtz, had met Sophie Wurtz several times, though only in her old age. He was able to use Adolphe's sister, also named Sophie, and a cousin, likewise named Adolphe (son of Uncle Adolph), as sources of information on the early years of his friend's life.

church and next door to—or perhaps even in the same building as—her husband's parents' home), where Sophie had gone to take advantage of better medical care in the city.

Wolfisheim was (and is) a small village of a few hundred mostly Protestant souls, set in a rich agricultural landscape of wheatfields and vineyards, in the shadow of the Vosges Mountains to the west and the Black Forest hills to the east. The medieval village church was tiny—though it served both Lutheran and Catholic congregations—but the parsonage was comfortable for the family of five.[8] Adolphe's maternal uncles were frequent visitors and often enlivened the dinner-table conversation with historical, literary, or theological topics. His early education was provided by his father.

Adolphe was described by boyhood friends as merry, charming, and lively, with bright eyes and curly brown hair—the personification of the popular image of the Alsatian personality.[9] His childhood was happy and active, and he retained a love for the countryside and for regular physical activity. It is also said that his rural activities, including much farm work,[10] contributed to his robust health, which lasted until the last few years of his life. After his family moved to Strasbourg, Adolphe sorely missed the country life, for which gymnastics, swimming, and rowing provided a poor substitute.

Linguistically and culturally, Alsace was in transition at this time.[11] Although by 1800 their region had been a part of France for more than 100 years, the Alsatians did not much feel themselves to be French. The language of daily life was German—more particularly, Alsatian, a distinctive Alemannic dialect closely related to Swabian and Swiss German and quite distinct from Hochdeutsch. This was the family idiom of Adolphe Wurtz's childhood. Wurtz and his family, like most other Alsations, were Lutherans, whereas the rest of France was largely Catholic. The Vosges Mountains formed a natural geographic barrier that accentuated the sense of isolation from the rest of the country. The language of high culture was Hochdeutsch,

8. I thank M. Freddy Sarg, the current pastor of the Eglise Luthérienne de Wolfisheim, for current and historical information about his church.

9. "Der Elsässer ist beweglich, heiter, aufgeweckt," in *Meyers Lexikon*, seventh edition (Bibliographisches Institut, 1925), vol. 3, p. 1567.

10. Lutheran country pastors in the nineteenth century derived their income from an agricultural tax on parishioners. Often a pastor's family worked on a small farm at the parsonage, though the situation of the Wurtz family in Wolfisheim is not known.

11. See Leuilliot, *L'Alsace*, esp. vol. 3, "Religions et culture."

and Alsatian intellectuals tended to look eastward to German peers and institutions, rather than westward to France. Alsace remained outside of the French tariff union of the ancien régime, and commerce was conducted mostly with states of the obsolescent Holy Roman Empire. Germanic language and culture pervaded the eighteenth-century university, Protestant seminary, and Gymnasium in Strasbourg; indeed, Strasbourg was the only important French city to have a Gymnasium, a typically German institution distinct from the French lycée.

However, the Napoleonic years heralded both cultural and political change, for it was then that the authorities in Paris began to get serious about assimilating their Germanic eastern province. In the 1790s, as part of a program of forced Frenchification of Alsace, French was established as the national language of instruction. Ironically, it was a relaxation of such policies in the following decade that removed tensions and led to gradual voluntary cultural assimilation. Simultaneously, administrative reforms and economic standardization accelerated integration, and Alsatians began increasingly to consider themselves a part of France. The growth of German nationalism across the Rhine and the increasing association of language with nation during the early decades of the nineteenth century increased the desire of French patriots, both in Paris and in Alsace, to continue this assimilative process.

The Alsatian language gradually came to be viewed as déclassé, both socially and politically. The decisive transition occurred during Adolphe Wurtz's childhood and adolescence, when the younger generation of Strasbourgeois began to become thoroughly Frenchified.[12] But there was always a limit to assimilation. Even today Alsace is notably Germanic in many respects, and one can still hear Alsatian spoken, especially by the elderly, in the cities and villages of the départements of Haut-Rhin and Bas-Rhin.[13]

---

12. Ibid., vol. 3, pp. 318–329.

14. Even today, it is commonly said that ethnic Alsatians say that they are going "to France" when they are about to cross the Vosges to the west, just as they say they are going "to Germany" when crossing the Rhine to the east. I was told this by a Strasbourgeoise in 1998. On the previous day, I had spoken with an elderly Alsatian couple in the village of Mittelhausen, not far from Strasbourg. They had a linguistic division of labor: he could speak excellent French, she had mastered Hochdeutsch. When we were speaking German, I could understand his Alsatian only with difficulty; she repeatedly turned to him when the conversation became French.

Wurtz's Education in Strasbourg

In 1826 J. J. Wurtz was appointed junior pastor at his father-in-law's church, and the family moved to the city of Strasbourg. In July of that year Adolphe began attending the Strasbourg Protestant Gymnasium, then located in a medieval monastery (since destroyed). This institution, older than the University and in fact its direct predecessor, served for years as an outpost of German language and culture against the encroachments of the French.[14] As late as 1820, French was taught there as a foreign language, little differently from Gymnasien in Munich or Berlin; by this date the university in Strasbourg and the Alsatian legal system were already fully Frenchified. But the end was already in sight: at the time Adolphe joined the classes, the Gymnasium was in the middle of its transition to the French language. However, the process was a slow one. Although the language of instruction was now French, German textbooks were used for several more years in many classes, and as late as 1838 the Director of the Gymnasium addressed his pupils in German for publication in the French-language Festschrift celebrating the tricentennial of the institution.[15] It was from this bilingual experience that Wurtz acquired his mastery, not only of his native idiom, but also of Hochdeutsch and Parisian French. Hofmann, a fine judge of verbal elegance, reported that Wurtz wrote and spoke perfect German, with only the merest hint of a South German drawl.[16]

Wurtz was a strong but not outstanding student, doing well in all subjects but standing out in none in particular. He developed a lively interest in many fields, including literature, art, music, and botany (even, it is said, reading the Naturphilosoph Lorenz Oken). His dyspeptic father, disturbed by the son's apparent lack of focus, predicted more than once that nothing very special would ever come of him. Adolphe's interest in the natural and physical sciences gradually grew, and he began to try out chemical experiments in the wash house of the parsonage for himself and small audiences of siblings and schoolfellows, assisted by small monetary grants from his soft-hearted

14. Leuilliot, *L'Alsace*, vol. 2, pp. 186–191.

15. Cited in Hofmann, "Wurtz," p. 183.

16. Hofmann, "Wurtz," p. 206. Hofmann's judgment was seconded by that of Wurtz's Austrian student Alexander Bauer, who could detect no trace of any accent ("Erinnerungen," *Oesterreichische Chemiker-Zeitung* 22 (1919), p. 117).

mother. His father was less happy, both because his son appeared not to be headed toward the pastorate and because he did not know that a chemical career was even possible. These misgivings seemed all the more warranted after Adolphe was severely burned when a goodly quantity of phosphorus burst into flames under his hands.

There is a durable but probably faulty assumption in the literature that Wurtz made friends at this time with Charles Gerhardt (1816–1856), a fellow student at the Gymnasium who later studied with Liebig and became a professor at Montpellier. At Montpellier and later in Paris, Gerhardt developed revolutionary new chemical theories—ideas that Wurtz would adopt in the 1850s. However, Gerhardt was 15 months older than Wurtz, and two Gymnasium classes ahead of him; moreover, he left the Gymnasium at a younger age than was usual, when he was barely 15, to attend the Karlsruhe Polytechnic. They may have known of each other at the Gymnasium, but it is likely that they did not become well acquainted until Gerhardt returned to Strasbourg in 1837.[17]

When Adolphe was awarded the bachelier ès lettres (equivalent to the German Abitur), in September of 1834, the question of a career became unavoidable, and the conflict between father and son caused serious strife in the family. Further pyrotechnics were avoided when a compromise was struck: Adolphe would study medicine in Strasbourg, at a branch of the national French university system.

After the dissolution of higher education in the French Revolution, the Université de France was reestablished by Napoleon in 1808. The Université was not located on a single campus or in a single city; rather, it was an administrative structure consisting of a national network of Facultés of letters, sciences, theology, law, medicine, and pharmacy, and in addition embracing secondary education (lycées and municipal collèges), all directed by a central bureaucracy in Paris. Ten Facultés des Sciences and three Facultés des Médicine (in Paris, Montpellier, and Strasbourg) were established in this first wave of decrees.

It is important to understand the outlines of this administrative structure. Each Faculté was directed by a dean (doyen) chosen from the professors.

17. M. Tiffeneau, ed., *Correspondance de Charles Gerhardt* (Masson, 1918–1925), vol. 2, p. 302; E. Grimaux and C. Gerhardt Jr., *Charles Gerhardt* (Masson, 1900), pp. 13–15, 29.

The various Facultés in a given city were collectively known officially as an Académie—l'Académie de Paris, de Bordeaux, etc. The Faculté des Sciences and the Faculté des Lettres de Paris were, and are, informally known as "the Sorbonne," after their physical location on the Left Bank of the Seine. Each Académie was led by a rector (recteur), who also was the administrative officer for secondary education in the same district. The various rectors reported to a Ministère de l'Instruction Publique, to a Conseil Royal [National, Impérial] de l'Instruction Publique, and to local Conseils Académiques for the respective Académies. But the Université de France and the various Académies were bureaucratic entities more than material or psychological ones; professors and students formed primary allegiances to their respective Facultés, and the various Facultés in a given city had little institutional or psychological cohesiveness.

The Facultés were also often rather insubstantial entities themselves, especially those in the provinces. Most consisted of four or five teachers, frequently borrowed from the lycées, instructing small groups of students. The chief function of French higher education was to examine students for degrees (the baccalauréat, the licence, the agrégation, and the doctorat), and thus reproduce their positions and functions for lycées and Facultés. Since there was no requirement to study formally at a Faculté in order to sit for the examination, student enrollments were low, sometimes vanishingly small. Facultés were also poorly funded, both for salaries and for facilities; laboratories and research budgets were poor in Paris and next to nonexistent in the provinces.

The faculty system of the Université was also overshadowed by the presence of the grandes écoles (such as the École Polytechnique and the École Normale) and the explicitly research-oriented institutions of Paris (such as the Collège de France, the Muséum d'Histoire Naturelle, and the Observatoire). In the face of this competition and implicit division of labor, the faculties of sciences were for decades not designed to fulfill a research function at all, but rather were considered purely as institutions for didactic instruction and examination. Further working against any establishment of a research ethic—particularly at the Sorbonne—was the custom of cumulating positions, which ensured that virtually every science professor at the Sorbonne had another appointment in one of the grandes écoles or in a research institution, so that his research could be done there.

In the first half of the nineteenth century, Strasbourg had perhaps the best set of Facultés outside of Paris. There were a number of distinguished names among the professors, and one of only three medical schools in France. From its origin in 1808, the language of instruction, by statute and practice, was French. At the time Wurtz was a student there, the medical faculty was the largest unit of the Académie de Strasbourg.[18] The Faculté des Sciences was also distinguished by the presence of the prolific chemist Jean François Persoz (1805–1868), a student of Louis Jacques Thenard and from 1852 a professor at the Conservatoire des Arts et Métiers in Paris. Persoz was a professor of chemistry at Strasbourg from 1833, and in 1835 he added the positions of professor of pharmacy and director of the Strasbourg École de Pharmacie. In 1838 Wöhler described him as "a most significant man . . . a small, dark Frenchman, still quite young, mais qui connait ses mérites, as the Parisians say. . . . He was quite full of chemistry and full of theories, and not without suffisance. . . . In any case, I think he has an excellent mind, works and thinks hard, lives alone in his science, but he is also a theory-maker, full of ambition, vanity,—a real Frenchman."[19]

In 1835 Persoz published a long article on "the molecular state of compounds," and 4 years later he published an "introduction to the study of molecular chemistry."[20] In these works he proposed standardizing formulas by taking them at the same number of volumes, and he pursued the theories of substitution and types; however, Persoz continued to maintain a dualist-electrochemical interpretation of reactions, which later reformers rejected. It is also important to note, in view of the generally dismal state of French provincial academic laboratory facilities, that Persoz's laboratory was, in Wöhler's words, "quite lovely and very well equipped."

18. In 1822, the medical faculty enrolled 277 of 757 total students. In the late 1830s, 15 to 20 MDs were graduating per year, and only about one doctor of sciences. See Leuilliot, *L'Alsace*, vol. 2, p. 280, and O. Berger-Levrault, *Annales des professeurs des académies et universités alsaciennes* (Berger-Levrault, 1892), pp. ccxii–ccxvi, 296, 299.

19. Wöhler to Berzelius, 14 October 1838, in *Briefwechsel zwischen J. Berzelius und F. Wöhler*, ed. O. Wallach (Engelmann, 1901), vol. 2, pp. 60–63.

20. Persoz, "Mémoire sur l'état moléculaire des corps composés," *Annales de chimie* [2] 60 (1835): 113–151; "Propositions faisant suite au mémoire sur l'état moléculaire des corps composés," *Annales de chimie* [2] 61 (1836): 411–419; *Introduction à l'étude de la chimie moléculaire* (Mathias, 1839).

In order to study medicine, aspirants had to present not only a diploma of bachelier ès lettres but also the bachelier ès sciences. After nearly 2 years of study, Wurtz obtained this second degree after examination by a committee from the Faculté des Sciences; he then began the serious study of medicine.[21] (Since Gerhardt studied at the Faculté in 1837–38,[22] he and Wurtz probably became acquainted during that year, if the had not done so earlier.) It would appear that in 1835, even before earning his science baccalaureate, Wurtz acquired his first position, as aide-préparateur in pharmacy at the Faculté de Médecine.[23]

In addition to his medical courses, Wurtz began to specialize early in chemistry. For this purpose he had available at the Faculté de Médecine a capable mentor by the name of Amédée Cailliot (1805–1884). Son of a navy physician, Cailliot had begun by studying medicine, but then switched to chemistry. In February of 1835 he was named agrégé en exercice (laboratory teaching assistant) at the medical faculty, and in 1838 he succeeded G. Masuyer in the chair of medical chemistry. Cailliot never attained a wide reputation, but he earned the lifelong gratitude of his student Wurtz, and subsequently he made at least one significant discovery: terephthalic acid.[24]

21. The requirement of a science baccalaureate was only added in 1836. It is not known with whom Wurtz studied at the Faculté des Sciences. Wurtz's mentor Cailliot had a poor relationship with Persoz, making it likely that Wurtz likewise avoided him. In a private communication to me, Natalie Pigeard cites F17/2269, "Extrait du registre des délibérations pour le concours aux chaires," in support of this information.

22. Grimaux and Gerhardt, *Gerhardt*, p. 29.

23. The bachelier ès sciences was awarded 10 May 1836; the first examination by the Medical Faculty on 3 August of the same year was passed with the mark "très-distinguée": "Feuille d'Examen de M. Wurtz, Charles-Adolphe (Fac. Med.)" 1T Sup., 496a, Wurtz, Archives du Bas-Rhin, Strasbourg. I have not been able specifically to document his first position. Wurtz later noted that he had begun studying medicine in 1835 (printed notice of works published, 1852, in AJ16/188), and in another document he stated that he was first appointed aide-préparateur de chimie at the Faculté de Médecine in 1835 (F17/21890). Friedel ("Wurtz," p. vii) cites the same date, perhaps relying on the same sources.

24. The *Royal Society Catalogue of Scientific Papers* and the *National Union Catalog, Pre-1956 Imprints* list fourteen papers by Cailliot, published between 1821 and 1847 (the last paper announcing the discovery of terephthalic acid). On Cailliot, see *Dictionnaire de biographie française* (Letouzey, 1956), vol. 7, p. 867; Leuilliot, *L'Alsace*, vol. 3, pp. 277–278; Berger-Levrault, p. 34; Grimaux and Gerhardt, *Gerhardt*, p. 25n.

Many years later Wurtz was to have the opportunity to return the favor of professional hospitality. After Strasbourg was annexed by Germany in 1871, Wurtz was pleased to be able to invite his old mentor into his Paris laboratory.

On 27 December 1836 Wurtz was examined by a jury of three for the position of aide-préparateur of chemistry, pharmacy, and physics. Although the competition was open (as was customary), Wurtz was the only candidate who applied. The examination consisted of a single problem: to explain the preparation and purification of nitric acid, pure potassium, and oxide of mercury. The candidate was given 10 minutes to ponder the question and 20 to answer orally. He succeeded and was duly appointed. Thereafter he was the préparateur for all the courses in the Faculté in chemistry, physics, pharmacy, and toxicology.[25] When Cailliot became professor of medical chemistry in 1838, he created a laboratory for practical work in chemistry, for which Wurtz provided the instruction; this may have been tied with Dumas's Rue Cuvier laboratory as the first such in France.[26]

In 1839, as the result of another competition, Wurtz was promoted to chef des travaux chimiques. His annual salary for the next 5 years was 1500 francs. In 1840 Wurtz passed the degree of licencié ès sciences.[27] Wurtz kept his mother from worrying by not telling her until each of these six examinations and competitions was safely past; this entailed donning the required formal attire elsewhere than at home. The last hurdle was of course the doctoral examination itself, which Wurtz did not pass until August of 1843, presenting a thesis entitled "Études sur l'albumine et sur la fibrine." Wurtz's total stint as a medical student, 8 years, was longer than normal; this was due not only to his pressing professional duties in Cailliot's laboratory but also to a 5-month visit to Giessen, the site of Justus Liebig's famous teaching and research laboratory.

This was an unusual step for a French student, so Wurtz had to prepare the conditions carefully. He made the formal request for a paid leave through the dean of the Faculté de Médecine and the rector of the Académie de Strasbourg, who on 4 March 1842 sent it with a favorable recommen-

25. Summary of jury report in F17/21890; Wurtz to Richard Lepsius, 7 March 1837, in Hofmann, "Wurtz," pp. 195–196.

26. Grimaux and Gerhardt, *Gerhardt*, p. 25n.

27. Printed bio-bibliographic notice of 1852, AJ16/188.

dation to the Ministry of Public Instruction in Paris, commenting that Wurtz could ill afford to do without a salary for this period. The Ministry was concerned about the work that would be left undone in Strasbourg, but the dean, J. B. R. Coze, responded that Wurtz had already arranged for a capable substitute, whom he had agreed to pay a portion of his salary. The Ministry then turned for advice to the dean of the Paris Faculté de Médecine, who happened to be a competent judge of the question: the prolific professor of toxicology M. J. B. Orfila (1787–1853). Orfila (the first functionary in this long series of memoranda to spell Liebig's name accurately) commented that Wurtz was quite right to be eager to spend a semester with "one of the most famous men of the age" and recommended that the request be approved. Wurtz's name was already known to Orfila; he had cited Wurtz's first chemical publication, an analysis of illuminating gas, in his textbook on legal medicine.[28] A ministerial decree followed, granting Wurtz 5 months' leave with pay.[29]

Why Wurtz resolved to go to Giessen is not known. He was by no means the first French chemist to go there: he was preceded by Jules Gay-Lussac, Victor Regnault, Jules Pelouze, and eight others,[30] including the Alsatian Charles Gerhardt. According to one account, J. J. Wurtz, who had become reconciled to his son's ambitions, consulted with Gerhardt regarding the best course of study for his son; Gerhardt recommended Giessen, where he himself had gone.[31] An even more probable source of advice was Charles Oppermann (1805–1872), another Giessen alumnus, who was at the time a professeur adjoint at the Strasbourg École de Pharmacie. In fact, Oppermann, a native of Strasbourg, had the distinction of being the first

28. Wurtz, "Analyse du gaz de l'éclairage," in G. Tourdes, *Relation médicale des asphyxies occasionnées par le gaz de l'éclairage* (Derivaux, 1841); Orfila, *Traité de médecine légale* (Béchet jeune, 1836), vol. 3, pp. 841–842.

29. F17/21890; the decree was dated 12 April 1842 and was effective from the 15th, but there are indications that Wurtz may have already arrived in Giessen by that time. Friedel placed Wurtz's stay in Giessen in 1843, an error that has been propagated through much of the Wurtz literature.

30. Armin Wankmüller, "Ausländische Studierende der Pharmazie und Chemie bei Liebig in Giessen," *Deutsche Apotheker-Zeitung* 107 (1967): 463–467.

31. Tiffeneau, *Correspondance*, vol. 2, p. 303. Gerhardt's maternal grandfather was legal administrator of the Lutheran church of Saint-Pierre-le-Vieux in Strasbourg (Grimaux and Gerhardt, *Gerhardt*, p. 9), and so the two families may have known each other.

of Liebig's students to publish the results of an independent investigation using the new apparatus for organic analysis invented by Liebig late in 1830; as was noted in chapter 2, Oppermann personally carried the first Kaliapparat from Giessen to Paris.[32] Moreover, as will be described below, Wurtz most likely had met Liebig during the latter's visit to Strasbourg in September of 1838, when Wurtz was directing Cailliot's teaching laboratory. The possibility of a conversation between them at this time, and conceivably even an invitation by Liebig, cannot be excluded.

Giessen

A quick summary from chapter 1: In 1824, only 2 weeks after his 21st birthday, Justus Liebig was called to a junior professorship in Giessen, site of the tiny state university of the Grand Duchy of Hesse, and in 1825 he was promoted to full professor. He gradually built his laboratory into a little empire of science. During his first few years in Giessen he was viewed by many as, in effect, a non-resident member of the Parisian chemical community. This tie to France began to weaken as Liebig's allegiances shifted toward the end of the 1820s. He formed an extremely close professional and personal friendship with Friedrich Wöhler, embraced as his second mentor Jacob Berzelius (who, although Swedish, was a principal founder of German organic chemistry), abandoned the French system of atomic weights and formulas for the Berzelian-German system, and became editor of the *Annalen der Pharmacie*. Around the same time, he devised the Kaliapparat, which made analysis far easier to perform and far more precise. As I have argued, Liebig's new analytical method fit well with his institutional context. Since his student days, Liebig had planned a career in which he would instruct *groups* of students in pharmaceutical and chemical laboratory operations.

32. For a discussion of Oppermann's work with Liebig in 1829–1831, see F. Holmes, "The Complementarity of Teaching and Research in Liebig's Laboratory," *Osiris* [2] 5 (1989), pp. 137–141. On at least one visit home to Strasbourg after he moved to Paris, Wurtz worked at the lab of the École de Pharmacie, presumably with Oppermann's permission (Wurtz to Dumas, 9 November 1848, Dossier Wurtz, Archives de l'Académie des Sciences). In 1852 he married Constance Oppermann, the daughter of a Paris banker and apparently a childhood friend of Wurtz's; indications are that she was a relative Charles Oppermann, perhaps his niece, but I have not verified the relationship. On Oppermann, see Grimaux and Gerhardt, *Gerhardt*, pp. 28–29n.

The years Wurtz spent as a student of medicine at Strasbourg were the very years during which Liebig rose to the preeminent position among chemists that he was to occupy for the rest of his life. Several events and trends contributed to Liebig's rise. His teaching lab, begun in 1826 as a private venture paralleling his official university instruction, attracted what were at the time relatively high numbers of students—from eight to thirteen at a time during the period 1830–1835. In 1835, when the University of Giessen finally began to finance and expand the laboratory, the number of Praktikanten increased to about 20. In the fall of 1837 Liebig began to use all those hands in a concerted and organized fashion to further his own research agenda to a degree he had not hitherto attempted. Two years later the laboratory was renovated and expanded.[33]

At first Liebig attracted students by promoting his institute as a training school for pharmacists, but with time he tended more and more to tout practical instruction in science as a necessary component of education in a wide variety of fields. The German chemical community, though geographically dispersed, was well interconnected, and Liebig's innovations were adopted elsewhere, such as in the Göttingen laboratory of Friedrich Wöhler and in the Marburg laboratory of Robert Bunsen. Liebig's new style of group laboratory teaching was also eventually adopted outside this network, both in other fields of science (such as physiology and physics) and in other countries (including France), but Liebig and his friends were the earliest pioneers.

Once one had a lab filled with budding chemists, one could make use of them as assistants for one's research program, and this was precisely what the German chemists did—but none better or more aggressively than Liebig. Armed with a simple and convenient analytical technique, and with a respected journal at his command, Liebig was able to provide constant employment for his typesetters. Liebig's fame was further promoted by the publication in 1837 of his detailed monograph on his analytical method (soon to be translated into French and English), by a long trip to England

33. Holmes, "Liebig's Laboratory," pp. 146–156; W. Conrad, *Justus von Liebig und sein Einfluss auf die Entwicklung des Chemiestudiums und des Chemieunterrichts an Hochschulen und Schulen* (Ph.D. dissertation, Technische Hochschule Darmstadt, 1985), pp. 124–127; J. Fruton, "The Liebig Research Group," *Proceedings of the American Philosophical Society* 132 (1988): 1–66; J. Volhard, *Justus von Liebig* (Barth, 1909), vol. 1, pp. 7–85.

that same summer, and by a short-lived but crucial collaboration with Dumas which began that fall. In England Liebig was given a contract to write a summary of the state of organic chemistry, an assignment that resulted 3 years later in the publication of his classic work *Die Organische Chemie in ihrer Anwendung auf Agrikultur und Physiologie*, and then in the publication of *Traité de chimie organique* (his only textbook). Liebig's turn from pure science to agricultural and physiological chemistry only increased his fame, in Germany and abroad; he seemed to be realizing the great hopes that academic chemistry might make an important difference to practical life.

The net effect of all this is that in the late 1830s Giessen was rapidly becoming something of a Mecca for chemistry students from around the world. By 1841 Liebig had 50 students working in his lab. Two years later, the number now at 68, Liebig opened a branch laboratory under the direction of his assistant, Heinrich Will.[34] Many of these students were foreigners, especially from Great Britain, Switzerland, Austria, the United States, Russia, and also France (especially Alsace). By the end of 1841 Liebig had had 12 French students; this number represented almost a third of his foreign students to that time.[35]

In March or April of 1842, Adolphe Wurtz traveled north up the Rhine and the Lahn to Giessen, where he joined the Liebig community. One of the most prominent younger members of that community was August Wilhelm Hofmann, the man who eventually would inherit the large mantle of his mentor. Already promoted to the doctorate but not yet an Assistent, Hofmann was nearly the same age as Wurtz. Like the Frenchman, he was sociable, articulate, cultured, and broad-minded. They soon became good friends. Hofmann described his first encounter with Wurtz (at a Sunday dinner with a group of the other Praktikanten at the Liebig home in Giessen) as follows:

Even at first glance one recognized in him the good comrade and lively companion. From his robust, slightly tanned face, framed by curly dark hair and full beard, peered so merrily his clever shining dark eyes, looking out into the world so confident of success! But it was no longer just youthful exuberance that animated his features. The powerful brow that arched over those fiery eyes, further revealed by

34. Volhard, *Liebig*, vol. 1, pp. 83–84; Holmes, "Liebig's Laboratory," p. 162.

35. A. Wankmüller, "Ausländische Studierende," p. 464.

Figure 3.1
Justus Liebig in 1853, photographed by Franz Hanfstaengl. Note combustion
train at left. Source: Photograph Collection, Deutsches Museum, Munich.

a slightly receding hairline, gave unmistakable evidence of habitual reflection and the gift of deep absorption. But this more serious look was only revealed in those rare instants when his mobile face paused a moment in its rapid play of expressions. The man's physical liveliness harmoniously matched his facial expressiveness. Sitting still for a time was difficult for him, and when forced to this position, his arms and hands found themselves in constant activity. Indeed, movement was a condition of well-being for his sturdy figure. For this reason he also loved lively conversation; he enjoyed talking, and always had much to say.[36]

After dinner Hofmann and Wurtz walked home together. They spent much of the afternoon and the evening hiking and getting acquainted. Hofmann reported that Wurtz quickly became the favorite of the laboratory. He roomed next door to Hofmann, near the lab. During lab hours Wurtz was always among the most industrious, but when the institute closed on Saturday afternoons and Sundays the chemists often went on walks and excursions in the beautiful Hessian countryside; Hofmann described no fewer than ten of these destinations.

The scientific life of the Liebig laboratory was a rich one. Present in Giessen that summer semester, in addition to Wurtz and Hofmann, were such students and junior staff members as Heinrich Will, Hermann Kopp, Remigius Fresenius, Adolf Strecker, Franz Varrentrapp, Carl Ettling, Friedrich Knapp, Friedrich Zamminer, and Ernst Dieffenbach. (The famous lithograph by Trautschold depicting an interior view of the laboratory with a dozen students working at the benches is probably from the 1841–42 winter semester. All but two of the twelve likenesses have been identified; Wurtz is not among them.)

Wurtz's research in Giessen, a project of his own devising, was on the formula and constitution of hypophosphorous acid. Berzelius regarded this substance, discovered a quarter-century earlier by P. L. Dulong, as a simple phosphorus oxide united with two molecules of water. According to Wurtz's own account, he started the project with the intent of refuting this view (which seemed unlikely to him, since these putative water molecules could not be separated out) and of demonstrating instead the truth of Heinrich Rose's alternative view that the compound consisted of phosphine united with phosphoric acid. Wurtz concluded from a study of its chemical behavior that Rose's hypothesis likewise failed the empirical test; rather, the compound was best formulable as $H_4P_2O_3$ in its anhydrous state, and was

36. Hofmann, "Wurtz," pp. 205–206.

Figure 3.2
A rare portrait of Adolphe Wurtz as a young man. Source: E. F. Smith Collection,
University of Pennsylvania Library.

a monobasic acid. Oxidation of this compound (replacement of H_2 by O) would thus yield dibasic phosphorous acid, $H_2P_2O_4$, and further oxidation would result in tribasic phosphoric acid, P_2O_5. The compound could therefore be viewed as the trioxide of the binary radical P_2H_4. Alternatively, Wurtz wrote, if one were to adopt Liebig's hydracid theory of 1838, the substance could be regarded as the monobasic hydracid of a triple radical, $H_2 + H_4P_2O_4$.[37]

Electrochemical-Dualist Theory and the Positivist Turn

Wurtz's hypophosphorous acid, $H_4P_2O_3$, looks little like its modern formulation, H_3PO_2, and the suspicion that Wurtz must have made analytical errors increases when one notes that Wurtz was using the same relative atomic weight values for hydrogen, phosphorus, and oxygen that are used today. However, the formula is indeed substantively the same as the one used today. In the custom of his day, Wurtz was using a "four-volume" formula—that is, twice the number of atoms as in such "two-volume" formulas as water = H_2O and phosphine = H_3P. Moreover, following the theory of "electrochemical dualism" (which Berzelius had constructed in the second decade of the nineteenth century), acids were always formulated in their anhydrous state. Double the atomic coefficients of the modern formula for hypophosphorous acid, then subtract two hydrogens and an oxygen (water, H_2O), and the result is Wurtz's formula.

The nomenclature and the logical derivation of nineteenth-century two-volume and four-volume formulas is intricate.[38] Suffice it to say that what nineteenth-century chemists referred to as two-volume formulas are essentially those used today, but that four-volume formulas, in which all the atomic coefficients were doubled, were used for many compounds, especially in the organic realm. Berzelius was convinced that all compounds consisted of molecules held together by polar (coulombic) attraction, and that they could schematically be dissected into paired component parts

37. Wurtz, "Ueber die Constitution der unterphosphorigen Säure," *Annalen* 43 (1842): 318–334; "Sur la constitution de l'acide hypophosphoreux," *Annales de chimie* 3 [7] (1843): 35–50.

38. Rocke, *Chemical Atomism in the Nineteenth Century* (Ohio State University Press, 1984), pp. 52–54, 171–172.

possessing opposite charges. For complex reasons,[39] he assumed atomic weights for many metals that were double the modern values, and this also had implications in the formulation of salts. For instance, he formulated lead acetate as $PbO + C_4H_6O_3$. The doubled weight of the lead atom fit nicely with the doubled size of the acetic acid formula; moreover, the dualistic formulation of the salt—lead oxide united coulombically with anhydrous acetic acid—tended to confirm the nonreducibility of the four-volume organic formula, for halving it would produce fractional atoms. In this view, the real acid entity was $C_4H_6O_3$; the hydrated acid was formulated as $H_2O + C_4H_6O_3$, regarded as an electrochemical-dualist formula exactly analogous to that of the salt. According to this way of looking at chemical composition, acids consisted of oxidized radicals, bases consisted of oxidized metals, and salts were additive combinations of the two. The scheme was simple and appealing; moreover, it was empirically satisfying, since the presumed acid and base components of salts could be isolated as free-standing chemical compounds.

For these more-than-sufficient reasons, electrochemical dualism was widely accepted in European chemistry during the 1830s and the 1840s, but it was not the only viewpoint to be found in the chemical community. Ever since Davy argued that muriatic acid consisted only of hydrogen and chlorine, the latter being "undecompounded" or elemental, an alternative view of the nature of acids and salts was possible.[40] Between 1814 and 1816, Gay-Lussac added other substances, such as hydriodic and hydrocyanic acid, to hydrochloric (calling the new class "hydracids"), and Pierre Dulong outlined a theory of acidity based on hydrogen content rather than on oxygen content. One could, of course, simply suggest that halogens were negative principles, like oxygen, thus maintaining the spirit if not the letter of the oxyacid theory. But the plot became thicker when Gay-Lussac found that hydrocyanic acid could combine not only with metals (to form, e.g., potassium cyanide, analogous to potassium chloride) but also with chlorine. The traditional oxygen-based theory of salts could also be maintained by supposing that chlorine and other bases of hydracids were not elemental at all, but rather contained oxygen. Many chemists explored this idea in

39. Ibid., pp. 76–77, 156–167.

40. J. Brooke, "Davy's Chemical Outlook," in *Science and the Sons of Genius*, ed. S. Forgan (Science Reviews, 1980).

the early years of the century, but in 1820 the last holdout, Berzelius, capitulated to the elemental character of the halogens, and that line of retreat was closed off.

In the meantime, it began to appear that chlorine could substitute for hydrogen in more than just hydrocyanic acid. In the years after 1820, Faraday, Gay-Lussac, Dumas, and others found that chlorine could substitute "volume for volume"—that is, atom for atom—for hydrogen in a wide range of organic compounds. After 1830, a number of chemists, including Dumas, Charles Gerhardt, and Auguste Laurent (1808–1853), explored a legion of such chlorine-substitution reactions. The phenomenon of substitution of intensely electropositive hydrogen by intensely electronegative chlorine was regarded as a fundamental problem for the theory of electrochemical dualism. For a dualist, all chemical reactions proceeded by addition or separation of two substances, not substitution of one for another, and the glue that held compounds together was electrical polarity. No dualist could countenance, say, a chlorinated ether whose molecules seemed to be held together with the same degree of affinity and to have the same essential properties as ordinary ether, for the two compounds were, to speak literally and metaphorically at once, poles apart. And yet such instances seemed increasingly common in the years after 1830.

At first, this opposition was played out between certain French chemists (especially Dumas, Laurent, and Gerhardt) and the dualist proponents of the theory of organic radicals (led by Berzelius but also including such German chemists as Liebig, Wöhler, and Bunsen). To oppose the radical theorists' electrochemistry and their focus on the chemical identity of constituent parts of molecules, the French substitutionists began, quite understandably, to focus on molecular arrangements. From this emerged their notion that there were chemical "types"—that is, pattern formulas that were maintained through whole groups of related compounds.

But the lines of battle had not yet even fully formed when a small potential breach appeared in the radical theory. In Liebig and Wöhler's classic 1832 paper on the oil of bitter almonds, regarded then and now as one of the cornerstones of the emergent dualistic radical theory, the authors suggested that their "benzoyl" radical seemed empirically to consist of $C_{14}H_{10}O_2$, to which could be added, e.g., hydrogen (to form benzaldehyde), chlorine (to form benzoyl chloride), oxygen (to form benzoic acid), and several other elements to form other related compounds. But this schema is entirely analogous to a hydracid and its salts, or a series of chemical types, a point to which Liebig made veiled allusion; as the formulas were written, the schema appeared to imply that substitution rather than electrochemical addition was taking place—moreover , substitution of electrochemically opposite atoms.   Through the 1830s Liebig could never seem to decide whether to fight the French substitutionists or to join them. Neither could Dumas. In his first major paper on chlorine substitution (1834), he attempted to explore this relatively new phenomenon while simultaneously using it to build support for his dualistic interpretation of alcohols, ethers, and esters. Dumas continued to try to find a via media between dualism and substitutionism for the next several years, while his former student Laurent was developing a more thoroughgoing version of the latter.

After his graduation from the École des Mines (1830), Laurent worked for 2 years in the École Centrale, then took a succession of short-lived industrial positions. In these years Laurent began to place arrangements on a higher plane than electrochemical identity; in his view, a chlorine atom could behave chemically just like a hydrogen atom, as long as it occupied the same position in the molecule. Dumas denied that this was possible.[41] In 1838 Laurent was appointed a professor at Bordeaux. Seven years later he moved to Paris, paying a substitute half his salary to teach in his place; in 1847 his leave was not renewed, and from that time on Laurent had to get by almost entirely on his own slender resources. Since he had quarreled with Dumas over substitution and types in the late 1830s, this talented chemist had no real supporter or protector, and he languished in poverty.

In August of 1836, Jules Pelouze, a protégé of Gay-Lussac and Liebig's best friend among the younger Parisian chemists, came to Giessen for several weeks of collaborative work, one subject of which was mellitic acid (today formulated as a benzene ring with six carboxyl groups on it). Citing the precedent of Dulong's view of oxalic acid, they formulated it in their joint paper as a dibasic hydracid.[42] Later that year, Liebig began to unfold

---

41. "Réponse de M. Dumas à la lettre de M. Berzelius," *Comptes rendus* 6 (1838), p. 699.

42. Liebig and Pelouze, "Vermischte Notizen," *Annalen* 19 (1836), p. 252.

his emergent hydracid theory of organic acids privately to Berzelius.[43] Such a theory ought to have brought him closer to the French substitutionists, especially in view of his hints in the 1832 benzoyl radical paper. However, he and Dumas, not completely in either camp, could agree on nothing in their correspondence or in their polemical articles.

As was mentioned earlier in this chapter, in the late summer of 1837 Liebig spent 2 months in Great Britain; in October, on his way back to Germany, he stopped for 17 days in Paris, where he had long conversations with Dumas. Apparently as a result of these conversations, Dumas capitulated to the ethyl-radical formulations of Liebig. Dumas and Liebig issued a joint manifesto declaring that organic chemistry was no more and no less than the chemistry of compound radicals.[44] Back home in Giessen, Liebig wrote to Dumas on 11 November to outline his hydracid theory of organic acids; Dumas responded favorably on the 24th. Upon receiving this reply, Liebig wrote out the theory for publication. It appeared 3 weeks later, with some additional experiments by Dumas.[45]

The unstable alliance of Liebig and Dumas did not last out the spring of 1838, but it marked a crucial stage in the thinking of both men. Dumas had second thoughts about some of the implications of hydracids, and backed away. In the meantime, Berzelius attacked both men for their apparent abandonment of electrochemical-dualist precepts. Liebig then published an expanded version of the hydracid paper in his own journal.[46]

And then Dumas suddenly joined the substitutionists in a whole-hearted fashion. Three months after having energetically denied, in an open letter printed in the *Comptes rendus*, that he had ever believed that chlorine could play the same chemical role as hydrogen, he discovered a remarkable new

43. J. Carrière, ed., *Berzelius und Liebig*, second edition (Lehmann, 1898), pp. 121–123, 134, 142–145, 158–159, 165–168.

44. Dumas and Liebig, "Note sur l'état actuel de la chimie organique," *Comptes rendus* 5 (1837): 567–572.

45. Dumas and Liebig, "Note sur la constitution de quelques acides," *Comptes rendus* 5 (1837): 863–866. The chronology is that stated later by Liebig in "Lettre de M. Liebig à M. le Président," *Comptes rendus* 6 (1838): 823–829, on 825–826. Dumas's undated letter to Liebig (Liebigiana IIB, Bayerische Staatsbibliothek, Munich) is accompanied by an annotation, possibly in Liebig's hand, of 29 November.

46. "Lettre de M. Berzelius à M. Pelouze," *Comptes rendus* 6 (1838): 629–644; "Ueber die Constitution der organischen Säuren," *Annalen* 26 (1838): 113–189.

substance: a fully chlorinated acetic acid. In 1839 and 1840 he developed a detailed theory of chemical types, the leading proposition of which was that "in organic chemistry there exist certain 'types' which are conserved even when one has introduced, in the place of the hydrogen which they contain, equal volumes of chlorine, bromine, or iodine."[47] He summarized his new position, making reference to the older orthodoxy to which until recently he had subscribed:

We have two systems before us: the one ascribes a principal role to the nature of the elements, the other to their number and arrangement. Pushed to extremes, each of these systems, in my opinion, can lead to absurdity. . . . The nature of the elements, their weight, form, and arrangement, all must exert a real influence on the properties of the body. The influence of the nature of molecules was well defined by Lavoisier; that of their weight was characterized by Berzelius's immortal work; Mitscherlich's discoveries relate to the influence of their form; and only the future can tell whether the current work of the French chemists is destined to yield the key to the role which their arrangement plays.[48]

In making this move, Dumas was really doing no more than following his former student Laurent, who had been making similar statements for several years. But Dumas regarded his theory as distinct from Laurent's, and the two men did not reconcile.

The final act of this drama had an unusual twist. Throughout their careers, from their first meeting in Paris early in 1823 until the rise of type theory at the end of the 1830s, Liebig and Dumas had followed curiously similar paths: organic analysis; chemistry of nitrogenous organic compounds; the relations of alcohols, ethers, esters, and aldehydes; radical theory; hydracid theory; metabolic chemistry of plants and animals; organic substitution and types. And now, independently and almost simultaneously, both gave up all their efforts at elucidating fundamental chemical theories.

Dumas, burdened with multiple positions, financially well off, and increasingly tapped for administrative assignments, became ever more caught up in political activities, both in the academic world and in the wider

47. Dumas, "Acide produit par l'action du chlore sur l'acide acétique," *Comptes rendus* 7 (1838), p. 474; "Mémoire sur la constitution de quelques corps organiques et sur la théorie des substitutions," *Comptes rendus* 8 (1839), pp. 621–622.

48. Dumas, "Mémoire sur la loi des substitutions et la théorie des types," *Comptes rendus* 10 (1840), pp. 177–178.

world. Liebig, worn down to the point of physical and mental illness by the strain of heavy research and teaching, and also by the vitriolic polemics in which he had been engaging for years, decided that he had "a real fear of theoretical discussions"[49] and threw it all over for more applicable areas of research, such as physiology, agricultural chemistry, and consumer products. Simultaneously, as I have mentioned, he expanded his laboratory, and the flood of students after 1839 facilitated certain kinds of research but discouraged others. Hofmann's description of Liebig's working patterns in 1842 reinforces the impression that Liebig had bid adieu to the theoretical dialectic after about 1840.

Liebig's turn away from theories took years to consummate, though the beginning point of the process can be fixed with precision. In September of 1838, Liebig, Friedrich Wöhler, Leopold Gmelin, Heinrich Rose, and Gustav Magnus traveled together to Freiburg to attend the annual meeting of the Gesellschaft Deutscher Naturforscher und Aerzte. There they met, among others, J. F. Persoz and Charles Oppermann (both of Strasbourg, and the latter a former student of Liebig's), who cordially invited the whole group to visit them in Strasbourg, 40 miles away. Since the meetings were boring and the weather and scenery magnificent, and since they had Wöhler's carriage at their disposal, they went. In Strasbourg, Oppermann served as tourist guide for the group. They also sampled some chemistry in Persoz's lab. Oppermann and Persoz even accompanied the five Germans on to Bouxwiller, where they toured one of the largest chemical factories in France. Wöhler reported that the champagne improved the difficult personality of Persoz.[50] He did not mention meeting other chemists in Strasbourg; however, Cailliot, Gerhardt, and Wurtz were all there at the time,[51] and it would be surprising if they were not at least introduced. Wurtz later reported that Gerhardt, despairing of an academic career, was at the point of accepting an industrial position when Liebig persuaded him to go to Paris and ally himself with Dumas. If the story is true, the conversation

49. Carrière, p. 191.

50. All of this is related in Wöhler's long, interesting, and humorous letter to Berzelius of 14 October 1838 (Wallach, *Briefwechsel*, vol. 2, pp. 59–65).

51. The five Germans were in Strasbourg during the last week in September, and departed home for Germany on the 30th. Wurtz was then still a medical student, but also first préparateur in the Faculté.

must have occurred during this visit to Strasbourg, for Gerhardt departed for Paris within a few days after the Germans departed.[52]

Wöhler commented that their joviality during the journey did not exclude long, earnest conversations on chemistry, "from which I think we derived more mutual benefit than from all the learned papers in Freiburg." Leopold Gmelin, something of a father figure to German chemists, persuaded the other four that atomic weights necessarily entailed atomic theory, a group of speculative notions about the world of the invisibly small. Chemical equivalents, on the other hand, Gmelin urged, are purely empirical entities that emerge directly from measured combining quantities of substances that react together. If chemists were uniformly to adopt equivalents, they not only would be rejecting speculation; they would be creating a uniform system that would never again change. The five chemists entered into a pact to do just that, though it would be several years before this would come to pass. Liebig was, however, sufficiently inspired by the conversation to write to Pelouze and to Berzelius, seeking their support for this organized movement toward equivalents.[53]

Unfortunately, Gmelin's reasoning was flawed: operationally considered, the system of chemical equivalents that developed in the nineteenth century was nothing more than a rival set of atomic weights, fully as theoretical as Berzelius's system and no more empirical.[54] This was not, however, noted at the time, and Liebig's resolve must have been strengthened by his upsetting polemics with the French chemists. After announcing his hydracid theory with such fanfare in the spring of 1838, Liebig abandoned it almost immediately and returned to the dualistic additive oxyacid formulations. His reasons are unclear, but one likely possibility is that he associated the hydracid theory with substitution and type theory (which were then giving

---

52. Wurtz ("Éloge de Laurent et de Gerhardt," p. 17) stated that Gerhardt arrived in Paris on 21 October 1838. Grimaux and Gerhardt Jr. (*Gerhardt*, p. 30) reported that Gerhardt left Strasbourg "in the first few days of October."

53. Liebig to Pelouze, 14 October 1838, Dossier Pelouze, Archives de l'Académie des Sciences (R. Fox, *The Caloric Theory of Gases* (Oxford University Press, 1971), pp. 319–320); Liebig to Berzelius, 5 September 1839 (Carrière, pp. 201–202). Either Pelouze did not respond to Liebig's appeal or (more likely) his reply was not saved, for nothing about this matter is mentioned in Pelouze's surviving letters to Liebig in the Bayerische Staatsbibliothek.

54. See chapter 10 below.

him so much polemical grief) and sensed that attacking the latter entailed abandoning the former. He may also have simply been more comfortable with the oxyacid formulations, regarding them as more empirical and less theoretical. And now Liebig's flirtations with chlorine substitution and types ceased. He returned wholeheartedly to the comfortable orthodoxy of radical theory. In 1840 he confessed to Berzelius "an unconquerable revulsion and aversion to what is happening in chemistry these days." "The battle over substitution theory," he continued, "is bringing it to a head, everything we have done and worked on is being used for the gilding of personalities. I have become absolutely sober, colder and more rational that you can imagine, after I read Persoz's thick book on our theories. . . ."[55]

For years thereafter, Liebig was inconsistent; as late as 1843 he was still writing essays in praise of atomic theory. In 1844, however, Liebig made a decisive and permanent break, after which he employed only the purportedly empirical equivalents in his chemical formulas. But 6 years was a long time to have wavered, and his vacillation on such central issues must surely have caused consternation among his students, who must have had difficulty in discerning a Giessen party line to follow. Wurtz's presumed first meeting with Liebig (in Strasbourg, in September of 1838) occurred within days of the beginning of this period of vacillation, and his semester in Giessen came toward the end of it, after Liebig pledged to leave the uncertain world of theory but before he fully capitulated to what has been called "textbook positivism."[56] Wurtz may therefore be forgiven for proposing two alternative constitutions for hypophosphorous acid—an oxyacid and a hydracid formulation—without attempting to decide between them.

Transition to Paris

On 21 August 1842, at the end of his semester in Giessen, Wurtz wrote the following to his sister Sophie: "What made life in Giessen so pleasant was the inner bond that prevailed among the chemists. United by the same scientific interest, suffused with the same devotion to the teacher, we worked the whole day together and supported each other at every instant with word

55. Liebig to Berzelius, 26 April 1840 (Carrière, pp. 210–211).

57. B. Bensaude-Vincent, "Atomism and Positivism," *Annals of Science* 56 (1999), p. 90.

and deed. . . . This wonderful life is now ending, or rather has already ended, for most of us have already departed. On the other hand, I very much look forward to returning to all of you." Then came a prophecy, soon to be fulfilled: "Next year I must accomplish something significant. Then I will go to Paris! That is my plan, that is my hope! I hope that God will give his blessing to it!"[57]

When the time came to bid farewell to his mentor, Wurtz was overcome with emotion and could hardly speak. He dearly wanted to request permission to write to Liebig from time to time, but lost heart at the last moment; he was all the more grateful when Liebig granted the privilege without prompting.[58] He also received from Liebig strong letters of recommendation to the Paris elite, especially Dumas. After a tour through Dresden, Prague, Vienna, Salzburg, and Munich, he returned home to Strasbourg in the fall.

Wurtz still had to finish his doctoral work at the Faculté de Médecine and write a thesis, and these tasks took him the better part of a year. In August of 1843, he passed the final defense of his thesis on soluble albumin and fibrin and was awarded a prize by the medical faculty. Wurtz's progress was slow partly because both Jacob Berzelius and Heinrich Rose had challenged his conclusions on the constitution of hypophosphorous acid, and he felt that in order to respond effectively he needed to repeat and expand upon the work he had done in Giessen.

Wurtz also had taken on translation assignments. Three months after his departure from Giessen, he received a letter from Liebig inquiring whether he would be willing to translate the final volume of the latter's organic chemistry textbook into French. Wurtz was thrilled, and hastened to accept, although for unknown reasons Liebig later reneged on the agreement.[59] Almost at the same time, on a visit to his home in Strasbourg in August of 1842, Gerhardt inquired of Cailliot and Oppermann whether

57. Hofmann, "Wurtz," pp. 214–215.

58. Ibid., p. 215.

59. Wurtz to Liebig, 14 December 1842 and 29 July 1843, Liebigiana IIB, Bayerische Staatsbibliothek. Gerhardt was the sole translator of Liebig's *Traité de chimie organique* (Masson, 1840–1844); it is probable that the uneven relations between the two men led Liebig temporarily to consider Wurtz as an alternative translator. See Grimaux and Gerhardt, *Gerhardt*, pp. 29, 44, 77, 461–463.

they could recommend a bilingual chemist to translate his new *Précis de chimie organique* into German. Cailliot and Oppermann recommended Wurtz; Gerhardt subsequently offered him the job, and Wurtz accepted.[60] Apparently, Wurtz took the work not because he was a disciple of Gerhardt (as is sometimes implied in the literature) but simply because it was a useful task that he could do well and get paid for. Wurtz, now 26 years old, needed all the income he could get. This translation must have occupied a great deal of his time during his last 18 months in Strasbourg.

After passing his dissertation defense, Wurtz had more free time to follow up two ideas that had spun off from the aforementioned chemical projects. He thought he was able to discern the generation of butyric acid in decomposed fibrin, a physiologically interesting observation, but it proved tricky to work through the details. A second and more significant discovery came from a novel reaction he encountered in his follow-up work with hypophosphite salts. He discovered a compound of copper and hydrogen, one of a very small number of metal hydrides then known to exist. He worked on these projects during the fall and winter of 1843–44 in Strasbourg, where he also continued in his duties as chef des travaux chimiques in the medical faculty. He kept Liebig informed on the details of his research,[61] and in March of 1844, following the advice given him by Gerhardt, he introduced himself by letter to Dumas.[62] Writing to Liebig in German and to Dumas in French, he tactfully indicated to each man his high respect and devotion as a past and a future protégé, respectively.

In his letter to Dumas, Wurtz enclosed three short papers, "which I hope," he wrote, "will not be unworthy of your attention." One summarized his doctoral work on soluble albumin, one the formation of butyric

60. Gerhardt to Wurtz, 9 January 1844, in Grimaux and Gerhardt, *Gerhardt*, pp. 76–77, 440–441, 461–464; Tiffeneau, *Correspondance*, vol. 2, p. 303; *Grundriss der organischen Chemie, aus dem Französischen von Adolph Wurtz* (Schmid & Grucker, 1844–1846).

61. Wurtz to Liebig, 17 October 1843, 28 November 1843, and 1 May 1844, Liebigiana IIB, Bayerische Staatsbibliothek. In his letter of 29 July 1843 he indicated his desire and intention to move to Paris that fall, but clearly that plan fell through.

62. Gerhardt to Wurtz, 9 January [1844], in Tiffeneau, *Correspondance*, vol. 2, p. 306; Wurtz to Dumas, 14 March 1844, Dossier Wurtz, Archives de l'Académie des Sciences.

acid from decomposing fibrin, and one the discovery of copper hydride. Less than a month later, all three appeared in the *Comptes rendus* , apparently having been inserted by Dumas, who must already have known the young chemist's name from his hypophosphorous acid paper.[63] On about 14 May 1844, Adolphe Wurtz arrived in Paris, where he would spend the remainder of his career.[64]

63. Wurtz, "Sur l'albumine soluble," *Comptes rendus* 18 (1844): 700–702; "Sur l'hydrure de cuivre," *Comptes rendus* 18 (1844): 702–704 (reprinted in *Annales de chimie* [3] 11 [1844]: 250–252); "Sur la transformation de la fibrine en acide butyrique," *Comptes rendus* 18 (1844): 704–705.

64. In his letter to Liebig of 1 May 1844 (Bayerische Staatsbibliothek, Liebigiana IIB), he indicated his intention to go to Paris: ". . . wohin ich nächsten Montag [6 May] über acht Tage abreisen werde."

4

Making a Career

Wurtz rented student digs near the Faculté de Médecine, on the Left Bank. Soon after his arrival (May 1844), he must have presented himself and his Alsatian and Giessen credentials to Dumas. At this time Liebig and Dumas were in one of the rockier periods of their tumultuous relationship, but there is no reason to believe that a strong word from Liebig had less than a strong effect on Dumas; moreover, Wurtz's name was by no means unknown to Dumas, for he had already published four papers, two of which were important by any measure. The two earliest Wurtz biographers state that Dumas wanted immediately to place Wurtz in his own private research laboratory, but that all the places were then occupied. Instead, Dumas recommended the young man to Antoine Balard, who was professeur adjoint at the Sorbonne, giving the chemistry lectures in place of the titular professor, Dumas. (In 1844 Dumas had three professorships, as well as serving as Doyen of the Paris Faculty of Sciences and directing his own private research laboratory; it was obviously too much for one person.)

Wurtz had no remunerative employment during this period, and was forced to live frugally in the extreme. He maintained a small income from his work on translating Gerhardt's *Précis de chimie organique* and from giving private lessons to medical students. Friedel related an anecdote from this first period in Wurtz's Paris career. Wurtz was a hard worker, and one day he was later to leave the Sorbonne than anyone else. He found the doors locked, could not raise anyone to let him out, and panicked. He finally found succor by throwing pebbles against the residence windows on the second floor of the courtyard. An old man appeared and asked "What do you want, my child?" It was the famous philosopher Victor Cousin, rector of the Académie de Paris, who then rescued the prisoner.[1]

This period in Balard's lab lasted only a few months, and probably ended in the early autumn of 1844; Dumas then took Wurtz into his private research laboratory, across from the Jardin des Plantes. Thus began Wurtz's long association with Dumas, one of the most powerful men in nineteenth-century French science.

Giessen on the Seine; or, Dumas in Search of a Laboratory

In 1836, Dumas was a professor at the École Polytechnique and at the École Centrale and a professeur adjoint at the Sorbonne. What he did not yet have, even with these multiple positions, was a good research laboratory.[2] Gay-Lussac inherited what had been Lavoisier's lab at the Paris Arsenal, but there was not much else in Paris, the decent equipment at the École Polytechnique having disappeared after the militarization of the institution (1805–1815) and its removal from the Palais Bourbon to the buildings of the Collège de Navarre near the Panthéon in the Latin Quarter.[3]

1. C. Friedel, "Notice sur la vie et les travaux de Charles-Adolphe Wurtz," *Bulletin de la Société Chimique* [2] 43 (1885), p. ix; A. Hofmann, "Erinnerungen an Adolph Wurtz," in *Zur Erinnerung an vorangegangene Freunde* (Vieweg, 1888), vol. 3, pp. 218–221.

2. On Dumas's research laboratories, see L. Klosterman, "A Research School of Chemistry in the Nineteenth Century," *Annals of Science* 42 (1985), pp. 7–11. As sketchy as Klosterman's information on this subject is, it has been invaluable in the absence of other more detailed documentation, and it is based upon careful examination of primary documents. Klosterman was much more interested in (and successful at) investigating the research school of Dumas, than the material resources at Dumas's disposal. I focus on the latter in the following text.

3. M. Bradley, "The Facilities for Practical Instruction in Science during the Early Years of the École Polytechnique," *Annals of Science* 33 (1976): 425–446. The 1820 report on the school's physical facilities, reproduced by Bradley (pp. 444–446) from the school's archives, emphasizes the decrepit nature of the building and its contents and the lack of appropriate laboratory space. However, an 1827 plan of the institution shows a "laboratoire de chimie" of about 50 by 6 meters in the eastern side of the complex, near the corner of the Rue d'Arras and the Rue Traversine: Général Alvin, *L'École Polytechnique et son quartier* (Gauthier-Villars, 1932), pp. 134, 231. On the decline of chemical instruction at the École, see also W. Smeaton, "The Early History of Laboratory Instruction at the École Polytechnique, Paris, and Elsewhere," *Annals of Science* 10 (1954): 224–233; J. Langins, "The Decline of

Upon his appointment as Thenard's répétiteur at the École Polytechnique in January of 1824, Dumas found the laboratory risible: in A. W. Hofmann's words, it consisted of "a sort of kitchen for the preparation of lectures, and a small room without a fireplace, possessing cabinets for preparations [but] no balance, no barometer, no thermometer, no graduated cylinder, no precision instruments of any kind for research." Indeed, it was not intended for research at all, but only for the preparation of lecture demonstrations.[4] Still, Dumas later averred that the acquisition of this laboratory was one of the greatest events of his lifetime, for it made it possible for him to pursue his science.[5] Dumas gradually improved the conditions, and by 1826 he could collaborate with a student there. Jules Pelouze arrived at the École Polytechnique as Gay-Lussac's répétiteur in 1831, replacing César Despretz, and occupied a neighboring laboratory space. Initially, Pelouze and Dumas got along well in these close quarters, even though their respective superiors, Gay-Lussac and Thenard, were no longer friendly with each other.[6]

Dumas made certain that there were laboratories at the new École Centrale after its founding in 1829; however, neither of his facilities was really adequate. And what of the famous Sorbonne, where he lectured from 1832 on? Very little is known about the laboratory facilities there during the Restoration and the July Monarchy. No chemical laboratory was provided

---

Chemistry at the École Polytechnique, 1794–1805," *Ambix* 28 (1981): 1–19; B. Belhoste, "Un modèle à l'épreuve," in *La formation polytechniennne, 1794–1994*, ed. Belhoste et al. (Dunod, 1994), pp. 19, 23–24.

4. Hofmann, "Zur Erinnerung an J. B. A. Dumas," *Erinnerung*, vol. 2, p. 242. This must refer to a laboratory occupied by the répétiteurs, and available also (in principle) as a place for personal scientific research, as opposed to the student labs where exercises were done. The exact location of this laboratory in the École Polytechnique campus is in doubt, nor can Hofmann's description (perhaps derived from an oral communication from Dumas) be even qualitatively confirmed from independent evidence.

5. Dumas to Liebig, n.d. [ca. June 1851], Liebigiana IIB, Bayerische Staatsbibliothek, Munich.

6. Pelouze wrote Liebig on 19 June 1832 that, despite his initial presumption to the contrary, Dumas "travaille avec un soin extrême et possède un grand savoir. C'est un des hommes les plus laborieux de Paris." On 30 April 1834 he told Liebig "il n'est personne pour la science duquel il [Dumas] professe plus de respect que pour vous" (Liebigiana IIB).

Figure 4.1
The École Polytechnique was housed in the former Collège de Navarre, near the Panthéon, from 1805 until 1976. Photograph taken in March 1999 by the author.

for the Faculté des Sciences upon its establishment in the Sorbonne buildings in 1821, and the leading Sorbonne chemists of the 1820s and the early 1830s, Gay-Lussac and Thenard, did no personal research there.[7] In an official document from 1837 (see below), Dumas referred to the Sorbonne's chemical laboratory as "tout à fait insuffisant." (But by then it must at least have existed, perhaps cobbled together on the initiative of Dumas.) Thirty years later it was described just as dismissively in Victor Duruy's massive report *Statistique de l'enseignement supérieur*. In 1871 a student of Wurtz's described it as "a dark and dank room, more than a meter lower than the Rue Saint-Jacques." A floor plan of the old Sorbonne (1892) shows it as a

7. Gay-Lussac conducted his research at the Arsenal, and Thenard did essentially no research at all after about 1810. Recall from the previous chapter that the Paris Faculté des Sciences was founded under the assumption that no research would be performed there. The state research institutions were exemplified by the Collège de France, and, since cumul was operating well, no administrator felt pressure to create laboratories at the Sorbonne. Instead, makeshift research facilities were slowly cobbled together by interested professors.

small annex tacked on to the mezzanine level, on the Rue Saint-Jacques side opposite the Église de la Sorbonne.[8] The research carried out by the young Wurtz in Balard's lab in the summer and fall of 1844 was apparently some of the earliest chemical research conducted within the precincts of the Sorbonne.

Dumas found the dearth of facilities frustrating in the extreme. After their marriage, Dumas and his wife Herminie lived with her brother and father, Adolphe and Alexandre Brongniart, at the official residence of the Muséum d'Histoire Naturelle. Alexandre, who did not lack for funds, was eventually persuaded to buy a plot of vacant land at 35 Rue Cuvier[9] (across from the Muséum and the Jardin des Plantes) for the planned purpose of building his son-in-law a chemical laboratory. Until this lab was available to him (as described later in this chapter), Dumas's work life was logistically difficult. The École Polytechnique was within a reasonable distance of his residence, but the École Centrale was much further, and his duties required his presence at each location every day. Dumas, who resented the loss of time entailed by his commutes and who envied the German academic tradition of combined residence-laboratories,[10] desired a single laboratory in or near his residence. Late in 1832 he was able to set up a small private lab in an annex of the École Polytechnique, but it was still far from what he wanted.

Dumas's struggles to improve the material conditions of his workplace must be viewed, as he himself clearly did, in comparison to the far more favorable situation of his foreign rival. Liebig lived in a small town, with no

8. *Statistique de l'enseignement supérieur 1865–1868* (Imprimerie Impériale, 1868), pp. 58, 424–425; F. Papillon, "Les laboratoires en France et à l'étranger," *Revue des deux mondes* [2] 94 (1871), p. 601; O. Gréard, *Nos adieux à la vieille Sorbonne* (Hachette, 1893), p. 393; Gréard, *Éducation et instruction*, second edition (Hachette, 1889), pp. 17–25, 245 (Dumas's 1837 report), and Plans 4 and 5; H. Nénot, *La nouvelle Sorbonne* (Colin, 1895). On these questions see also Klosterman, "Dumas," pp. 7–11. All these details are imprecise and are in need of verification and elaboration.

9. Actually at 14 Rue de Seine-Saint-Victor until the street was renamed around 1840. The house no longer exists. After the Revolution of 1848, Dumas and his family moved to the Brongniart family home at 3 Rue Saint-Dominique, in the Seventh Arrondissement. See L. de Launay, *Une grande famille de savants, les Brongniart* (Rapilly, 1940), pp. 139–144.

10. Liebig (in Giessen) and Stromeyer and Wöhler (in Göttingen) lived above their labs; Bunsen (in Marburg) lived directly across the street from his.

commutes or urban distractions, in modest but extremely convenient lodgings directly upstairs from his institute. He was continually improving his laboratory there, and by 1835 he was accepting 20 students at a time. In that year his institute was taken over by the state, and the building underwent a significant renovation. Some of Liebig's Praktikanten were advanced students or collegial guests, so that the Giessen institute was gradually coming to operate as a virtual factory of new science. Pelouze, who visited Giessen for a few weeks in the late summer of 1836, ever after viewed the undistinguished little German town as no less than heaven on earth[11]; on his return, he may have painted Liebig's situation in rosy terms to Dumas, with whom he was still friendly.

In the spring of 1837, Dumas accepted a proposal Liebig had made 6 years earlier for a collaborative project.[12] It was not a good example, Dumas wrote, for the two leaders—the two creators, in his view—of the science of organic chemistry to feud with each other; they ought to combine their forces. "And moreover, Liebig," he wrote, "even if we were to separate ourselves, posterity would reunite us, if it concerns itself with us. Until now, we have made nearly equal efforts to found organic chemistry, and if we continue, it is possible that one day it will be definitively established by our efforts, whether they remain separate or united, each of us retaining a nearly equal part in the success achieved."[13] After paying a visit to Dumas in Paris in October of 1837, Liebig suggested organic acids as the topic of their joint

11. "Giessen, Giessen, ah!," Pelouze exclaimed, almost overcome with passion; "jamais matelot n'a demandé la terre avec plus d'impatience." Later, in response to the engagement of Liebig's daughter Agnes to the Giessen professor P. M. Carrière, he wrote: ". . . te donnera l'occasion de revoir plus souvent ta vieille patrie, le théâtre de tes plus brillants exploits, les bois charmants, les passages délicieux de la célèbre ville chimique." (Pelouze to Liebig, 25 January 1838 and 18 December 1852, Liebigiana IIB)

12. In his letter to Dumas of 23 October [1831] (Fonds Dumas, Archives de l'Académie des Sciences, Paris), Liebig suggested that Dumas propose a subject for a collaborative project, just as he had recently done with Wöhler.

13. Dumas to Liebig, May 1837, Liebigiana IIB: "Et d'ailleurs, Liebig, nous aurions beau nous séparer, la postérité nous unirait, si elle s'occupe de nous. Jusqu'à présent, nous avons fait des efforts à peu près égaux pour fonder la chimie organique et si nous continuons, il est possible qu'un jour elle s'établisse définitivement par nos efforts, soit qu'ils demeurent séparés, soit qu'ils se réunissent, chacun de nous conservant une part à peu près égale dans le succès obtenu."

project. Liebig sent Dumas a detailed prospectus of his emerging hydracid theory, and proposed, if he agreed with the content, that Dumas publish it in Paris under both their names.[14] However, Dumas was well aware of his disadvantageous position in such a collaboration. "How very fortunate you are to have a battalion of eager chemists at your disposal," he replied. "I hope someday to offer you something comparable, but for the moment I am far from that."[15] Apparently one of their conversations in Paris was on laboratory facilities; it seems that Dumas, looking to the future, wanted to pick Liebig's brain.[16] Beyond professional collaboration, Dumas and Liebig each had complicated reasons for pursuing their truce and brief alliance.[17]

About this time the new Minister of Public Instruction, Narcisse de Salvandy, asked the dean of the Faculté des Sciences, Thenard, for advice on how it might be possible to raise the overall level of the science faculties of France. Thenard turned to his protégé Dumas, then professeur adjoint at the Sorbonne, and asked him to draft a report. Dumas's draft, dated 15 November 1837, is highly revealing.[18] He proposed increasing the number of science chairs and salaries in the provincial faculties, increasing the faculty size and physical size of the Sorbonne, adding a proper library, renovating and improving the physical collections, and adding courses in mathematical physics and experimental mechanics.

Most significant for the present story, Dumas also proposed improvements at the Sorbonne that were modeled on Liebig's facilities in Giessen.

14. Liebig to Dumas, 19 November 1837, Fonds Dumas, Archives de l'Académie des Sciences.

15. Dumas to Liebig, n.d. (with annotation, possibly by Liebig, as received on 29 November [1837]), Liebigiana IIB: "Que vous êtes heureux de pouvoir ainsi disposer d'un bataillon de chimistes zélés. J'espère vous en offrir autant quelque jour; mais pour le moment je suis loin de là."

16. This presumption of such a conversation between Liebig and Dumas, and the archival evidence for it, is nicely laid out on p. 9 of Klosterman, "Dumas."

17. See F. Holmes's biography of Liebig in *Dictionary of Scientific Biography*; see also Mi Gyung Kim, "Constructing Symbolic Spaces," *Ambix* 43 (1996): 1–31.

18. In Gréard, *Éducation* it is reproduced from the Procès-verbaux of the Faculté des Sciences of 15 November 1837 (pp. 236–252), along with faculty comments on the draft (pp. 252–255). On the final draft of 6 December 1837, which I sought but failed to find in the Archives Nationales, see also H. Paul, *From Knowledge to Power* (Cambridge University Press, 1985), p. 18; Paul, *The Sorcerer's Apprentice* (University of Florida Press, 1972), p. 5.

He argued for a new chemical lecture theatre "analogous [to those] recently constructed in other countries." "In the foreign institutions to which we refer," he noted, the laboratory where demonstration experiments were prepared was located immediately adjacent to the lecture hall, which made preparation of demonstrations convenient. Dumas also proposed a new arrangement whereby two agrégés devoted exclusively to laboratory preparations and scientific research would be added to the staffing of the chemical section. This intimate confluence of theory and practice "would perfectly prepare men to render great services to science."[19] Further details in the description of the physical facility desired and of the two new positions (essentially a French version of German academic Assistenten) leave no doubt as to the single model in Dumas's mind, notwithstanding the vague references to "foreign institutions" and use of the plural number. Simply stated, Dumas wanted to transform the Sorbonne chemical laboratory into an organization patterned on the Giessen laboratory. Indeed, the details of the recently concluded renovation of the Giessen lab are clearly recognizable in Dumas's text.[20] Unfortunately, Dulong objected to the plan to add two new agrégés as too expensive. Despite Dumas's rejoinder proclaiming the advantages that "foreign countries" gained from such an arrangement, the proposal was dropped from the draft.[21]

Meanwhile, a new vacancy at the Faculté de Médecine had been announced. It is said that the French response to deficiencies in higher education has always been to invent a new institution rather than to reform existing ones. There was an analogous canon in nineteenth-century French academic career strategies: the response to a personal deficiency, in money

19. Gréard, *Education*, pp. 245–249. Dumas, like Liebig, always stressed this dynamic relationship. For those who might complain that his textbook contained too much atomism and other theory, Dumas preemptively responded: ". . . this book is addressed to young men and not to manufacturers who are already educated; that my intention has not been to describe technological practice, but to clarify its theory; and that the scientific details that intimidate middle-aged manufacturers will be but child's play for their children, when they will have learned in their secondary schools a little more mathematics and a little less Latin, a little more physics and chemistry and a little less Greek." (*Traité de chimie appliquée aux arts* (Béchet, 1828), vol. 1, p. viii)

20. On the Giessen renovation of 1835, see J. Volhard, *Justus von Liebig* (Barth, 1909), vol. 1, pp. 78–79.

21. Gréard, *Education*, p. 252.

or perquisites, was often simply to add a new position. Dumas was already a professor at three institutions, but there was no reason why he could not apply for the new opening if he wished. However, there was no chemical laboratory at the École de Médecine, and in December of 1837, when his proposals for renovating the Sorbonne lab were still being considered, Dumas declared to Liebig his intention not to apply:

I tell you . . . that I do not want the School of Medicine. I prefer to remain at peace in my mediocrity and totally engaged in science. I have a very nice theatre for teaching [at the Sorbonne]. There I shall be the master of chemistry in France, if I am allowed to do what is needed to improve the material means. Our ideas will be proclaimed near and far, and they will triumph. But in order to make no compromises I want no new occupations, I want to be completely in my laboratory.[22]

But Dumas's lingering dissatisfaction with his small lab at the École Polytechnique was apparently increased by the extraordinary cold spell that gripped Paris the next month; the building became uninhabitable.[23] Liebig was just then beginning to make use of his scientific "battalions" in a concerted and organized fashion to further his own well-articulated research agenda on organic hydracids. If Dumas was to be able to keep pace with Liebig in a collaborative project, something clearly had to change. Soon after writing the letter quoted above, he learned that his report to Salvandy would have no positive issue, so that even his Sorbonne position would prove worthless as far as material resources were concerned. Faced with all this, he suddenly changed his mind. He applied for the professorship in the Faculté de Médecine an hour before the registration period was to end.[24]

22. Dumas to Liebig, n.d., but probably 19 December 1837, Liebigiana IIB: "Je vous dis de plus que je ne veux pas de l'école de médecine. J'aime mieux rester tranquille dans ma médiocrité et tout à la science. J'ai un assez beau théâtre pour l'enseignement. J'y serai le maître de la chimie en France, si on me laisse faire ce qu'il faut pour améliorer le matériel. Nos idées seront proclamées haut et ferme et elles triompheront. Mais pour ne rien compromettre, je ne veux pas de nouvelles occupations, je veux être tout entier au laboratoire."

23. Described in Dumas's letter to Liebig, n.d., postmarked 21 January 1838, Liebigiana IIB. It was 17 degrees Celsius below zero in the lab, which made even the simplest operations impossible. Dumas simply was forced to wait out the cold weather.

24. This at least was the report in Pelouze to Liebig, n.d., postmarked 15 January 1838, Liebigiana IIB.

In three letters written in the first half of 1838, Dumas wrote to Liebig to say that he (actually, his father-in-law) was building a new laboratory near his residence, and that he had decided to compete for the position at the Faculté de Médecine after all, relinquishing his professorship at the École Polytechnique. These events were related, for Dumas's plan was to use the extra income from the more lucrative medical professorship to enable him to finance the private lab.[25] (Ironically, Pelouze inherited Dumas's professorship at the École Polytechnique, and used the increase in salary to open *his* private lab at same time; by this time Pelouze and Dumas had become enemies and rivals.) In the first of these letters Dumas told Liebig that he was planning to use this facility, after renovation, to devote the entire year to their substantial collaborative project on organic acids.[26] At that time he was already in the middle of the series of examinations that constituted the competition for the new chair.

The Paris medical faculty had always had a chair of pharmacy (held successively by Antoine Fourcroy, Louis Nicolas Vauquelin, and Nicolas Deyeux) and a separate chair of medical chemistry (held for many years by M. J. B. Orfila, who from 1831 was also dean of the faculty). When Deyeux died in 1837, a proposal was floated to change the definition of the chair of pharmacy, in order to give greater emphasis to the burgeoning field of organic chemistry, to "pharmacy and organic chemistry." The Ministry of Public Instruction and its national Conseil approved the plan, and that summer the Ministry announced that a competition for the new chair would be held.[27]

There were five candidates for the chair. The competition required no fewer than 27 sessions, conducted over a period of nearly 2 months (February and March 1838) in front of a jury of 13 scholars selected from

25. "Par là [taking the position at the Faculté de Médecine] je me procurais l'argent nécessaire pour fonder un laboratoire, comme je l'entendais et comme je vais le faire." (Dumas to Liebig, n.d., postmarked 21 May 1838, Liebigiana IIB)

26. Dumas to Liebig, n.d., postmarked 25 February 1838, Liebigiana IIB. Liebig had declared his intention thenceforth to publish separately, and Dumas was attempting to keep their collaborative alliance together. See letters of Liebig to Dumas, 19 February and 10 March 1838, Fonds Dumas, Archives de l'Académie des Sciences.

27. AJ16/6311; A. Corlieu, *Centenaire de la Faculté de Médecine de Paris (1794–1894)* (Imprimerie Nationale, 1896), pp. 242–243, 279–285.

the Académie de Médecine and the Faculté de Médecine. Among other tests, Dumas was required to present extemporaneous lectures on topics drawn by lot: one on "the chemical and pharmaceutical relations of sugars" with a day's preparation, and one on "the chemical and pharmaceutical relations of milk" with 3 hours' preparation. Dumas won in every category of the competition, and was named to the chair on 9 April 1838.[28] This was the first academic chair for organic chemistry in the world, predating any other by decades. It was the chair that Wurtz would eventually inherit.

After the decision was announced, Dumas wrote to Liebig attempting to excuse his inconsistency in changing his mind about the position:

I find myself in contact with people who do not understand how someone could devote his life to science and who have not the slightest idea of the requirements of a position such as yours, or of that which I want to acquire, regarding the means of work [i.e., laboratory facilities]. I was unable to obtain a thing when I tried to develop my material means, and I have had to procure resources myself. . . . Since I came to Paris, I have been seeking a way to create a laboratory broadly constituted under my direction. I think I have finally succeeded in this, and that gives me some consolation. In two or three months I will be able to put ten selected students to work in my house, and I will be able to devote four or five thousand francs per year to their experiments. Only then will I be in a position to resume my experiments in competition with yours. At the moment I can't keep pace with you.[29]

Dumas spent the summer of 1838 converting the Brongniart property into a suitable facility and moving his equipment from the École Polytechnique to the Rue Cuvier. By autumn he was able to begin accepting students. Among those accepted were Auguste Cahours, Charles Gerhardt, and an Italian political refugee named Faustino Malaguti. Others who

---

28. AJ16/6311.

29. Dumas to Liebig, n.d., postmarked 21 April 1838, and n.d., dated in unknown hand May 1838, Liebigiana IIB: "Je me trouve en contact avec des gens qui ne comprennent pas que l'on dévoue sa vie à la science et qui n'ont pas la moindre idée des exigences d'une position comme la vôtre ou celle que je veux acquérir sous le rapport des moyens de travail. Je n'ai rien pu obtenir quand j'ai voulu développer mes moyens matériels et j'ai dû me faire des ressources. . . . Depuis que je suis à Paris, je cherche un moyen d'avoir un laboratoire largement constitué sous ma main. Je crois y être enfin parvenu et c'est là ce qui me fait quelque consolation. Je pourrai dans deux ou trois mois faire travailler dix élèves choisis chez moi et je pourrai consacrer à leurs expériences quatre ou cinq milles francs par an. Alors seulement, je serai en mesure de reprendre des expériences en concurrence avec les vôtres. Je ne puis pas aller votre pas dans ce moment."

Figure 4.2
Jean-Baptiste Dumas at about 50, as shown in J. Sheridan Muspratt, *Chemistry, Theoretical, Practical, and Analytical, as Applied to the Arts and Manufactures*, volume 1 (London: MacKenzie, 1853). Source: Photograph Collection, Deutsches Museum, Munich.

worked prior to Wurtz's arrival in this lab, perhaps then the best chemical research facility in France, included Jules Bouis, Eugène Chevandier de Valdrome, Z. Delalande, F. Hervé de la Provostaye, Félix Leblanc, Eugène Péligot, Édouard Saint-Èvre, and Francis Scribe. Dumas even accepted such foreigners as the Dane Bernhardt Lewy, the Belgians Louis Melsens and Jean Servais Stas, the Italian Rafael Piria, and the Pole Filip Walter. Those still there at the time of Wurtz's arrival (late in 1844) included Bouis, Cahours, Leblanc, Lewy, Saint-Èvre, Scribe, and Walter.[30]

As Dumas's letter to Liebig indicates, the cost of his students' research was borne by Dumas, whose newly enhanced income could now accommodate such an expense. Here his model diverged from Liebig's. Dumas apparently was intent on keeping pace with his rival while also taking action to ameliorate what he considered the lagging fortunes of French chemistry. The lab was, in essence, an amalgamation of the Liebigian model and the traditional French model. Dumas now had a proper "research group," but it was predicated on expanded personal patronage rather than the environment of a university or a private school.

Dumas's new position on the medical faculty had the effect of directing his research more toward physiologically relevant subjects, and so he drifted in a course curiously parallel to that of Liebig. Not coincidentally, their alliance did not survive the spring of 1838, and disputes concerning substitution theory and chemical theories of metabolism brought their relationship to its lowest ebb ever. Not everyone in the Paris collegial community was on Dumas's side in this confrontation. Gay-Lussac wrote the following to Liebig at the time of the split:

Now, my dear Liebig, I congratulate you on your escape from the straits you were in. I never understood your marriage, in particular I never imagined that it could

---

30. See Klosterman's painstaking work on the identity and evolution of the Dumas research group, summarized in "Dumas," pp. 7–25. Both Hofmann ("Wurtz," p. 221) and Friedel ("Wurtz," p. ix) state that it was the departures of Stas and Piria that created room for Wurtz, but Klosterman shows that both had departed by 1840. (Hofmann, who took much from Friedel's biography, may have repeated this from Friedel.) It is more likely that Melsen's return to Belgium in 1844 created the space. It should also be noted that Leblanc departed the next year, and that Cahours's presence in the lab after 1842 is probable, even if undocumented. Otherwise, the Dumas group remained nearly constant until his laboratory was forced to close in the revolutionary conditions of 1848. See also M. Chaigneau, *J. B. Dumas* (Guy le Prat, 1984), pp. 201–207.

last, and your divorce did not surprise me. Your two personalities were too opposed—great honesty and sincerity can never go well with craft and guile [une fine finesse]. You knew him so well, this brother Ignatius, this Jesuit, as you and others used to call him. You must have been hypnotized [fasciné], and in a way that I never would have believed you could have been; you must recognize in him, as in a serpent, a great hypnotic power. Don't think that there is another such in France![31]

Gay-Lussac's animus against Dumas is surprising but not impossible to understand. Liebig was always Gay-Lussac's protégé, and Dumas was Thenard's (indeed, it is possible to see Thenard's hand in some of Dumas's advancements during the 1830s), and when a split occurred between the younger men the older men chose sides.[32]

Upon Thenard's move to the vice-presidency of the Conseil Royal de l'Instruction Publique in 1841, Dumas was promoted from professeur adjoint to professeur at the Sorbonne, and in 1842 he succeeded Thenard as Doyen of the Faculté des Sciences. Thenard, now the de facto academic czar, relied heavily thereafter on Dumas's recommendations for chairs and promotions, and in this way Dumas became the most powerful academic scientist in France. Parisian chemical students began to refer to Dumas jocularly as "l'être suprême."[33] It is no wonder that his private laboratory became the French analogue to Liebig's. In the 1840s, the route to an academic chemical career in France almost necessarily led through the Rue Cuvier.

31. Gay-Lussac to Liebig, 12 June 1838, quoted directly (in French) in Liebig's letter to Berzelius of 28 June, in *Berzelius und Liebig, Ihre Briefe von 1831–1845*, ed. J. Carrière, second edition (Lehmann, 1898), p. 171. Liebig also showed the letter to Wöhler during their trip to Strasbourg. Wöhler described his gleeful reaction to Berzelius (*Briefwechsel zwischen J. Berzelius und F. Wöhler*, ed. O. Wallach (Engelmann, 1901), vol. 2, p. 61).

32. Unlike Gay-Lussac, neither Thenard nor Dumas knew the German language: "As I [Thenard] do not know German, he [Gay-Lussac] did me the favour of translating it."(M. Crosland, *The Society of Arcueil* (Harvard University Press, 1967), p. 375, referring to a paper of 1807) "Voilà la première fois de ma vie où je [Dumas] puisse me féliciter de ne pas savoir assez d'allemand pour vous [Liebig] lire" (Carrière, *Briefe*, p. 277).

33. A. Lieben, "Erinnerungen an meine Jugend- und Wanderjahre," in *Adolph Lieben Festschrift* (Winter, 1906), p. 5. Lieben was a student in Wurtz's lab in Paris in the period 1856–1858 and in 1862.

Gerhardt was one of the first who followed this path.[34] Born and raised in a Protestant[35] family in Strasbourg, he was a graduate of the Protestant Gymnasium there. After attending technical colleges in Karlsruhe and Leipzig, Gerhardt studied with Liebig in Giessen (1836–37) and, less satisfactorily, with Persoz in Strasbourg (1837–38).[36] Apparently Liebig persuaded him to go to Paris to seek out Dumas when Liebig visited Strasbourg in late September of 1838; Liebig knew directly from Dumas about the opening of the Rue Cuvier lab. Gerhardt spent 3 years in Paris, taking Dumas's courses and performing research (at first in Dumas's lab and later in the private lab of Henri Sainte-Claire Deville, who had set up a modest facility in a garret in the Rue de la Harpe, and in Pelouze's private lab). During this time Gerhardt established a somewhat distant but respectful relationship with Dumas, and exchanged warm letters with Liebig. Liebig continued to urge his former student to trust Dumas ("He has a magnificent character," Liebig wrote in August of 1839), and to avoid advocating theories lest Dumas and other bigwigs in the Académie

34. On Gerhardt, see J. Brooke, "Chlorine Substitution and the Future of Organic Chemistry," *Studies in the History and Philosophy of Science* 4 (1873): 47–94; "Laurent, Gerhardt, and the Philosophy of Chemistry," *Historical Studies in the Physical Sciences* 6 (1975): 405–429.

35. In *The Quiet Revolution* (University of California Press, 1993), p. 86, I stated that Gerhardt came from a Jewish family. The definition of "Jewish" has always been problematic; many historians include in this category Jewish converts to Christianity, or even sons of Jewish converts, which sometimes causes problems of correct identification. Gerhardt's name appears regularly in the philosemitic literature, whence my statement, but there has never been any question that his parents raised him as a Lutheran, at least conventionally speaking. In a letter of 21 April 1843 to his half-brother, Gerhardt accused the rector at Montpellier of discriminating against him "parce que je suis protestant" rather than Catholic (E. Grimaux and C. Gerhardt Jr., *Charles Gerhardt* (Masson, 1900), p. 75n.). I have continued to seek evidence of genuine Jewish background in his family, but in vain; I now believe that the philosemitic literature is in error. This would also negate my speculation that there may also have been Jewish background in Laurent's family. I thank Georges Bram for a stimulating discussion on this question.

36. "M. Persoz does not concern himself with organic chemistry, or rather he does not do organic analyses" (Gerhardt to Liebig, 18 May 1837, in Grimaux and Gerhardt, *Charles Gerhardt*, p. 29); "I believe that [Persoz] has never performed a good combustion [analysis]" (Gerhardt to Laurent, 17 May 1845, in *Correspondance de Charles Gerhardt*, ed. M. Tiffeneau (Masson, 1918–1925) (hereafter "Tiffeneau"), vol. 1, p. 52).

des Sciences turn against him as they had already turned against Persoz and Laurent. In 1841 Gerhardt was named to the Faculté des Sciences of Montpellier, a position which Gerhardt attributed (incorrectly) to Liebig's rather than Dumas's influence. He spent seven unhappy years there.[37]

The Liebig-Dumas "divorce" of spring 1838 produced acrimony and indignation in Giessen, but Liebig could never maintain anger against Dumas. By the summer of 1842, Liebig's wrath had cooled once more, partly because Liebig was then named foreign member of the Académie des Sciences. In 1850, Liebig again met Dumas in Lille, and they reconciled fully; thereafter their relationship was not only collegial but genuinely warm.[38] This reconciliation could not but have been gratifying to their joint student Wurtz, who no longer had to worry about divided loyalty.

Early Years at the Faculté de Médecine

Wurtz had the dubious distinction of having three of his four earliest contributions attacked shortly after their appearance, but he also had the pleasure of emerging victorious—or at least unscathed—in all three instances.

Jean Louis Lassaigne, a respected physiological chemist, wrote a critical report on Wurtz's summary of his doctoral work; having failed to extract soluble albumin by the specified procedure, he claimed that all the stated results were incorrect. But Dumas must have had reason to distrust Lassaigne, for Wurtz's paper was published whereas Lassaigne's report was not.[39] The second critique was by Friedrich Wöhler, who had repeated Wurtz's new reaction that was said to generate copper hydride and had failed to replicate the synthesis; he wrote to Berzelius that Wurtz's putative new compound "appears to be an error," but he did not publish this

37. Grimaux and Gerhardt, *Charles Gerhardt*, pp. 28–49.

38. F. Holmes, *Claude Bernard and Animal Chemistry* (Harvard University Press, 1974), p. 47 and chapters 2–4 passim; Carrière, *Briefe*, pp. 277–278.

39. J. Lassaigne, "Note sur l'extraction de l'albumine par le procédé de M. Wurtz" (undated manuscript in miscellaneous correspondence folder of Dossier Wurtz, Archives de l'Académie des Sciences); Wurtz, "Sur l'albumine soluble," *Comptes rendus* 18 (1844): 700–702.

Figure 4.3
Charles Gerhardt. Source: E. F. Smith Collection, University of Pennsylvania Library.

opinion.[40] (In fact, the reaction is quite difficult to manage. Although today the compound is fully recognized, a generation ago a leading textbook of advanced inorganic chemistry stated only that "Copper hydride, CuH, seems to be a definite compound," possessing the properties reported by Wurtz,[41] and the compound's entry in a recent edition of the *CRC Handbook of Chemistry and Physics* is followed by the annotation "(exists ?)." All the more reason to admire Wurtz's chemical artistry, even at the very start of his career.)

As for the work on hypophosphorous acid, it was the critical responses by Jacob Berzelius and Heinrich Rose to Wurtz's first paper of 1842 that led him to repeat and expand the work, first at the medical faculty in Strasbourg and then in Balard's and Dumas's labs in Paris. The expanded paper was completed by October of 1845, though it did not appear in print until early in 1846.[42] In 1842 Wurtz had formulated the hydrated form of hypophosphorous acid as $H_6P_2O_4$, using four-volume atomic weights as was usual in Germany at that time. Over the intervening 3 years, however, conventional equivalents had been generally adopted throughout Europe, and so (in a two-volume formulation) the same formula for hypophosphorous acid now had to be written H_3PO_4.[43] Wurtz argued that this formula still was the best, most logical, and most empirical formula for the compound. One could

40. Wöhler to Berzelius, 3 October 1844, in Wallach, *Briefwechsel,* vol. 2, pp. 497–498. Thirty-three years later, Marcellin Berthelot also claimed, this time in the pages of the *Comptes rendus,* that copper hydride does not exist. Wurtz replied effectively to the charge.

41. F. Cotton and G. Wilkinson, *Advanced Inorganic Chemistry,* second edition (Wiley, 1966), p. 206.

42. Wurtz, "Recherches sur la constitution des acides du phosphore," *Annales de chimie* [3] 16 (1846): 190–231; Dumas, reporting also for Pelouze and Regnault, "Rapport sur un mémoire de M. Wurtz," *Comptes rendus* 21 (1845): 935–943. Wurtz commented (p. 191) that he had worked on this subject an entire year in Dumas's private laboratory, which suggests (contra Hofmann, "Wurtz," p. 221) that he arrived in the lab before the end of 1844. The report by Dumas, Pelouze, and Regnault is extremely favorable.

43. The conventional equivalent for oxygen is 8 rather than the atomic weight of 16, so the number of oxygen atoms must be doubled; then halve all of the resulting atomic coefficients to go from a four-volume to a two-volume formula. Wurtz actually wrote the formula PH_3O_4, using the superscripts that were typical of French chemical formulas. In what follows, I silently translate superscripts to subscripts.

speculate as to the parent radical, but it was not reasonable to assert confidently that the radical was a simple oxide, as Berzelius and the dualists did, for there was simply no evidence for this in the reactions of the compound. Purely empirically, all that could be said was that the acid was monobasic.

Throughout most of his paper Wurtz retained the dualistic anhydride formulations, but he argued that the view of Dulong and of Davy was more consistent with the reactions of the compound, especially as "molecular questions are still surrounded by so much obscurity." Davy's theory, Wurtz said, was superior to the dualistic formulas precisely because it avoided "any hypothesis regarding the intimate arrangement of the molecules." Without naming it, Wurtz was advocating the abandoned Liebig-Dumas hydracid theory. In the very act of touting a purely empirical interpretation of chemical events, Wurtz was willy-nilly taking theoretical sides.[44]

Wurtz's paper on copper hydride had exactly the same character: purportedly merely a chemical description of a novel compound, it carried an important theoretical message.[45] Wurtz noted that decomposing the new compound thermally generated only half as much hydrogen gas as using hydrochloric acid; clearly in the latter case the acid must also have been losing its hydrogen, so a double decomposition had to be occurring. The energy with which this reaction proceeded seemed curious, even anomalous. Copper itself was attacked by hydrochloric acid only with difficulty, and copper hydride, one would naturally think, must add an additional barrier to reactivity, since the hydrochloric acid would have to overcome the added affinity of the copper for the hydrogen. And yet hydrochloric acid seemed to have more, not less, affinity for copper hydride than for copper. In his 1844 paper Wurtz addressed the anomaly only by vague reference to "contact action," but 11 years later he noted that the hydrogen of copper hydride must have affinity for the hydrogen of hydrochloric acid, more than compensating for the weak affinity of hydrogen for copper in copper hydride.[46]

44. Wurtz, "Recherches," pp. 205, 230–231.

45. Wurtz, "Sur l'hydrure de cuivre," *Annales de chimie* [3] 11 (1844): 250–252.

46. Wurtz, "Sur une nouvelle classe de radicaux organiques," *Annales de chimie* [3] 44 (1855), pp. 301–302. At this time and ever after, Wurtz used this reaction as evidence for the assertion that hydrogen molecules have two atoms of hydrogen (see also his *Leçons de chimie professées en 1863* [Paris, 1864]), but it cannot be ascertained whether he interpreted the reaction this way before 1855.

The notion that an element can have affinity for itself, so antithetical to electrochemical dualism, would be central to the theories of valence and structure that would develop over the next 20 years. Amedeo Avogadro had argued in 1811 that only by proposing molecules of hydrogen and oxygen that consisted of pairs of like atoms could one construct a molecular theory that was also consistent with chemical (stoichiometric) data, a point that was independently seconded by André Marie Ampère 3 years later. The rise of electrochemical dualism about the same time tended to suppress these ideas, since, for an electrochemist, like atoms must have like charges and hence must repel rather than attract one another. But the ideas did not vanish: in the first volume of his *Traité de chimie appliquée aux arts* (1828), Dumas proposed reactions which, in modern notation, could be written $H_2 + Cl_2 = 2HCl$ and $2H_2 + O_2 = 2H_2O$. In his famous published lectures of 1836 there is similar support for an Avogadrian view of molecular chemistry, though there it is ambiguous and hedged.[47] In traveling down this road, therefore, Wurtz could well have had a sense that he was following his mentor, rather than blazing a new path.

There is every indication that Dumas had the highest regard for Wurtz, at this time and ever after.[48] With Dumas's help, Wurtz attained his first two Parisian positions late in 1845: préparateur at the Faculté de Médecine, and chef des travaux chimiques for the second and third years at the École Centrale.[49] His salary for the first position appears to have been 900 francs,[50] and probably about the same for the second; the two together constituted a bare living wage. Two years later Wurtz won a competition for the agrégation in chemistry in the Faculté de Médecine. As was customary, the competition involved a variety of tests, such as analyzing an unknown substance (Wurtz needed 25 minutes to determine that his substance was oxalic acid), as well as the preparation of oral and written papers on subjects drawn by lot (for example, Wurtz was required to speak on "the

47. Dumas, *Traité*, vol. 1, pp. xxxviii–xxxix; *Leçons de philosophie chimique* (Gauthier-Villars, 1837), pp. 162–165, 178–179.

48. Dumas, reporting also for Pelouze and Regnault, "Rapport sur un mémoire de M. Wurtz," *Comptes rendus* 21 (1845): 935–943.

49. AJ16/6565; AJ16/188, printed bio-bibliographical notice, 1852.

50. At least, the préparateur who succeeded Wurtz was paid 900 francs in 1852 (Rapport à M. le Ministre, n.d. [ca. 1865], AJ16/6556).

production of heat in organized bodies" after 24 hours' preparation). The competition required 19 formal sessions, held in July and August of 1847; there were seven candidates for three positions, three judges, and a jury of eleven, including Dumas and Orfila. It is said that Wurtz's lecture on pyrogenic compounds made a vivid impression on the jury. The decision for Wurtz was ratified by a ministerial decree on 23 September. Wurtz's salary rose to 1000 francs, and he was now certified to teach in the Faculté, though there was no immediate opportunity to do so.[51]

All this time Wurtz continued to work in Dumas's private laboratory, but after becoming agrégé Wurtz began to scout around for a space in which he could establish his own lab, even if on a modest scale. Modest was a generous word for the facility Wurtz cobbled together, in the summer or fall of 1847, in the attic of the Musée Dupuytren. This institution, located in the courtyard of the École Pratique de Médecine, was a museum of pathological anatomy that in 1835 had been established in the former refectory of a medieval Franciscan monastery. (The fifteenth-century refectory, known simply as the Couvent des Cordeliers, is today the only building predating 1877 at the Faculté de Médecine.) Wurtz always insisted, within his financial constraints, on bright and pleasant surroundings and "a certain elegance in his work," and his first act, assisted by his préparateur A. Rigout, was to paint the smoky walls of the dark and narrow garret.[52]

Even after setting up this personal lab, Wurtz probably continued to work in Dumas's lab; Auguste Cahours, with a equally modest facility at his disposal as a guest in Deville's private lab in the Rue de la Harpe, probably did the same. Wurtz also worked occasionally in the facilities of the École de Pharmacie in Strasbourg, during his visits home in the "long vacation" of late summer and early fall.[53] Wurtz was referring to this period in

51. *Moniteur universel*, 18 August 1847, p. 2445; AJ16/6556 and AJ16/6321; Corlieu, *Centenaire*, p. 185; Friedel, "Wurtz," p. x.

52. Friedel, "Wurtz," p. x; Wurtz to Ministre de l'Instruction publique, "Rapport à M. le Ministre de l'Instruction publique sur l'état des bâtiments et des services matériels de la Faculté de Médecine," 1 February 1872, AJ16/6357. The courtyard, off the Rue de l'École de Médecine, still exists, but the space is now surrounded by newer buildings of the Université de Paris VI.

53. Documented by Wurtz's letter to Dumas of 9 November 1848, written from that laboratory (Dossier Wurtz, Archives de l'Académie des Sciences).

Figure 4.4
The refectory of the fifteenth-century Franciscan monastery on the Rue de l'École de Médecine, on the Left Bank, as it appeared in the middle of the nineteenth century. This was the site of Wurtz's first research laboratory. Watercolor by Hubert Clerget (1818-1899), private collection. © Photothèque des Musées de la Ville de Paris (photo 99CAR0414NB).

Figure 4.5
The Couvent des Cordeliers as it appears today. Wurtz's lab was in the attic.
Photograph taken by the author, March 1999.

his life when he wrote: "[Dumas] appeared to us as a courageous athlete, a conqueror, when we gathered round him in his modest laboratory in the Rue Cuvier, which he had set up at his own expense, and from which flowed many students and memoirs."[54]

Just at this time, Wurtz found occasion to visit England, where his sister had moved. Sophie Wurtz had spent the winter of 1844–45 in Rome, nursing her ill father, who died in the spring. In Rome she met and was successfully wooed by the prominent German engraver Ludwig Gruner (1801–1882). Gruner was subsequently commissioned by the Prussian government to engrave a number of famous art works and decorations in Hampton Court Palace and Buckingham Palace. This activity brought him in contact with Prince Albert, who was German by birth and a connoisseur of great art. The upshot was that Sophie settled in London with her husband for several years, and they were visited nearly every year by her brother Adolphe. By coincidence, the Gruners established a residence on Fitzroy Square virtually next door to A. W. Hofmann, who moved there in 1845 to direct the new Royal College of Chemistry. Hofmann met Sophie fortuitiously at a dinner party a year or two after his arrival, and was pleased to establish this unexpected connection. Thereafter, Hofmann and Wurtz had frequent occasion to see each other, Wurtz when he was in London visiting his sister and Hofmann when he was in Paris preparing British exhibits for the Crystal Palace Exhibition of 1851.[55] Wurtz's first such visit occurred in September of 1847; he used the time not only to visit his sister and his old chum from Giessen but also to learn English. Thereafter, Wurtz translated Hofmann's papers into French in exchange for Hofmann's translating Wurtz's papers into English.[56]

---

54. "Discours de M. Wurtz" at Dumas's funeral, *Comptes rendus* 98 (1884): 940–944, on 942.

55. Hofmann, "Wurtz," pp. 216–219.

56. Wurtz to Dumas, 24 September 1847, Dossier Wurtz, Archives de l'Académie des Sciences; Wurtz to Hofmann, 25 November 1848, Chemiker-Briefe, Berlin-Brandenburgische Akademie der Wissenschaften. Wurtz's early letters to Hofmann are in German, but this first surviving letter has a passage in French in which Dumas asked Wurtz to tell Hofmann, regarding such translations of his papers, that "tant qu'il [Hofmann] voudra, . . . il se considère comme un des rédacteurs des Annales [de chimie]." Wurtz's close friendship with Hofmann is indicated by his use of the pronoun "du."

The First Efforts at Institutional Reform and Their Failure in Revolution

Dumas was of course painfully aware of the deficiencies of instructional and research labs in the French educational system, and well familiar with the superior facilities across the Rhine; furthermore, owing to his position as dean of the Faculté des Sciences, his was a powerful voice. After a 2-year stint as Minister of Public Instruction in the 1830s, Narcisse de Salvandy was reappointed in 1845, and the next year he commissioned Dumas (again) to study the state of the country's post-secondary science instruction. Dumas's report recommended a general reorganization of the Faculté des Sciences, and a reconstruction of the Sorbonne with the addition of proper laboratory facilities. When the Faculté was founded at the beginning of the nineteenth century, Dumas wrote, it took but little apparatus to make major discoveries:

> Armed with a few flasks and test tubes, Gay-Lussac in France and Dalton in England paved the way for exact science. Some furnaces and vessels of the most ordinary kind enabled Thenard, Vauquelin and Davy to win immortal fame with the most brilliant chemical discoveries. . . . But how everything has changed today for anyone who aspires to make discoveries in the physical or natural sciences. . . . Chemical laboratories, having subsequently become, by the close alliance of chemistry and physics, veritable precision workshops, require of anyone desirous of completing research for a thesis at one of the Facultés an expenditure of as much as three or four thousand francs.

Dumas noted that private efforts could only go so far to rectify this unfortunate situation:

> As regards myself, for the last twenty years[57] I have opened my laboratory to all young people who appeared to me to be capable of benefiting, and I have only regretted the necessary expense on those occasions when I have reflected that its utility would have been an order of magnitude greater had the University given the least support to it.

Dumas further averred that for a mere 6000 francs of capital investment and an annual outlay of 3000 francs a perfectly adequate teaching and research laboratory for experimental sciences could be installed in the Sorbonne, intended for licenciés preparing for the agrégation or the

57. Dumas was obviously counting the time that he accepted occasional students into his laboratory at the École Polytechnique, which presumably began around 1826.

doctorat. He thought that a dozen student places would be about the right number, meeting the minimum size to have a real impact and produce beneficial rivalry among the group, but without outrunning the potential demand:

The Faculté [des Sciences], which has allowed itself to be overtaken by Germany and England, would soon regain its rightful place, and would no longer fear the competition of foreign laboratories, if it could direct a competition of well-organized efforts toward the solution of some of the problems of science, as it is practiced on the other side of the English Channel and the Rhine. Today it is necessary for a Faculté to found a school; to direct the movement of minds; not to be forced to wait for a question to be resolved by the individual work of one of its professors extended over several years, when it can do so in a few weeks under his direction by the collective effort of a dozen beginners in science; above all, it is necessary that it not leave to others this high honor of serving to guide youth, reserving for itself the thankless and difficult role of criticism and control.[58]

Press reaction to this published report was mixed. One writer argued that, were this report's recommendations to be implemented, the Faculté des Sciences would be transformed into nothing more than an industrial school, teaching "manual arts, joinery, carpentry, locksmithing . . . and awarding diplomas of doctor of mechanical science, agricultural science, etc. Was this one of the chimerical projects that sprout from the brain of M. de Salvandy like weeds in an abandoned field, or that escape from the evaporating liquids in the flasks in M. Dumas's laboratory?"[59]

Other reactions were more positive, and in any case Salvandy was convinced. His immediate response was bureaucratic: he appointed a commission, headed by Thenard, to recommend and oversee the process of renovating and expanding the Sorbonne; costs for the project would be shared by the state and by the city of Paris. He also commissioned a second report by a team led by Dumas, this time on primary and secondary education. This report, submitted in April of 1847, proposed weakening the strong classicist orientation of early education, introducing scientific sub-

58. Dumas, "Rapports adressés à M. le Ministre de l'Instruction publique par M. Dumas, Doyen de la Faculté des Sciences," 20 June 1846, in Carton 18, Fonds Dumas, Archives de l'Académie des Sciences, on pp. 7–10. The report was also printed in *Moniteur universel* (28 October 1846, pp. 2448–2450).

59. Undated newspaper clipping, circa November 1846, Carton 18, Fonds Dumas, Archives de l'Académie des Sciences.

jects earlier, and tracking science students (what would later be introduced as "bifurcation").[60]

As I noted above, Dumas was motivated to found his Rue Cuvier laboratory by competition with Liebig, and he may even have solicited Liebig's advice in planning the lab. Similarly, his suggestions for reforming France's state educational system had a number of features that were notably present across the Rhine, as he explicitly remarked. Strikingly reminiscent of Giessen was his model of group laboratory research well organized and articulated by a scientific director, of the resulting beneficial tension between cooperation and competition among members of the group, and of the efficient use of large numbers of only semi-skilled scientific workers. The influential biologist Henri Milne-Edwards, a friend of Dumas, lobbied the government along similar lines; French higher education, he wrote, was far too exclusively oral and far too theoretical.[61]

There is no doubt that Salvandy was impressed by these ideas or that he attempted to implement them. In 1847 the Ministry's architect, Alphonse de Gisors, an ex officio member of the Thenard Commission, presented his plans for reconstructing some of the Sorbonne's buildings to include laboratories and didactic collections. On 10 February 1848, Salvandy notified the dean of the Faculté de Médecine of his approval of Dumas's proposal for a double laboratory there for chemistry and microscopy.[62] Unfortunately, just 2 weeks later Louis Philippe abdicated in the wake of the first insurrection of what was immediately named "the Revolution of 24 February" or "the Revolution of 1848." The Ministry's plans and proposals all became victim of this event, and

60. *Rapport sur l'enseignement scientifique dans les collèges, les écoles intermédiaires et les écoles primaires* (Ministère de l'Instruction Publique, 1847). See Gréard, *Éducation*, pp. 256–270, for the Dumas Commission's specific recommendations for the Sorbonne renovation, and Salvandy's decree establishing the Thenard Commission (18 September 1846).

61. Milne-Edwards, article in *La Presse*, 3 January 1847, cited in *Exposé fait au Conseil académique de Paris sur l'enseignement supérieur des sciences* (Imprimerie Impériale, 1868), p. 3.

62. Ministre de l'Instruction publique to Dean, Faculté de Médecine, 10 February 1848, AJ16/6556; B. Bergdoll, "Les projets de Léon Vaudoyer pour une reconstruction sous le Second Empire," in *La Sorbonne et sa reconstruction*, ed. P. Rivé (La Manufacture, 1987), pp. 55–56.

nothing substantial was done for French laboratory science for another 20 years.

In Paris, the February insurrections were followed by a bewildering and terrifying 4-month period of political lurches to the far left, pulls toward the center, and class warfare. Within days of Louis Philippe's abdication the republican Provisional Government appointed Hippolyte Carnot (son of Lazare, the "organizer of victory" for Napoleon, and brother of the engineer and heat theorist Sadi) to succeed Salvandy as Minister of Public Instruction. Carnot—a democrat and a republican, and much influenced by Saint-Simonian socialism—attempted to design and implement republican educational reforms, including a sensitivity to technical, trade, and scientific education which Dumas and many of his colleagues would have applauded. Unfortunately, the bloody June days resulted in a shift toward much more moderate policies and evoked a visceral dread in most bourgeois Parisians of what were perceived as radical schemes. Carnot was forced to resign after serving only 4 months; what little he managed to accomplish was mostly concerned with primary education. After the election of December 1848, which brought Louis Napoleon to the presidency, the new Minister of Public Instruction was Armand de Falloux of the antirepublican "party of order."[63]

In the more centrist and anxious political climate of late 1848 and 1849, an "extraparliamentary commission" worked out new provisions for primary and secondary education that retreated from the republican agenda of removing the Church's influence from the educational arena. The "loi Falloux" of 1850 consolidated this turn to the right and toward the influence of Catholic clergy. It proved to be a harbinger for the future of educational policy in the Second Empire. Louis Napoleon himself was essentially uninterested in this sphere of public administration; he left the actual development and implementation of policies to his ministers, at the same time giving educational funding a low priority. His Minister of Public Instruction after the coup d'état of December 1851 was Hippolyte Fortoul, a once liberal but now conservative Bonapartist intellectual who became the principal instrument of political reaction in the administration of the Université de

63. P. Carnot, *Hippolyte Carnot et le ministère de l'instruction publique de la IIe République* (Presses Universitaires de France, 1948), pp. 5–63; R. Anderson, *Education in France, 1848–1870* (Oxford University Press, 1975), pp. 39–50.

France. He was much disliked by the members of the Facultés for his dogmatic authoritarianism, and his reforms removed the last vestiges of corporate autonomy from the French system of higher education. The system was now under the firm control of the Ministry and of Louis Napoleon.[64]

Meanwhile, like the nation, the chemical community was also sorting itself into new political constellations. In March of 1845 Laurent had moved to the capital, forsaking his position at Bordeaux and attempting to make up with Dumas. This proved difficult: not only had Laurent accused Dumas of pilfering substitution and the theory of types, but Laurent's republican politics grated on Dumas's moderate conservatism. To make matters worse, Laurent had recently allied himself with the fellow outcast and fellow republican Gerhardt (who was still at Montpellier). Since 1842, Gerhardt had been advocating an atomic weight system largely synonymous with Berzelius's old atomic weights of 1826 and had been urging the non-Berzelian idea that all formulas, organic and inorganic, should be expressed in the two-volume convention. But in the immediate years after 1842, led by Liebig and Gmelin, an opposite shift took place in European chemistry, toward the "equivalents," which were widely considered to be non-theoretical.[65]

The February 1848 revolution was warmly welcomed by Laurent, Gerhardt, and a few others in the Parisian chemical community, including the ardent republicans Arago and Pelouze and the liberal Cahours. Laurent was named assayer at the Paris Mint 2 weeks after the revolution, and it looked like there were further spoils soon to be gathered; however, patience would be required. Laurent wrote to his friend in the provinces: "Our cumulards of science are not rushing to abandon their positions." Laurent (whose new job paid 5000 francs) thought it particularly unfair, even "revolting," that Eugène Péligot, formerly essayeur and now vérificateur des essais at the Mint, was also drawing salaries from the Conservatoire des Arts et Métiers and from the École Centrale, for a total of something like 17,000 francs.[66]

64. Anderson, pp. 50–57; G. Weisz, *The Emergence of Modern Universities in France, 1863–1914* (Princeton University Press, 1983), pp. 32–36.

65. Examples of conventional equivalents are H = 1, C = 6, O = 8, S = 16, N = 14, Cl = 35, and so on. Gerhardt wanted to substitute H = 1, C = 12, O = 16, S = 32, N = 14, Cl = 35, and so on. See Rocke, *Chemical Atomism*, pp. 200–210.

66. Laurent to Gerhardt, 13 March 1848, in Tiffeneau, vol. 1, p. 265.

Even richer than Péligot was the newly appointed Président de la Commission des Monnaies and Directeur du Laboratoire des Essais, Jules Pelouze, who also held the chair of chemistry at the Collège de France. Pelouze, though a leading cumulard, was probably behind Laurent's appointment to Pelouze's old position as essayeur at the Mint (and Laurent acquired Pelouze's old assay lab there, at 8 Rue Guénégaud). Gerhardt and Laurent agreed that Pelouze would probably have to leave the Collège de France and thought that Cahours would be his likely successor there. The old and infirm Gay-Lussac would have to be replaced at the Muséum, probably by Edmond Frémy; Frémy would then have to resign at the École Polytechnique, and that chair would then be available. Even Dumas might be forced to yield at least one of his chairs.[67]

Such prospects dismayed and frightened some in the community, including, presumably, Thenard and his protégé Dumas, who had both been closely associated with the government of the July Monarchy. Less than 3 weeks before the revolution began, a group of students carrying a petition protesting the political suspension of the historian Jules Michelet interrupted Dumas's lecture (to 1200 auditors), but were firmly rebuffed by both professor and audience.[68] In the ensuing weeks and months Dumas displayed some of his highly touted "fine finesse" in this turbulent political period. "But what disgusts me," Gerhardt wrote to Cahours from Montpellier 10 days after the abdication of Louis Philippe,

is to see people who were rabid conservatives scarcely a week ago, to see these people rally for the Republic and act more fanatically than anyone. This is just what our friend Dumas is doing in Paris; I was scandalized to read his speech in favor of the Republic the other day, he who less than a month ago attacked in his lecture the young people who protested the suspension of Michelet. But Dumas should watch out, we now have real freedom of the press and nothing can stop anyone from telling him the truth.

Cahours could not help but agree, noting how all the "gros bonnets" were plotting carefully how to keep their positions. "How long the faces are, how full of curses for the Republic are the hearts, when lips twist their smile of satisfaction and open to make lovely speeches in favor of the new order.

67. Ibid.; Charles Gerhardt to Jane Gerhardt, 27 March 1848, in Grimaux and Gerhardt, p. 175. Gerhardt thought that the only serious candidates for chemistry positions other than Laurent and himself were Cahours and Malaguti—and Malaguti was excluded as being a foreigner.

68. *Moniteur universel*, 5 February 1848, p. 284.

Figure 4.6
The Paris Mint, at Quai de Conti and Rue Guénégaud. Pelouze, Laurent, and others had laboratories here. Photograph taken by the author, March 1999.

Good God, what hypocrisy!"[69] On 26 March, Gerhardt "told Dumas the truth" in a long conversation, in which (as Gerhardt related the incident to his wife) Dumas flattered and cajoled Gerhardt in order to avoid being denounced. In April, both Gerhardt and Laurent lobbied a sympathetic Carnot for positions in Paris, and were optimistic about the eventual outcome. Carnot told Gerhardt that the sort of discrimination he had experienced under Louis Philippe would never have happened in the Republic. "If there is justice still in the world," Gerhardt wrote to his wife Jane, "I will succeed this time; now or never."[70]

Jane Gerhardt's Scots mother was shocked by the politics of her son-in-law, who attempted to justify himself to her as follows:

There are but two parties in France, the men of the past and those of the future. The latter have been called, under the Restoration, liberals; under Louis Philippe, republicans; and today they are called socialists. . . . The liberals have been guillotined,

69. Gerhardt to Cahours, 6 March 1848; Cahours to Gerhardt, 18 March 1848, in Tiffeneau, vol. 2, pp. 47–48.
70. Grimaux and Gerhardt, pp. 175–180.

the republicans shot, and, logically, the same treatment will be accorded to the socialists. It is the history of the first Christians, it is the history of the Reformation, it is the history of humanity. It is the battle of the weak against the strong, of justice against brutal force. . . . The socialists have been depicted to you as the ruin of society; I proclaim to you that they are its guardians.[71]

One may discern a touch of paranoia and grandiosity here. There is no question that the community had closed ranks against Gerhardt and Laurent in the late 1840s, or that the powerful elite of the field could and certainly should have dealt with them more generously than they did. However, Gerhardt (and to a lesser extent Laurent) shared at least flashes of both intransigence and arrogance with their opponents. Moreover, Gerhardtian chemistry was by no means free of anomalies, which both men clearly recognized and agonized over; nor could either man point to much positive evidence to support his ideas. They offered the community a wholesale alteration of the time-tested orthodoxy based only on thoughtful rationalization and systematization, not on experimental proof. The old guard wanted more, and that requirement was not unreasonable.

There had in fact been peace offerings from the "gros bonnets." For one thing, Gerhardt did not realize that he owed his position in Montpellier to Dumas. In 1844, in a published referee's report for the Académie, Dumas highly praised Gerhardt's *Précis de chimie organique*, and indicated that he agreed with Gerhardt's reforms[72]; in 1845 he invited Gerhardt to a soirée (where, according to the latter, he was treated with the greatest imaginable kindness, which Gerhardt imputed solely to "pure politique").[73] Thenard also hosted Gerhardt at dinner in 1845, and was treating him with respect and kindness the next year. In 1849 Thenard asked Gerhardt to edit the new edition of his *Traité de chimie*, and in 1855 it was Thenard's influence that procured a position for him at Strasbourg. It would appear to have

71. Gerhardt to Mrs. Saunders, 24 February 1850, Grimaux and Gerhardt, pp. 184–185.

72. "The treatise of M. Gerhardt thus represents the system of ideas which is connected to work executed in France for the last few years, and it will certainly greatly contribute to propagate ideas which, speaking for myself, I have sought to spread for a long time in my public instruction." (*Comptes rendus* 18 (1844): 809–810) Gerhardt thought, probably rightly, that this was a subtle way of arguing for priority for some of Gerhardt's ideas (Gerhardt to Wurtz [summer 1844], in Tiffeneau, vol. 2, pp. 307–308).

73. Chaigneau, *Dumas*, p. 152.

been only Gerhardt's actions during the radical republican phase of the Revolution that turned Dumas away from Gerhardt definitively.[74] Similarly, Gay-Lussac tried (unsuccessfully) to have Laurent appointed répétiteur at the École Polytechnique in 1836, Dumas asked Laurent to substitute teach for him in the Sorbonne in the summer semester 1847, Thenard offered him the deanship at Caen, and it was probably due to Thenard's as well as Pelouze's influence that he was named to the Mint.

Dumas was clearly unhappy with Gerhardt's campaign to deprive him and others of some of their income. Immediately after the February revolution, a short-lived Association pour l'Abolition du Cumul quickly put out a petition, which was signed by no fewer than 200 scientists and physicians, asserting that 212 professorships were being occupied by 57 people averaging 15,460 francs per chair holder, whereas if 212 different scholars had held these functions the mean salary would have been 4152 francs. The petition was presented to the National Assembly on 12 April 1848, and the same day an article appeared in the semi-official journal of the Republic, *Le National*. Scientific dynasties, the paper opined, should fall just as political ones should; "the merchants should be chased from the temple of science; the temple should be restored to the workers and no longer be abandoned to the jackals of the professoriate." A commission was established to study the question, headed by the journalist Ferdinand Flocon, a member of the ruling junto and a friend of Laurent's.[75]

However, cumul was never abolished. Carnot was gone by July of 1848, and the energy of the reform movement declined. The republican journalist Gustave Augustin Quesneville, writing in August, deplored the loss of momentum:

The economies in scientific establishments that have been carried out have been taken out on the smaller employees, the important people have been protected. To be sure, the abolition of cumul has been timidly advanced, but the chair holders have been so moved, have so well had their friends maneuver for them, have cried "vive la République" so high and loud and have sung the Marseillaise with such gusto, that all has remained the same. . . . In the days following 24 February the high and mighty barons of science looked the picture of misery; they and their courtesans trembled for the precious privileges that they had won by the flexibility of

74. Grimaux and Gerhardt, pp. 98, 150, 190–192, 259.

75. *Moniteur universel*, 23 April 1848, p. 877, regarding a decree announced the day before.

their backbone [à la souplesse de leur échine] . . . [but now] those who were trembling have become just as arrogant as they were humble and groveling. . . .[76]

The language of Gerhardt's letters began to move from patient optimism to frustration and anger. In August he could see, as did Quesneville, that cumul would not be abolished, and in September he was "disgusted" by politics. An anonymous article arguing militantly against cumul was published in *La patrie* in the spring of 1849; probably written by Gerhardt, it was the last gasp of this hopeless campaign.[77] Dumas was elected to the Legislative Assembly in 1849. "If you could see him, you would laugh at his air of importance," Gerhardt wrote to Chancel; ". . . the doors could not have been wide enough when he made his entrance [at the Assemblée]."[78] Frémy was appointed at the Muséum but chose to retain his position at the École Polytechnique. Gerhardt was a candidate for supervisor of the state Manufacture des Tabacs, but that job was given to Cahours.[79]

Dumas was forced to close his laboratory on the Rue Cuvier permanently, owing to his loss of income and other disruptions of the revolution. Later he described it this way:

Political events had dispersed my own collaborators; after twenty-five years of activity, my laboratory of advanced chemical studies had closed. Henri Deville was my successor; but, more fortunate than I, he opened his laboratory with the cooperation of the state, which I never had. . . .[80]

In the summer of 1848 Dumas was visited in the Sorbonne by a "surly but kind" stranger who pulled out a large roll of banknotes and offered to

76. Quesneville, "Du Cumul," *Revue scientifique* 32 (1848): 533–540; Grimaux and Gerhardt, pp. 178–179; *Le National*, 12 April 1848.

77. Gerhardt to Chancel, 4 August 1848, 9 September 1848, and 5 June 1849, in Tiffeneau, vol. 2, pp. 73, 76, 93–94, 94n.

78. Tiffeneau, vol. 2, p. 93.

79. Gerhardt to Chancel, 12 October 1850, in Tiffeneau, vol. 2, p. 104. Gerhardt was especially bitter about this, for Cahours had been his good friend and had promised him that he would not compete for the job.

80. J. Dumas, éloge of Henri Sainte-Claire Deville, in *Discours et éloges académiques* (Gauthier-Villars, 1885), vol. 2, p. 307. Dumas was speaking figuratively, for he had never been at the École Normale; Deville's predecessor was Balard. When Deville came to the École Normale, he was fortunate that a new building had recently opened (in 1847) to house it (at 45 Rue d'Ulm, where it has been ever since). The new building, in contrast to the old lodgings in the Collège du Plèssis, included laboratories.

finance another lab. It was a wealthy bachelor physician named Jecker, who years earlier had attended Dumas's lectures, who was "passionately devoted" to organic chemistry, and who had only a short time left to live. Dumas declined the offer and counseled him to endow a prize instead. When Jecker died, 2 years later, he left the Académie des Sciences an endowment to provide an annual award of 5000 francs for the best work in organic chemistry.[81] But this gift, munificent though it may have been, could do little to arrest the decline in the fortunes of French laboratory science. The bloody June days were a terrible shock, and thereafter the Republican tide waned. All of Gerhardt's and Laurent's schemes started to evaporate.

As Gerhardt correctly foresaw, in 1850 Pelouze resigned his position in the Collège de France. (In addition to his new appointment as director of the Mint, Pelouze had replaced Gay-Lussac on the counsel of the great Saint-Gobain chemical works, after the latter's death; the two appointments constituted full-time employment, and a very rich income.) J. B. Biot, an eccentric and liberal member of the French scientific establishment, became concerned when Balard, who then held a chair at the École Normale as well as at the Sorbonne, announced his candidacy for the position. Parisian wags said that Balard, long inactive in research, was "discovered by bromine," and Gerhardt cattily remarked that many would be surprised to learn that he had not died long ago.[82] Laurent wrote to Alexander Williamson about this situation:

I told Gerhardt to put his name in, competing with Balard. After testing which way the wind was blowing Gerhardt saw that the devil himself had a better chance than he. Father Biot, seeing that the chair would go to Balard, sought me out to enjoin me to put my name in. I told him NO! and suggested Gerhardt. "Impossible," he replied, "and if you refuse to be a candidate in friendship to Gerhardt, you will lose the opportunity to have a chair, and thus to be of service, sooner or later, to him."[83]

And so Laurent was persuaded. Biot, the senior professor at the Collège, strongly urged his colleagues to accept Laurent's nomination, arguing that Balard had accomplished little over the last decade with two labs at his

81. Chaigneau, *Dumas*, p. 204; M. Crosland, "The Emergence of Research Grants within the Prize System of the French Academy of Sciences," *Social Studies of Science* 19 (1989), pp. 85–86. In its first two decades, Wurtz, Cahours, Marcellin Berthelot, and Charles Friedel each won the Jecker prize twice.

82. Gerhardt to Liebig, 21 April 1851, in Grimaux and Gerhardt, pp. 213–214.

83. Laurent to Williamson [December 1850], in Tiffeneau, vol. 1, p. 54.

Figure 4.7
Antoine Balard, 1857. Source: E. F. Smith Collection, University of Pennsylvania
Library.

disposal, whereas Laurent had done a great deal of fine research with virtually no material help at all; giving the chair at the Collège to Balard would add nothing to his means, whereas it would make a world of difference for Laurent. Laurent won the vote by professors at the Collège itself, by 13 to 9, but Louis Napoleon required a second présentation by the Académie des Sciences, where Balard was a member and Laurent was not. The vote there, 35 to 11 in favor of Balard, was a foregone conclusion, and the chair went to Balard.[84]

Laurent retained his position at the Mint, but the lab which he was given there, Pelouze's old facility in the Rue Guénégaud, was cold, damp, and dark.[85] About the time of Biot's visit Laurent contracted a case of tuberculosis; he lingered on for a little over 2 years before succumbing in the spring of 1853. The injustice over the Collège position was said to have hastened Laurent's death.[86] Gerhardt once more hoped for a position, either as Laurent's successor as assayer, or as Balard's successor in the École Normale, but again in vain: Cahours succeeded Laurent, and Deville succeeded Balard. Since neither Cahours nor Deville had had proper positions to that time, the chain of successions was then closed, and Gerhardt was locked out once more.

The noted historian of French chemistry Jean Jacques trenchantly characterized Dumas and Laurent, in language that could just as well apply to Gerhardt:

[Dumas's] political ideas—order and religion—are of a most reassuring conformism, his scientific philosophy, marked by a cautious positivism, forbade him from going beyond experience where he would be lost in a host of confutable theories. He is a fortunate man. Laurent is like the negative of this portrait. This scientific romantic is a polemicist who obstinately wants to be right; he insists on his rights, he antagonizes; he was given, to have peace, a quiet professorship at

84. E. Grimaux, "Auguste Laurent," *Revue scientifique* 58 (1896): 161–170, 203–209, on 162; Tiffeneau, vol. 1, pp. 292–293; Biot, open letter to the Académie des Sciences, in Grimaux and Gerhardt, pp. 588–591; *Moniteur universel*, 15 January 1851, p. 140.

85. This at least is Grimaux's judgment; Gerhardt described it as "charmant" in a letter of 8 April 1848 to his wife. See Grimaux and Gerhardt, pp. 177, 209.

86. See C. DeMilt, "Auguste Laurent—Guide and Inspiration of Gerhardt," *Journal of Chemical Education* 28 (1951): 198–204. The best recent work on Laurent is M. Novitski, *Auguste Laurent and the Prehistory of Valence* (Harwood, 1992).

Bordeaux, he wants a laboratory in Paris; against the immense majority of scientists of his day he believes in atoms, in molecules, in chemical structure of compounds; he is poor and sick. He is the star-crossed chemist, as one speaks of the star-crossed poet.[87]

When Laurent and Gerhardt both died young (within a span of 3 years), the chemical establishment attempted to make amends, at least to the families. As Laurent was on his deathbed, a delegation headed by Biot and supported by the entire chemistry section of the Académie des Sciences petitioned the Minister de l'Instruction Publique for a pension for Laurent's widow and two small children, who were otherwise without resources. Upon Gerhardt's death, Baron Thenard founded a Société des Amis des Sciences to support both families, and others in a similar condition.[88]

The tragedy of Laurent's last years of life symbolize, for the chemical community, the turn to the right after 1848. This was the difficult political climate in which Wurtz, 9 years younger than Laurent and a year younger than Gerhardt, and with just as insecure a career when the revolution began, was forced to navigate.

Wurtz's Navigation of the Revolution and the Second Republic

There is no evidence of Wurtz's reaction to the events of February and June 1848. The indications suggest only that he worked assiduously on his research, which was continuing to yield important discoveries, and that he was careful to maintain good relations with his patron Dumas. Two letters written in November 1848, one to Dumas and the other to Hofmann, contain only chemical details and a reference to the beginning of the new term at the École Centrale.[89] When the brilliant young physicist Léon Foucault published an article attacking Dumas for not sufficiently supporting Wurtz's candidacy to the Légion d'honneur, Wurtz wrote a letter

87. J. Jacques, "Auguste Laurent et Jean-Baptiste Dumas d'après une correspondance inédite," *Revue d'histoire des sciences* 6 (1953): 329–349.

88. Chaigneau, *Dumas*, pp. 145–147, 289–291.

89. Wurtz (in Strasbourg) to Dumas, 9 November 1848, Archives de l'Académie des Sciences; Wurtz to Hofmann, 25 November 1848, Chemiker-Briefe, Berlin-Brandenburgische Akademie der Wissenschaften, Berlin.

to Dumas disclaiming knowledge of the article and protesting his good faith and loyalty:

When I compare the position in which I found myself nearly six years ago, the day I saw you for the first time, with that which your benevolence has since attained for me, I have reason to rejoice and be gratified. It is to you, and, permit me to add, also to some extent to the good and excellent M. Balard, whom I owe any success I may have had.[90]

This contretemps had no effect on Wurtz's relationship with his mentor, and Wurtz was indeed named Chevalier of the Légion before the end of 1850.[91]

Wurtz was, as A. W. Hofmann put it, "not exactly thrilled" by Gerhardt's and Laurent's aggressive polemics; Gerhardt and Wurtz were at this time wary rivals, who maintained friendly surface relations but did not fully trust each other's loyalties and visited little together. To this was added jealousy on the part of Gerhardt, who saw the younger Wurtz repeatedly preferred over himself. Wurtz stated that he "scarcely knew" Laurent.[92] Gerhardt revealed his true feelings about Wurtz in an 1846 letter to his friend Jérome Nicklès:

If you have definitively decided to remain in Paris . . . do as our compatriot W, who well knows how to make his way in the world, in devoting himself body and soul and in giving his all, including his conscience, to some powerful patron whom he will enthusiastically praise down to the most insignificant details. You know that W translated my book; as long as he thought he still needed me he wrote and consulted me; this lasted until he was settled in Paris as Dumas's préparateur. Since then he has become one of my most ardent detractors.[93]

90. Wurtz to Dumas, 14 January 1850, Fonds Dumas, Archives de l'Académie des Sciences: "Quand je compare la position dans laquelle je me trouvais il y a bientôt six ans, le jour où je vous ai vu pour la première fois, à celle que votre bienveillance m'a faite depuis ce temps, j'ai lieu de me réjouir et de me contenter. C'est à vous, et en partie aussi, permettez-moi de l'ajouter, à ce bon cet excellent M. Balard, que je dois les succès que j'ai pu avoir."

91. For details of this affair, see A. Carneiro, The Research School of Chemistry of Adolphe Wurtz (Ph.D. dissertation, University of Kent/Canterbury, 1992), p. 83.

92. Hofmann, "Wurtz," p. 237; Tiffeneau, vol. 2, pp. 303–204; Wurtz, "Éloge de Laurent et de Gerhardt," *Moniteur scientifique* 4 (1862), p. 484.

93. Gerhardt to J. Nicklès, 6 October 1846, in J. Jacques, "Quelques lettres inédites de Charles Gerhardt," *Bulletin de la Société Industrielle de Mulhouse* 833, no. 2 (1994), p. 26. It must be noted that Gerhardt could only have been getting information on Wurtz's opinions indirectly.

In February or March of 1850, A. W. Hofmann and Thomas Graham traveled to Paris for 8 days. This was the first time Hofmann had been to Paris, though he and Wurtz had seen each other two or three times in England. In his biography of Wurtz, Hofmann described the visit in nearly 20 pages of detail. The visitors were treated to a lavish chemists' dinner, to which all the important names were invited—with the exception, Thenard had commented, of "les deux" (Gerhardt and Laurent).[94] Both Hofmann and Wurtz participated in, or at least countenanced, this ostracism of the two reformers, to their later regret. Hofmann had met Laurent as early as 1844, when the Frenchman visited Giessen, and had even collaborated briefly with him; soon thereafter, Laurent urged Hofmann to adopt Gerhardt's new equivalents, but Hofmann declined. In 1855 Hofmann wrote to Gerhardt: "I have long wanted to make your acquaintance, even though I found it necessary to forego that pleasure during a short trip I made to Paris, because you have often injured my sensibilities by the form of your journal articles." That was putting it mildly. In an 1847 letter to Liebig, Hofmann called Gerhardt a "despicable cur" ["infam gemeiner Hund"] for publicly implying that Hofmann had stolen a discovery from him.[95]

There is no question that Dumas favored Wurtz over Gerhardt. In October of 1849 the newly elected Prince-President Louis Napoleon appointed Dumas Minister of Agriculture and Commerce. France had been troubled by disastrously poor harvests, and Napoleon told Dumas on this occasion, alluding to his uncle's agricultural chemist, "You will be my Chaptal."[96] Dumas's new bureaucratic and political role required him to withdraw from teaching (though not from his positions). He asked Deville to substitute for him at the Sorbonne, and Wurtz at the Faculté de

94. Hofmann, "Wurtz," pp. 226–243. Hofmann stated that the visit took place in 1851, but this was clearly an error, for within days of his return to London Hofmann descibed what was clearly the same trip in a letter to Liebig (*Justus von Liebig und August Wilhelm Hofmann in ihren Briefen*, ed. W. Brock (Verlag Chemie, 1984), p. 93). This letter is unfortunately undated, but internal references make it certain that it was written in March 1850, plus or minus a month.

95. Tiffeneau, vol. 1, pp. 11, 292; Grimaux and Gerhardt, p. 264 (Hofmann to Gerhardt, 5 February 1855); Brock, *Briefe*, p. 67 (Hofmann to Liebig, 7 February 1847).

96. Chaigneau, *Dumas*, p. 210. Napoleon was no prophet: Dumas remained in this position only 2½ months.

Médecine. Wurtz was an obvious choice for the latter role, above Gerhardt and other alternatives, because, apart from questions of possible favoritism, Wurtz was a physician and already agrégé in the faculty. Thus began Wurtz's teaching career at the Faculté, which continued without interruption for 35 years.

Dumas's appointment was not a sinecure, for he had always had a strong interest in technical and agricultural chemistry. Almost immediately he convinced the President to create a new National Institute for Agronomy, essentially an agricultural college, located in Versailles. One of the new professorial positions was for chemistry, and Dumas engineered Wurtz's appointment there.[97] During the term, Wurtz lectured twice a week in Versailles and also taught at the Faculté de Médecine. The salary at the new institute, 6000 francs, enabled him to resign his position at the École Centrale, which in any event was inconveniently located for a resident of the Latin Quarter (being on the Right Bank); he continued to be paid at the Faculté in his function as agrégé.[98] The original appointment at the agronomy institute was only provisional, but it was made titular from the beginning of 1852. Unfortunately, Louis Napoleon, who by then was emperor in all but name, began to see the institute both as a financial drain and as a suspect republican institution, and he abolished it 9 months after Wurtz took formal possession of the chair.[99]

During the brief Second Republic, Wurtz also attempted to establish a private chemical teaching and research laboratory modeled on Liebig's. For this purpose he joined forces with François Verdeil, son of a noted Swiss historian, and Charles Dollfus, son of an Alsatian industrialist, both of whom had just finished a course of study in Giessen. They found a

97. The chair was created on 22 August 1849, but apparently Wurtz did not begin teaching there until the next year.

98. Wurtz to Liebig, 18 April 1851, Liebigiana IIB; AJ16/6618, Traitement du personnel, Faculté de Médecine; Rapport à Ministre de l'Instruction publique, n.d. [ca. 1865], AJ16/6556. The salary at the Faculté, 1000 francs, does not include the éventuel (capitation), which was 745 francs in 1850 and 953 francs in 1851; he was paid nothing additional for lecturing in place of Dumas. In 1852 Wurtz was paid 1200 francs by the Faculté. From June 1850 through the end of 1854, professors at the Faculté earned 6000 francs plus 3000 éventuel.

99. *Moniteur universel*, 26 December 1851, p. 3167; Chaigneau, *Dumas*, pp. 295–298.

promising site for the laboratory at Rue Garancière 8, between the Église St. Sulpice and the Luxembourg Garden (about 400 meters from the Faculté de Médecine), where Wurtz's colleague, Charles Robin, likewise agrégé in the Faculté, had recently set up a physiological laboratory. The building they moved into was (and is) known as the Hôtel de Sourdéac, a magnificent seventeenth-century mansion that until 1850 had housed the town hall of the former Eleventh Arrondissement.[100] Dollfus provided the not inconsiderable capital and constituted the technical department, while Wurtz and Verdeil represented pure chemistry. Wurtz was the director.

This venture, as Wurtz described it in a letter to Liebig dated April 1851, was not designed for personal profit; rather, "following the example of our teacher," it was directed toward advancing the science of chemistry—that is, intended for research as well as for teaching. It also included a stronger element of applied science than Liebig's laboratory ever had.[101] Wurtz also described it to Liebig as "the first laboratory of this type founded in Paris." Dumas, of course, had run his Rue Cuvier laboratory, in which Wurtz himself had worked, and Wurtz's Rue Garancière lab was also preceded by a variety of private chemical laboratories, such as Pelouze's in the Mint, Deville's in the Rue de la Harpe, and Wurtz's at the Faculté de Médecine. Wurtz may have regarded Dumas's lab in the same category, albeit a much larger version, as the other private labs, namely as institutions run purely on the basis of personal patronage. In contrast, Wurtz and his co-proprietors seem to have charged a standard fee for a standard syllabus of "practical chemistry," modeled, as Wurtz professed, on that of Liebig at Giessen. Still, as we will see below, this model was not unique in Paris.

There seems to have been some demand for Wurtz's product, for he had a dozen customers by the time the lab school officially opened (probably toward the end of 1850). Among his students during the brief lifetime of this laboratory were Eugène Caventou, Philippe de Clermont, an American named Phillips, William Marcet (an Englishman), the Spaniards Echevarria and Ramon Torres Muñoz y Luna, the Swiss Adolphe Perrot, and the

100. On the history of the Hôtel Sourdéac, see A. Jacob and J.-M. Léri, *Vie et histoire du VIe arrondissement* (Hervas, 1986), pp. 87, 108, 136.

101. Wurtz to Liebig, 18 April 1851, Liebigiana IIB; Hofmann, "Wurtz," pp. 222–225.

Figure 4.8
The Hôtel de Sourdéac (1640), at Rue Garancière 8. This was the site of Wurtz's private laboratory school in 1850–1853. It has been occupied by a publishing company since 1854. Photograph taken by the author, March 1999.

Alsatians Auguste Scheurer-Kestner and Eugène Risler.[102] Scheurer-Kestner later reminisced about the bad ventilation in the lab, where the notably foul and even lethal cyano and amino compounds were being investigated. However, he noted,

we were richly compensated by our teacher's tireless readiness to help. Whoever was motivated by the sacred fire in the heart for science could expect to receive everything from him. His devotion to his students went so far that he often insisted on taking things in hand himself. Wurtz was an exceptional artisan in glass, and it gave him great satisfaction to blow the apparatus we needed. The whole laboratory used to gather in admiration around our congenial master when he sat at the glassblowing table, turning out one Kaliapparat after another. I myself used one given me by Wurtz for twenty years, and it almost broke my heart when it finally cracked in my hands.[103]

Scheurer-Kestner arrived there in the fall of 1852, and he studied with Wurtz for 2 years. Wurtz, "lively as gunpowder," was

distracted and preoccupied, talking to himself sometimes in a low voice, sometimes out loud, with his incredible mobility of expression, chewing spit-balls, then flicking them absent-mindedly onto the ceiling or onto his apparatus. When he was in a good mood he sang at the top of his voice, very nicely, too, the "Virginum proeclara" of Rossini's Stabat Mater. . . . *Bonus filonus! Bonus filonus!* he cried, while working on his compound ammonias.[104]

But the Rue Garancière laboratory had been operating only about 3 years[105] when their building was sold to the publisher Plon and the lease

102. I have not been able to identify Phillips. William Marcet (1828–1900), a physician active in physiological chemistry, was the son of the well-known Swiss-English chemist Alexander Marcet. Both Luna (ca. 1818–ca. 1889) and Echevarria were professors of chemistry in Madrid; Luna was back in Madrid at least by June 1852. Perrot (1833–1887) studied with Wurtz for a decade, became his préparateur in 1861, then moved to an industrial position in Geneva. Scheurer-Kestner (1833–1899), the son-in-law of Karl Kestner, an industrialist and the discoverer of racemic acid, took over the family company in Thann and later was a prominent politician in the Third Republic. Risler (1828–1905) became an agricultural chemist at the revived Agronomic Institute in the Third Republic. Caventou (1823–1912), Wurtz's unofficial assistant, was the son of J. B. Caventou, discoverer of several alkaloids. Clermont (1831–1921) became sous-directeur of the chemical laboratory at the Sorbonne.

103. Scheurer-Kestner to Hofmann, in Hofmann, "Wurtz," pp. 224–225.

104. Scheurer-Kestner, *Souvenirs de la jeunesse* (Charpentier, 1905), p. 33.

105. Not just one year, as reported by Hofmann (p. 225). Its startup in 1850, as reported by Friedel ("Wurtz," p. xi), is supported by Wurtz's letter to Liebig of 18

was canceled. By this time Wurtz had been named professor at the Faculté de Médecine, so he did not join Verdeil and Dollfus when they moved the lab to an industrial concern which Dollfus purchased in a Paris suburb.[106]

About 5 months after Wurtz opened his laboratory, Gerhardt became a direct competitor. In the wake of the disaster of the Collége de France succession, Gerhardt decided (as he wrote Chancel on 2 February 1851) that only "a new 24 February [1848]" would provide him a state position; since no second revolution was in sight, private enterprise was his only way to make a living. He decided to open an "École de Chimie Pratique" collaboratively with Laurent (but not counting on any immediate aid from the invalid). The advertised novelty of the endeavor was its emphasis on laboratory manipulations rather than lectures, thus "completing the theoretical instruction that students receive in the public or private [chemistry] courses of Paris"—though Gerhardt did not neglect the opportunity to promulgate his own theoretical views.[107]

Gerhardt calculated that the enterprise would cost about 5000–6000 francs per year to run; charging students 100 francs per month would mean that five or six students would allow him to meet his costs, and twelve would give him 6000 francs annual profit. This enrollment would be easy to achieve, he thought, for Pelouze had sometimes had up to 30 students, with not much to offer. He found an appropriate building at 29 Rue Monsieur-le-Prince, on a street bordering the Faculté de Médecine.

April 1851, which refers to his "newly" established laboratory; I conclude from this that the facility must have opened late in 1850. A subsequent undated letter from Wurtz to Liebig refers to the prospective dissolution of the laboratory at the end of that year. This letter is datable by Wurtz's reference to his accession to Dumas's position at the Faculté de Médecine (beginning of February 1853) "a few months ago"; it was thus likely written in the summer or early autumn of 1853. The total lifetime of Wurtz's private laboratory was therefore about 3 years. The two letters are in the Bayerische Staatsbibliothek, Liebigiana IIB. This chronology is also supported indirectly by the dates in Jacob and Léri, *Histoire du VIe arrondissement,* which leave an unexplained gap in occupancy of the Hôtel de Sourdéac between 1850 and 1854.

106. Both Hofmann and Friedel cite financial exigencies as connected with the breakup of the lab, but there is no mention of this in Wurtz's explanation to Liebig, in the letter of summer 1853. Plon still occupies the building today.

107. Prospectus of the school, printed in Grimaux and Gerhardt, pp. 592–593.

He estimated it would cost 5000 francs for the renovation and a similar amount for fitting out. It could hold 25 or 30 students at a time, he thought.[108]

At this time Gerhardt was still on leave from his position at Montpellier, drawing his half-salary of 158 francs a month (while his former student Gustave Chancel was receiving the other half to teach the courses in Montpellier). His last extension of leave was due to expire on 1 April, so he hoped to open the lab at that time. Nonetheless, he wrote to the Minister of Public Instruction (Giraud) to request one more extension, while the enterprise was getting up to speed. Gerhardt's memo described the new school as "a scientific establishment new for France . . . organized on the model of [Liebig's] celebrated Giessen Institute and of [Hofmann's] Royal College of Chemistry in London," and he promised important services to science and to France to be rendered by the enterprise. Although the rector approved this request, Thenard and the Conseil Académique did not, and in the end it was denied.[109]

Gerhardt told Chancel that he hoped to get the necessary capital from a "rich industrialist," but in the event his rich mother-in-law lent him 10,000 francs (£400) and his half-brother another 5000.[110] The lab opened around 1 May 1851. Gerhardt thought it "the most functional and best situated in Paris" and "quelque chose de chique" in its interior details. He hired J. T. Silbermann (1806–1865) to teach physics and F. Hautefeuille (later E. Kunemann) to teach industrial applications.[111] Although in February Gerhardt claimed to have five prospective students and two more

108. The street constitutes, then and now, the back side of the campus of the Faculté de Médecine and the École Pratique de Médecine, but none of the present buildings date earlier than the 1870s. Gerhardt's "petite maison" consisted of ten rooms of varying sizes on two floors and a garret; the rent was 2500 francs per year. All these details are from Gerhardt to Chancel, 2 and 16 February 1851, in Tiffeneau, vol. 2, pp. 112 and 114–115.

109. Gerhardt to Ministre de l'Instruction publique, 18 March 1851, in Tiffeneau, vol. 2, p. 264; Grimaux and Gerhardt, pp. 214–215.

110. Grimaux and Gerhardt, pp. 210–215, 222–223, 244–245.

111. Gerhardt to Liebig, 21 April 1851, in Grimaux and Gerhardt, pp. 214–215; prospectus of school (from 1853 or 1854?), reprinted in Grimaux and Gerhardt, pp. 592–593.

promised,[112] it appears that the customers were slow to come. The young Liebigian August Kekulé arrived in Paris within days of the opening of the school and quickly struck up a close collegial friendship with Gerhardt; Gerhardt offered him an assistantship in his school, with the rumuneration of half his fees, but since there was then only one student (as Kekulé later related) he declined the generous offer.[113] We have indirect testimony of eleven or twelve students by the end of the year; Charles Gerhardt Jr. mentions Rogojski, Chiozza, and Ollivier as students and Chevreul, Pasteur, Malaguti, Graham, Williamson, Bernhard, Erdmann, and Kekulé as visitors to the lab.[114] The younger Gerhardt comments that the coup d'état of Louis Napoleon (2 December 1851) was devastating for his father's enterprise. By 1853 there were only four students; however, as the elder Gerhardt remarked in a letter to Chancel, Wurtz then had exactly the same number.[115]

It is curious to note that, within a few months of each other, two young Alsatians of nearly the same age, both former students of Liebig, founded rival private teaching laboratories in the immediate neighborhood of the Faculté de Médecine, explicitly modeled on Giessen, and that the two enterprises appear to have had similar trajectories. Neither could really be called successful, though it appears Wurtz made a better run than Gerhardt, for it seems that the latter had built up a sizable debt by late 1854.[116] Wurtz appears to have had more students, in general; moreover, he was able to capitalize the enterprise without taking on external debt (thanks to M. Dollfus's accounts).

112. Gerhardt to Chancel, 16 February 1851, in Tiffeneau, vol. 2, pp. 114–115.

113. When in 1892 Kekulé told the story publicly, he simply said that Gerhardt's conditions of employment had been unacceptable; he privately told his student Richard Anschütz the details. See Anschütz, *August Kekulé* (Verlag Chemie, 1929), vol. 1, pp. 26–27; vol. 2, pp. 943, 949–950.

114. Chancel to Gerhardt, 8 December 1851, in Tiffeneau, vol. 2, p. 117 (Chancel heard this from Richard Gordon, a mutual friend of his and Gerhardt's, who resided in Montpellier); Grimaux and Gerhardt, pp. 222–223, 244–245.

115. Gerhardt to Chancel, 16 May 1853, in Tiffeneau, vol. 2, p. 36.

116. Thenard to Ministre de l'Instruction publique, 24 November 1854, in Tiffeneau, vol. 2, pp. 297–298. Thenard stated that Gerhardt had sunk 20,000 francs into the enterprise and so could not accept an appointment at Strasbourg unless it were for two professorial chairs. Scheurer-Kestner stated flatly that the Wurtz lab in Rue Garancière "was not successful" (*Souvenirs*, p. 33).

There was even a third similar institution, founded earliest, about which almost nothing has ever been written: the laboratory of Jules Pelouze. In his éloge of Pelouze, Dumas commented that their destinies had been confounded for 40 years, and there is much truth to the statement.[117] The two men overlapped for 5 years as répétiteurs at the École Polytechnique, occupying neighboring lab benches, and both welcomed the occasional research student there; both also taught at the École Centrale, and both also provided substitute lectures for Thenard at the Collège de France. After a 2-year period when Dumas had become professor at the École Polytechnique while Pelouze had become assayer at the Mint, Pelouze then succeeded Dumas in his professorship (1838) and subsequently succeeded Thenard at the Collège (1846). When Dumas used his new professorship at the Faculté de Médecine to enable him to open a private lab in the Rue Cuvier near his residence, Pelouze simultaneously used his new professorship at the École Polytechnique to enable him to open a private teaching and research lab adjacent to his official residence in the Rue Guénégaud at the Mint (1838). In subsequent years, many fine chemists were trained there, or pursued their own original research, including Barreswil, Zinin, Claude Bernard, Millon, Gélis, Chancel, and Gerhardt.[118]

Late in 1845, Pelouze created a much larger private teaching lab at Rue Dauphine 24/26, near the Mint (and the Seine), about half a mile from Gerhardt's lab. The lab consisted of several commodious rooms on two floors at the back of a courtyard of a large corner house. This expanded enterprise lasted 12 years and was highly successful. He had

117. Dumas, *Discours et éloges académiques* (Gauthier-Villars, 1885), vol. 1, pp. 196–197.

118. Pelouze had a small laboratory installed in (or attached to) his new lodgings soon after taking the position of assayer in 1834. In his letter to Liebig of 22 August 1835, he wrote: "On m'a donné deux chambres à côté de mon logement. J'en fais dans ce moment un laboratoire et j'espère bien travailler plus que par le passé et d'une manière d'ailleurs moins fatiguante." Three years later (1 May 1838), however, he announced to Liebig the completion that day of "un magnifique laboratoire de chimie qui va remplacer le petit trou que j'occupais au bout de mon appartement" (Liebigiana IIB). Various witnesses rate the capacity of this lab (Rue Guénégaud) as being between six and twelve workers; it is apparently the lab that Laurent inherited in 1848. Many students and guest workers accommodated in Pelouze's laboratory can be documented prior to the opening of the better-known lab treated in the following paragraph.

Figure 4.9
Jules Pelouze, ca. 1860. Source: Photograph Collection, Deutsches Museum,
Munich.

enrollments of about 30 at a time, consisting mostly of young men from
the provinces and from abroad, preparing for future roles in their families' chemical businesses. He charged either 100 or 150 francs per month,
depending on the level of instruction, and he accommodated some pure
research in small rooms off the main teaching lab. On 2 February 1851,
Gerhardt wrote to tell Gustave Chancel that the "illustrious [Charles]
Barreswil ran the shop" for the owner Pelouze; for the previous 13 years,
Barreswil had served as préparateur for the Pelouze laboratory in the Rue

Guénégaud, where both Gerhardt and Chancel had worked in the mid 1840s.[119]

Pelouze's enterprise could only have been benefited when Dumas was forced to close his own lab in the midst of the insurrections and the far-left politics of the spring of 1848, and Gay-Lussac, his patron, was induced to retire. The apparent success of Pelouze's enterprise may help to explain why Gerhardt's and Wurtz's schools had trouble getting off the ground: Pelouze was older and much better established—he was, for instance, a member of the Académie des Sciences—and he may have siphoned off much of the customer demand. It is no coincidence that Pelouze, like Gerhardt and Wurtz, was closely associated with Justus Liebig; all three men had worked with the master in Giessen. Clearly the spirit of the "Giessen plan" was alive and well in Paris, even if its actualization proved difficult.

In the context of early 1852, Wurtz, with his position at the National Agronomic Institute and his private lab school, was able to think about marriage. On 17 March of that year he married the "lovely and amiable" Constance Pauline Henriette Oppermann (1830–1906), daughter of Chretien Guillaume Oppermann (a wealthy Paris banker) and Charlotte Constance Oppermann (née de Suze). Christian Wilhelm Oppermann was from Strasbourg, though his daughter had been born in Paris; Wurtz had been a family friend for some time. This family was almost certainly related to Wurtz's former professor, fellow Liebigian, and collegial friend in Strasbourg, Charles Oppermann, the hospitality of whose lab he enjoyed on

119. E. Jungfleisch ("Notice sur la vie et les travaux de Marcellin Berthelot," *Bulletin de la Société Chimique* [4] 13 (1913), pp. xvii–xviii) dates the origin of this lab to 1848. See also Grimaux and Gerhardt, pp. 122, 211. There is little information in printed form on this lab, and what exists is contradictory. C. Martius ("Nekrolog auf T. J. Pelouze," *Neues Reporatorium für die Pharmacie* 17 [1868], p. 507) dated the origin of the school to 1846 and commented that it was founded on German models; the same date was cited by A. Chevalier ("T. J. Pelouze," *Journal de chimie médicale* 3 (1867): 444–448). By far the best source of information on Pelouze's various laboratories (he had the two teaching labs mentioned here, plus no fewer than four other private research labs at various times in his career) is a 17-page anonymous document (probably written by Aimé Girard at Dumas's behest shortly after Pelouze's death) preserved in the Archives of the Académie des Sciences, in which it is stated that the large teaching lab in the Rue Dauphine was opened late in 1845.

occasion, even after his move to Paris.[120] In view of Wurtz's still relatively unestablished professional state, one can only assume that the bride's parents helped considerably with the finances of the new household. The couple moved to an elegant apartment, Rue Saint-Guillaume 27, in the Faubourg Saint-Germain (now just off the Boulevard Saint-Germain in the eastern part of the Seventh Arrondissement).

In September of 1852, Wurtz was without a secure position; however, he foresaw, probably even before his wedding, that one would fall into his lap within a few months. Dumas's increasing political activities had led to his decision to resign his professorship at the Faculté de Médecine, and he had assured Wurtz that the position would be his. This came to pass in February of 1853.[121]

Wurtz's successful navigation through the difficult waters of authoritarian monarchy, socialist revolution, republic, and authoritarian empire was attributable to foresight, tact (or, in Gerhardt's view, hypocrisy), and a cconsiderable amount of luck. Not yet considered here, but decisive, was the remarkable research he had completed during the first 8 years of his residency in Paris.

120. Gerhardt, then in Strasbourg, learned of a serious accident in Wurtz's Paris laboratory from a Mme. Oppermann (Gerhardt to Wurtz, 7 January 1856, in Tiffeneau, vol. 2, p. 10). I assume that this was the wife of Professor Charles Oppermann. The most likely assumption is that Chretien Guillaume Oppermann and Charles Frédéric Oppermann were brothers, though I have not been able to verify this.

121. A letter from Louis Pasteur's wife to his father dated 12 April 1852 (less than a month after Wurtz's wedding) related a recent conversation between Pasteur and Dumas in which Dumas stated that he intended to resign his position at the Faculté de Médecine and that Wurtz would be his successor. Dumas promised the position at the Agronomic Institute to Pasteur. (*Correspondance de Pasteur, 1840–1895* (Flammarion, 1940), vol. 1, p. 237)

5

Covering the Bases

Liebig and Dumas were intensively involved with nitrogenous organic bases during the 1820s and the 1830s, and it is not surprising that their student Wurtz would show such an interest himself. To understand Wurtz's early scientific work, we must begin with that of his mentors.

Cyanic Esters and Substituted Ureas

In 1830 Justus Liebig and Friedrich Wöhler published an article in which they reported the formation of ethyl cyanate by the action of cyanic acid (modern HOCN) on grain alcohol; in 1845 they repeated their observation and also produced methyl cyanate by an analogous reaction with wood alcohol.[1] A few months after the publication of the 1845 paper, Laurent and Gerhardt aggressively attacked Liebig's research on the "mellon" family of complex urea derivatives, declaring that his work was riddled with errors.[2] Inter alia, they asserted without further evidence or justification that Liebig's and Wöhler's recently announced "ethyl cyanate" was nothing of the kind, but rather urethane (modern H_2NCO_2Et), discovered by Dumas in 1833. Liebig's fury was limitless; his ties to Gerhardt were irrevocably broken.

Laurent and Gerhardt knew that it was Adolphe Wurtz who had discovered the error, though they did not cite him.[3] In fact, an article by Wurtz

1. Liebig and Wöhler, "Untersuchungen über die Cyansäuren," *Annalen der Physik* 20 (1830): 369–400; Wöhler and Liebig, "Cyansaures Aethyl- und Methyloxyd," *Annalen der Chemie und Pharmacie* (hereafter "*Annalen*") 54 (1845): 370–371.

2. Laurent and Gerhardt, "Recherches sur les combinaisons melloniques," *Comptes rendus* 22 (1846): 453–462.

3. Wurtz's discovery was mentioned in Laurent's letter to Gerhardt of 14 March 1846; 2 days later, Laurent presented their attack on Liebig to the Académie des

on this subject appeared in the very same issue of the *Comptes rendus*; it was Wurtz's first paper on organic bases, indeed, his first paper on organic chemistry proper, as opposed to biological or physiological chemistry. He studied a reaction similar to the one Wöhler and Liebig had studied, but using cyanogen chloride rather than cyanic acid as the reagent. His product, however, was not ethyl cyanate, nor any cyanic ester, but rather urethane. He then showed that Wöhler and Liebig had misidentified their product, which was the same substance. (By suggesting that their putative cyanic ester contained a molecule of water of hydration, Wöhler and Liebig were able to match the formula that they measured for the product; the purported hydrated cyanic ester is isomeric with urethane.) Wurtz wrote that this newly discovered fact "cannot but add new interest to Wöhler and Liebig's observation."[4]

In view of the eventful history of the cyanates and urea derivatives in the hands of Wöhler and Liebig during the preceding 20 years, this was a bold publication for a young chemist who was not yet professionally established. The curious thing is that Wurtz had apparently been working on this reaction for at least 4 years without publishing anything; moreover, it seems that Liebig and Wöhler had been aware of Wurtz's research since 1842.[5]

Sciences (M. Tiffeneau, ed., *Correspondance de Charles Gerhardt* (Masson, 1918–1925), vol. 1, pp. 163, 166). It was precisely because of a fear of being further anticipated by Wurtz that Laurent rushed the article into publication.

4. Wurtz, "Note sur la formation de l'uréthane par l'action du chlorure de cyanogène," *Comptes rendus* 22 (1846): 503–505.

5. "Wurtz has anticipated us with the second cyanic ester. He worked on the reaction of gaseous cyanogen chloride with alcohol, and found urethane, among other products." (Liebig to Wöhler, March 1842, in *Aus Justus Liebig's und Friedrich Wöhler's Briefwechsel in den Jahren 1829–1873*, ed. A. Hofmann (Vieweg, 1888), vol. 1, p. 191) The passage is curious, for it suggests that Wurtz had been working for some time on this reaction, and yet he did not arrive in Giessen until March or even April of 1842. Internal references confirm that Hofmann dated this letter correctly (at least appoximately). However, Hofmann often silently edited these letters; indeed, he sometimes compressed several from a similar time period into a single printed letter under a single date. It is therefore possible that these sentences were written by Liebig in April or even May of1842, and that the research referred to had been carried out in Giessen. It is also possible that Wurtz had been working on this reaction in Strasbourg, before coming to Liebig's lab. Other than this one reference, there is no evidence of Wurtz's interest in this subject until his first publication in March 1846.

But Wurtz's tactful approach to correcting his elders was obviously appreciated by Liebig. In sharp contrast to his explosion against especially Gerhardt, he gracefully accepted the correction by "this skilled chemist"; he commented that he and Wöhler had never made a positive identification, but had only stated that the formula of the product matched that of ethyl cyanate. A few months later he and Wöhler determined that their "ethyl cyanate" of 1830 had actually been allophanic acid. In 1846 Liebig published a translated abstract of Wurtz's article in his journal, a full year after it had appeared in French.[6]

Wurtz was evidently captivated by this family of substances, despite their disgusting odors and often dangerous properties. In early 1847 he published a second article containing three discoveries: a superior way to prepare cyanuric acid (the trimer of cyanic acid), a novel liquid form of cyanogen chloride, and a putative new compound consisting of cyanogen chloride combined with hydrogen cyanide (which Wurtz later showed to be a mixture rather than a compound). At the end of the note he apologized for the preliminary character and dilatory appearance of his communication; the work, he said, was terribly difficult and he had been forced to take frequent pauses due to fatigue. It was another year before he published his third article, but this time one could see major results on the way. Wurtz showed that it was indeed possible to prepare the true methyl and ethyl esters of both cyanic and cyanuric acid (by reacting potassium sulfovinate—i.e., potassium ethyl sulfate—with potassium cyanate). By preparing a triple ester of cyanuric acid, Wurtz demonstrated that it was tribasic, as Liebig had asserted in 1838.[7]

The new cyanic esters were interesting and scientifically fruitful substances, and Wurtz was sufficiently skilled and industrious to take full advantage of the opportunities. In 1848 Wurtz showed that both esters—ethyl and methyl cyanate—absorb ammonia gas and are thereby transformed into new

6. Liebig, "Beleuchtung einer Untersuchung von Laurent und Gerhardt über die Mellonverbindungen," *Annalen* 58 (1846), pp. 260–261; Liebig and Wöhler, "Ueber die Einwirkung der Cyansäure auf Alkohol und auf Aldehyd," *Annalen* 59 (1846): 291–300; Wurtz, German-language abstract of French paper, *Annalen* 60 (1847): 264–265.

7. Wurtz, "Mémoire sur les combinaisons du cyanogène," *Comptes rendus* 24 (1847): 436–439; "Recherches sur l'éther cyanurique et sur le cyanurate de méthylène," *Comptes rendus* 26 (1848): 368–370.

compounds, which he named ammoniacal cyanic ester and ammoniacal methylene cyanate, respectively.[8] Treated with water, the cyanic esters dimerized and lost carbonic acid to form "methylic and ethylic cyamethane." Curiously, the ethyl compound in the first case was isomeric with the methyl compound in the second (that is, ammoniacal cyanic ester was isomeric with methylic cyamethane).[9]

Wurtz later determined that his cyanic esters were not true cyanates after all, but rather an isomeric version subsequently named "isocyanates." He also eventually recognized that when ethyl and methyl isocyanate react with ammonia the products by intramolecular rearrangement are ethyl and methyl urea (in modern terms, $EtHNCONH_2$ and $MeHNCONH_2$), and that his second set of products (also due to rearrangements and loss of carbonic acid) were actually dimethyl and diethyl urea. Dimethyl urea is indeed isomeric with ethyl urea. But how did Wurtz interpret these reactions at the time of writing (August 1848)?

In 1830 Dumas had formulated urea as the double amide of carbonic acid; this was the origin of the word "amide," which Dumas used to refer to the moiety AzH^2. However, Victor Regnault argued a few years later that it was the isomeric substance carbamide that had this constitution.[10] (Urea and carbamide were first recognized as identical, rather than isomeric, in the 1850s.) Regnault's paper suggested that the accepted constitution of urea had to be altered, and it was thereafter generally agreed that urea must have the constitution of ammonium cyanate, C^2AzO,HO,AzH^3 (consistent with the conventions of the 1840s, this is a dualistic formula using conventional equivalents, i.e., C = 6 and O = 8). Wurtz formulated his cyanate esters in an analogous fashion, namely as C^2AzO,C^2H^3O and C^2AzO,C^4H^5O (or C^2AzO,MeO and C^2AzO,EtO). Once these theoretical commitments were made, it was straightforward to formulate the new ammoniacal esters; Wurtz regarded them as substituted ureas, namely C^2AzO,MeO,AzH^3 and C^2AzO,EtO,AzH^3.

8. These are direct cognate translations of Wurtz's French; the only alteration I have made is to translate "éther" as "ester." On p. 98 of *The Quiet Revolution*, I erroneously identified Wurtz's "ammoniacal cyanic ester" with modern ethyl isocyanate.

9. Wurtz, "Recherches sur les éthers cyaniques et leurs dérivés," *Comptes rendus* 27 (1848): 241–243.

10. Regnault, "Sur l'acide chlorosulfurique et la sulfamide," *Annales de chimie* [2] 69 (1838), pp. 180–183.

It was less easy to decide how to regard the constitutions of the dimerized products. Wurtz concluded that they were methyl cyanate combined with "méthylamide," $C^2H^3(AzH^2)$ = $MeAzH^2$, and ethyl cyanate combined with "éthéramide." He regarded the substituted ureas (he called them "acetic urea" and "propionic urea") as the true homologues of urea itself, in just the same way that acetic and propionic acids were homologues of formic acid. The dimers were, in contrast (he said), quite new compounds with no direct homologous relationships to known substances.

Conjugated Ammonias or Amides?

Wurtz placed this research not only in the context of recent work on cyanates and ureas but also in the context of a new program that was being successfully exploited by the Englishman Edward Frankland (1825–1899), a former pharmaceutical apprentice, and the German Hermann Kolbe (1818–1884), a student of Wöhler and Bunsen. Frankland and Kolbe met in London in 1845, and in 1847 they worked together for a summer in Bunsen's laboratory in Marbug. In the spring of that year they published an important paper showing that alkyl cyanides (also known as nitriles) could be hydrolyzed to corresponding organic acids: methyl cyanide to acetic acid, ethyl cyanide to propionic acid, amyl cyanide to caproic acid, and so on. Not only was this a convenient way to increase the size of an organic compound (for the ethyl cyanide could be prepared from ethyl chloride); it also gave important insight into the constitutions of the acids. Kolbe and Frankland argued, reasonably, that the homologous alkyl groups (methyl, ethyl, amyl, and so on) could be regarded as real preexistent radicals united to the rest of the molecule, namely C_2O_3,HO in the dualistic convention that is equivalent to the modern "carboxyl" group. Kolbe and Frankland regarded C_2O_3,HO as identical to oxalic acid. (In modern terms, the carboxyl group is to oxalic acid as a monomer is to a dimer.)

In the broadest sense, Kolbe and Frankland thought that this research supported, or even demonstrated, the truth of the theory of copulated or conjugated organic compounds. This theory, due primarily to Berzelius, was originally designed as a way of accommodating the new research on chlorine substitution while retaining the principles of electrochemical dualism. According to Berzelius, conjugated compounds are formed of parts that can have independent existence and are bound in a non-electrochemical fashion that does not affect the saturation capacity of the acidic or basic components. In the Kolbe-Frankland example, oxalic acid retains its full acidity even when "conjugated" to methyl or ethyl in acetic or propionic acid, respectively. Methyl and ethyl, the "copulas," may then be indifferently substituted by chlorine or any other element without materially affecting the properties of the compound.

Kolbe and Frankland, working independently in 1848 and 1849, drew further support for this conception by developing methods for generating the copulas as independent compounds, named methyl, ethyl, and so on. Frankland did this by chemical means—for instance, by reacting zinc and ethyl iodide to produce zinc iodide plus free gaseous ethyl. Kolbe did the same thing by physical means—for example, by electrolyzing potassium valerate to produce the free "valyl" (later named "butyl") radical. For Kolbe and for Frankland, the methyl that came bubbling out of Kolbe's electrolysis of potassium acetate was the unchanged copula of acetic acid, simply freed from its combination with oxalic acid and potash.[11]

After Gerhardt and Laurent began collaborating (in 1844), they started to see all sorts of reasons not to agree with these schemes, as logical and as empirically supported as they seemed to be. In their alternate vision of organic chemistry, electrochemical properties were regarded as unimportant as a means of discerning composition, so the phenomenon of chlorine substitution was not anomalous; all their suggested reaction mechanisms were substitutions, in contrast to the dualists' invariable addition mechanisms. Moreover, they urged a return to two-volume atomic weights and a sensitivity to polybasic acids and hydracid theory, which, they affirmed, were more than a change of indifferent conventions. Consequently, rather than use the dualists' monomers, they formulated oxalic acid and the Kolbe-Frankland free alkyl radicals in the dimeric form. For Gerhardt and Laurent, acetic acid was not methyl conjugated with oxalic acid and hydrated with a molecule of water, all three components being isolable as stable substances; rather, it was a unitary hydracid of the form $C_2H_3O_2 \cdot H$. There are no oxides in salts, and no water in acids, they affirmed.

11. On copulas and the discovery of the "alcohol radicals," see J. Partington, *A History of Chemistry* (Macmillan, 1964), vol. 4, pp. 372–375 and 504–510; Rocke, *Quiet Revolution*, pp. 55–68.

Liebig and Dumas had done much to prepare the way for these ideas. Since 1834 Dumas had championed the notion of substitution, at first from within the safe haven of dualism but later in a fundamentally different version known as the theory of types. During the same period, Liebig had developed a theory of polybasic acids and organic hydracids and had done much in an implicit fashion to promote substitutionist ideas. However, as we have seen, after 1840 both men retreated from these ideas into the security of more (apparently) empiricist and epistemologically cautious ideas, namely the use of "equivalents" and the older, comfortable dualistic formulas. Indeed, neither man was ever thereafter engaged with fundamental theories of composition in organic chemistry, and most of the chemical world, especially the older generation, followed them along this path.

The work of Kolbe and Frankland in 1847–1849 only reassured Dumas and Liebig that they had been right to reject the extreme notions of Gerhardt and Laurent. That work was focused on organic acids, but the same ideas were applied to the bases. In the spirit of the copula theory, all organic bases should be formulated as complexes of the prototypical base, ammonia, with organic copulas. This view was at first opposed by August Wilhelm Hofmann for the aromatic amines he had been studying in Giessen the summer he had become friendly with Wurtz. Hofmann preferred the amide formulations of his teacher Liebig. However, 6 years later, in the summer of 1848, while Wurtz was puzzling through the isocyanate reactions discussed above, Hofmann —probably under the influence of Kolbe's theoretical preferences—published a review article that indicated a strong preference for the Berzelian viewpoint. He collected all the evidence in favor of the "exceedingly probable" view "that the organic bases are indeed conjugated ammonia-compounds," aniline itself being considered to be $C_{12}H_4 \cdot NH_3$. Still, Hofmann appears to have thought, it ought to be possible to separate these components and isolate them independently, as Frankland and Kolbe were doing with the organic acids; "vainly have I hoped," he plaintively stated, to do just that.[12]

There was, however, an alternative way to view organic nitrogen bases. As far back as 1830, Dumas had proposed that urea had two NH_2 groups bound to C_2O_2 (or CO in atomic weights); he also proposed that oxamide,

12. Hofmann, "Researches on the Volatile Organic Bases," *Journal of the Chemical Society* 1 (1848), p. 317.

the double amide of oxalic acid, consisted of $(C_2O_2)_2(NH_2)_2$. In 1840 Liebig published an encyclopedia article in which he wrote:

When one considers the compound NH_2, namely amide, as a compound radical . . . it is clear that ammonia consists of the hydride of a basic radical. . . . We know that amide, the radical in ammonia, can replace the oxygen in many organic acids equivalent for equivalent, and we find that the new compounds that are formed in this manner entirely forfeit their acid characteristics. . . . Were we able to replace the oxygen in ethyl and methyl oxide [that is, in alcohols, esters, and ethers] . . . by one equivalent of amide, we would then without the slightest doubt have compounds that would behave entirely similarly to ammonia. Expressing this in a formula, a compound of $C_2H_5 + NH_2 = Ae + Ad$ would possess basic properties.[13]

Liebig's "amidogen theory" of organic bases was certainly influenced by Dumas and Laurent, both of whom had also proposed such a radical. It appears to differ little from the Berzelian copula theory, for the difference between NH_3 and NH_2 as a proximate constituent does not seem large. But in fact the two theories were as different from each other as electrochemical dualism was distinct from the Gerhardt-Laurent reforms. In proposing amidogen, Liebig was in effect allying himself with the latter, for NH_2 can only make sense in a substitutionist theory, whereas NH_3 requires additive mechanisms. However, his proposal was written just before he abandoned the theoretical dialectic, and so he published nothing more on the idea.

There is little that is remarkable about Wurtz's suggestions for the constitutions of his new compounds; his ideas reflected the dominant tone of theoretical chemistry of the 1840s: epistemological caution combined with an almost instinctive preference for electrochemical-dualist-additive formulations. Thus, each of his formulas for acids specified a water molecule within, and each of his salt formulas had an oxide subcomponent; his formula for urea had three independent additive components (oxidized cyanogen, water, and ammonia); and he thought nothing about proposing a compound consisting of cyanogen chloride and hydrogen cyanide. It is, however, interesting that his cyamethanes contained components that do not fit this model, namely "méthylamide" in his methylic cyamethane and

---

13. Liebig, "Basen, organische," in *Handwörterbuch der reinen und angewandten Chemie* 1 (Vieweg, 1840), pp. 697–699. The title page of the volume is dated 1837, whence Partington's date for this article (*History*, p. 437), but this is an error. The date on the title page is the publication date of the first fascicle of the volume; the fourth fascicle, in which this article appeared, was not published until 1840.

"éthéramide" in his ethylic cyamethane. This hint of reformist chemistry fits the pattern I have noted in Wurtz's early papers on inorganic chemistry (1842–1846).

One could explain this ambivalence in various ways. It is possible that Wurtz was insensitive to theory, and adopted an eclectic rather than a severely consistent approach. One might also suggest that he did not regard these points as evidence of inconsistency or ambivalence at all, although it is hard to see how they can be reconciled. A third possibility is that he had been convinced by the arguments of Laurent and Gerhardt but wanted to maintain his apparent orthodoxy in order to stay in Dumas's good graces and thereby advance his career. Finally, it is conceivable that, although he was a theory-driven personality, he was still uncertain as to which direction he or the field as a whole should or would go; maintaining his options would be a good strategy in such a case, and might lead to the use of ambiguous expressions.

Whatever was Wurtz's position of the 1840s regarding dualist versus substitutionist chemistry, there is no question that he was seeking a single theoretical framework for the whole science, organic and inorganic alike. Electrochemical dualism had arisen in the context of inorganic chemistry of the 1810s and the 1820s, whereas substitution theory had come of age with the newer organic chemistry of the 1830s and the 1840s. Should inorganic chemistry provide the basis for interpreting organic, or vice versa? The great conflicts of the age resulted from placing the analogical arrow in one direction or the other.[14] But chemistry itself was a single science, and Wurtz was convinced that one theoretical structure ought to suffice for both of its divisions, for, as Isaac Newton had put it, nature must be always be simple and conformable to itself.

This conviction is clearly revealed in a paper of Wurtz's published early in 1847. The subject was the novel compounds thiophosphoric acid and phosphorus oxychloride. The latter compound, Wurtz stated, should not be regarded as an oxide of PCl^3, "in the sense of the dualistic theory"; rather, it is analogous to a chloro-substituted organic compound. In PCl^5, two of the

14. J. Brooke, "Chlorine Substitution and the Future of Organic Chemistry," *Studies in the History and Philosophy of Science* 4 (1873): 47–94; Brooke, "Laurent, Gerhardt, and the Philosophy of Chemistry," *Historical Studies in the Physical Sciences* 6 (1975): 405–429.

chlorine atoms are held more loosely than the other three, and "occupy a different position" in the molecule, thus are extractable in the same way as stones may be removed from the top or sides of a building more easily than from the bottom. PCl^5 may therefore be regarded as the schematic generator of a series of compounds all derived "from the same type." This is a case of substitution in inorganic chemistry, Wurtz concluded—still a rare phenomenon, but one that "will perhaps contribute to eliminating one of the barriers that still separate mineral from organic chemistry."[15] A similar sentiment opens one of Wurtz's most famous articles, to which we now turn.

The Discovery of the Primary Amines

In the midst of his new teaching responsibilities at the Faculté de Médecine, Wurtz spent the fall of 1848 in his small lab in the Musée Dupuytren trying to follow up on the discovery of cyanic esters and alkyl-substituted ureas. Once a chemist has a new toy, one way to play the game was (and still is) to make new derivatives that were analogous or homologous to those one already had made, and to hydrolyze the derivatives to examine the pieces of the molecule. The first route is synthetic in character, the second analytic. In the synthetic direction, Wurtz proceeded to prepare amyl cyanate and amyl urea[16]; however, he encountered difficulties when he tried to hydrolyze his cyanic esters with potassium hydroxide solution. He could tell that free ammonia was liberated, for he could smell it as it bubbled out of solution, and he found potassium carbonate in the reaction vessel, which demonstrated that carbon had separated from the substance. However, he never was able to find the organic remainder of the starting material. He repeated the procedure a number of times over a few weeks, but the result was always the same.

In this experiment Wurtz's perceptions were being conditioned by his theoretical expectations. If alkaline hydrolysis of his ethyl cyanate proceeded

15. Wurtz, "Recherches sur l'acide sulfophosphorique et le chloroxyde de phosphore," *Comptes rendus* 24 (1847): 288–290 (290). A full version appeared under the same title in *Annales de chimie* ([3] 20 (1847): 472–481). The quoted passage is identical in the two articles.

16. Wurtz to Dumas, 9 November 1848, Dossier Wurtz, Académie des Sciences; Wurtz to Hofmann, 25 November 1848, Chemiker-Briefe, Berlin-Brandenburgische Akademie der Wissenschaften.

analogously to the Kolbe-Frankland hydrolysis of ethyl cyanide, then the compound should lose ammonia and form an organic acid with more carbon than acetic acid, since in addition to ethyl the radical would also contain the carbonaceous portion of cyanogen. However, the circumstance that potassium carbonate was found in solution after the reaction suggested that the compound had lost carbon, so Wurtz had expected to find an ethyl-derived remainder, such as a salt of acetic acid. However, no organic substance at all could be found in solution. Wurtz continued to be perplexed by this curious reaction until, by accident or by design, the escaping "ammonia" caught fire and burned brightly. This was not ammonia after all, but an organic compound with the olfactory character of ammonia. That was where the missing organic residue had gone.[17] The paper he rushed into print in the *Comptes rendus* for 12 February 1849 was written with the flair of a man who knows he is making a major announcement. Ammoniacal compounds, Wurtz argued, form "in some manner a transition" between inorganic and organic chemistry. He continued:

Assuredly, ammonia would have to be regarded as the simplest and most powerful of organic bases, and it would be for all chemists the type of this numerous class of bodies, if it were not for one characteristic, important without doubt, but to which we have perhaps attached an exaggerated value. Ammonia does not contain carbon. It appears that this difference in composition does not suffice to separate ammonia from the organic bases. Indeed, I have succeeded in making from this alkali a true organic compound by adding to it the elements of the carbonated hydrogen C^2H^2 or C^4H^4, without making it lose the characteristics of a powerful base, nor its most striking properties, such as for instance its odor.[18]

He then related how alkaline hydrolysis of cyanic esters, of cyanuric esters, and even of his "acetic and propionic ureas" result in "méthylammoniaque" and "éthylammoniaque," or, perhaps better named, "méthylamide" and "éthylamide," AzH^2,C^2H^3 and AzH^2,C^4H^5. Other than the obvious fact that a molecule of carbonic acid had departed and a molecule of water had been added, Wurtz did not speculate on a reaction mechanism.

In order to understand the great impact that Wurtz's discovery made on his colleagues, let us suppose that a wide variety of organic bases of all

17. The story is told in A. Hofmann, "Erinnerungen an Adolph Wurtz," in *Zur Erinnerung an vorangegangene Freunde* (Vieweg, 1888), vol. 3, pp. 341–342. Hofmann presumably got the anecdote directly from Wurtz.

18. Wurtz, "Sur une série d'alcalis organiques homologues avec l'ammoniaque," *Comptes rendus* 28 (1849), p. 223.

degrees of complexity had long been known, but that only a few large organic acids were known to exist—for instance, oleic, stearic, benzoic, and salicylic acid. Suppose further that some éminence gris had proclaimed the likelihood that simple organic acids corresponding to inorganic carbonic acid must in principle exist and would soon be found. Imagine the excitement that would have been generated by the subsequent announcement of the discovery of formic and acetic acids. In this hypothetical case, as in Wurtz's real one, a new route was discovered into a fertile area of investigation, and moreover one that strikingly illustrated the connections between inorganic and organic chemistry.

Dumas was lavish in praise of his protégé. The entire chemical community, he said, had been "powerfully struck" by the announcement, which revealed a "new and vast field at a stroke." He continued:

To discover a new series of the same nature [as the homologous acids], perhaps the most important series of all by the number and variety of its derivatives; to show, by a striking example, that science may enter with confidence in the pathways that synthesis offers her, is at once a high honor and a rare happiness. . . . [The paper] will remain a model in science for the authority of its views, the justice and rigor of its conclusions, as well as the abundance of facts. In our opinion it has been a long time since chemistry has been so enriched by a series of compounds so important, and by a theory so fecund. . . .[19]

After Wurtz's death, Hofmann wrote that no other paper in his memory had made such an impression on the collegial community.[20] We have no surviving letter of early 1849 from Wurtz to Hofmann telling him the news, but even if he did not do this directly Hofmann would soon have found out, for an English translation appeared in the *Chemical Gazette* just a month after the issue date of the *Comptes rendus*.[21] Hofmann wrote to Liebig on 24 March to tell him about Wurtz's paper; in a postscript, he congratulated Liebig on his prediction of Wurtz's compounds 9 years earlier. Liebig was as struck as Hofmann and Dumas, not only by Wurtz's paper but also by

19. Dumas (reporting as chair of a commission also including Thenard and Chevreul), "Rapport sur un mémoire de M. Wurtz, relatif à des composés nouveaux analogues à l'ammoniaque," *Comptes rendus* 29 (1849): 203–206. See also Dumas, *Comptes rendus* 28: 323–324.

20. Hofmann, "Wurtz," pp. 339–340.

21. *Chemical Gazette* 7 (15 March 1849): 115–117. Since we know that Hofmann was translating at least some of Wurtz's papers into English in this period, it is even possible that the translation was done by Hofmann.

Hofmann's comment. "Oddly enough," wrote Liebig, "it was only the post-script in your letter that reminded me that I had at one time so much as thought about the constitution of the organic bases. I was so enchanted by Wurtz's work that I wrote him and congratulated him on such lovely dis-coveries; I had completely forgotten the relationship of this work to my own thoughts."[22] Once reminded, however, Liebig made sure that his pre-diction of 1840 was republished in his journal when a translation of Wurtz's paper appeared there.[23] The editor (Hermann Kopp?) noted that the impor-tance of Wurtz's work was such that this preliminary communication should appear in translation, contrary to the normal policy of the *Annalen* to wait for the appearance of the definitive publication in French.

A second editorial note accompanied the German translation, correct-ing Dumas's statement that Wurtz had prepared four of the five theoreti-cally simplest amines: methyl, ethyl, butyl, and amyl. Wurtz had by this time published two follow-up articles on homologous bases, including announcing amylamine; however, the editor noted that Wurtz had pub-lished nothing on butylamine.[24] But Thomas Anderson had earlier pre-pared a substance which he called "petinine," with a formula which Gerhardt had corrected. This all happened just before Wurtz's discovery, after which Anderson's compound was recognized by everyone as buty-lamine. It is unclear whether Wurtz knew of Anderson's work before he published his paper, and Gerhardt's correction of the formula probably was not published until after Wurtz's publication.[25]

22. Liebig to Hofmann, 23 April 1849, in *Justus von Liebig und August Wilhelm Hofmann in ihren Briefen*, ed. W. Brock (Verlag Chemie, 1984), p. 84. Hofmann's letter to Liebig of 24 March is lost, but the date is cited in this letter.

23. *Annalen* 71 (1849), pp. 346–348.

24. Wurtz, "Recherches sur les ammoniaques composées," *Comptes rendus* 29 (1849): 169–172; "Note sur la valéramine ou l'ammoniaque valérique," *Comptes rendus* 29 (1849): 186–188; editor's note, *Annalen* 71 (1849): 345–346. (The unsigned note could have been written by Liebig but was more likely written by his managing editor, Hermann Kopp.)

25. Anderson's paper was "On the Products of Destructive Distillation of Animal Substances," read to the Royal Society of Edinburgh in April 1848 but first published in September in *Philosophical Magazine* ([3] 33 (1848): 174–186). Wurtz could also have seen a German translation that appeared that autumn in the *Journal für prak-tische Chemie*. For Gerhardt's correction of Anderson's formula and commentary on Wurtz's discovery, see *Comptes rendus des travaux de chimie* 4 (1848): 324.

Gerhardt, who had arrived in Paris about a year before Wurtz's discovery, was not napping as this was playing out. He was both envious and angered by what he regarded as Dumas's excessive praise of Wurtz's paper, especially since Wurtz had used elements of his and Laurent's ideas—homology, amide theory, substitution, and the example of butylamine—without acknowledgment. Dumas had praised the "fecund theory" of Wurtz, but Gerhardt could not help noticing that there was no explicit theoretical content at all in Wurtz's paper: "M. Wurtz, after having profited from my observations, without, of course, saying from whom he took them, keeps the bases for himself. . . ." Gerhardt's son and biographer went so far as to suggest that Wurtz lacked integrity in ignoring Gerhardt's contributions in order to advance his career and to curry favor with Dumas.[26]

It must be noted, however, that Gerhardt had corrected Anderson's formula not by analyzing "petinine" but rather by noting that Anderson's proposed formula would violate his and Laurent's "even-number rule." Moreover, Gerhardt did not attempt to investigate the constitution of the new base. He had come to the conclusion, against Laurent's advice, that the determination of molecular constitutions was inaccessible to science, and that only empirical formulas were permissible in chemistry. Wurtz, in contrast, was, from the time of his earliest chemical work, strongly oriented toward the determination of constitutions. Contrary to Gerhardt's recriminations, Wurtz deserved the credit he received for this major discovery.

The Multiple-Alkylation Strategy

Hofmann's research on aniline derivatives was, as he put it, "given a new direction" by Wurtz's paper. Since Wurtz had succeeded in inserting hydrocarbon radicals into ammonia, Hofmann attempted to do the same with aniline, at first by reacting this substance with methyl and ethyl bromide.[27] At this time both Hofmann and Dumas were still using Berzelian copula

26. Gerhardt, *Comptes rendus des travaux de chimie* 5 (1849), pp. 23 and 313; Grimaux and Gerhardt, *Charles Gerhardt*, pp. 197, 379–380.

27. Hofmann, "Recherches sur l'éthylaniline et sur la méthylaniline," *Comptes rendus* 29 (1849): 184–186. This article appeared in the same weekly issue of the journal as Wurtz's second and third communications on his new amines (13 August 1849), and one week before Dumas's highly laudatory referee report.

formulas to describe the new bases. (Hofmann had affirmed that they accurately represented the constitutions of the compounds; Dumas was silent on the theoretical question and may have used these formulas in a more conventional sense.) However, in November of 1849 Hofmann converted to the substitutionist amide formulations that Wurtz had used.[28] Hofmann placed all the weight for his conversion on a reaction he had investigated which was anomalous for the theory of copulated bases but which would have been predicted by amide theory. However, it is reasonable to believe that Wurtz's work may also have played a role in his thinking. At the end of this paper Hofmann announced that he had succeeded in replacing the amino hydrogens of aniline and the hydrogens of ammonia by methyl, ethyl, and amyl, using methyl, ethyl, and amyl iodide. A long paper thoroughly exploring the new reactions followed in January of 1850.[29] It was an instant classic. Hofmann may have been chagrined not to have been the first to prepare Wurtz's simple amines, but Wurtz must in his turn have envied his friend's fine work, which provided the first flexible synthetic route to an immense number of novel amines. (Their close friendship was, however, completely unaffected by their scientific rivalry.) Hofmann began with a masterly discussion of recent history. The copula theory, regarded until recently as so well established, was now shown to be untenable, and Liebig's "amidogen" theory (Hofmann did not mention the French chemists) was shown to be the only empirically supported view of the constitution of the organic bases. This new phase, he wrote, had been initiated by Wurtz's "splendid investigation" of the simplest organic derivatives of ammonia. "Now these compounds," he continued, "imagined in 1840 by Liebig in illustration of his views, have sprung into existence in 1849, with

28. Hofmann, "Researches on the Volatile Organic Bases," *Journal of the Chemical Society* 2 (1849): 300–335; "Recherches sur la série anilique," *Comptes rendus* 29 (1849): 786–788. Hofmann's evidence was the circumstance that aniline and benzoic acid do not react together with loss of two water molecules to form a nitrile, as benzoic acid does with ammonia. This would be expected only if the nitrogen group in aniline were NH_2 rather than NH_3, for the two hydrogens of amide plus the acidic hydrogen of benzoic acid provide only three rather than the four hydrogens necessary to form two molecules of water.

29. Hofmann, "Researches regarding the Molecular Constitution of the Volatile Organic Bases," *Philosophical Transactions of the Royal Society* 140 (1850): 93–101.

Figure 5.1
August Wilhelm Hofmann, ca. 1850. Source: E. F. Smith Collection, University of
Pennsylvania Library.

all the properties assigned to them by that chemist. . . . It would be difficult to imagine a more brilliant triumph for any theoretical speculation. . . ."[30]

Hofmann published a second major paper on this subject in 1851. He had succeeded in forming an enormous number of amines—not only Wurtz's "primary" amines, but also "secondary" and "tertiary" amines, such as diethyl-, dimethyl-, triethyl-, and trimethylamines, and ethyl, methyl, diethyl, and dimethyl aniline.

Another amine discovered around this time was a degradation product of narcotine; it was an intensely fishy-smelling substance which its discoverer, the Austrian chemist Theodor Wertheim, called "Oenylamin." The formula suggested to Wertheim that it was the next larger homologue of methyl- and ethylamine, that is, propylamine. However, Hofmann suspected that it was trimethylamine (isomeric with propylamine), and he demonstrated this in 1852. Wurtz told Hofmann after reading his paper that he had never believed that Wertheim's substance was propylamine, and that he had been about to investigate the subject himself.[31]

Wurtz published a long review article on amines in the December 1850 issue of the *Annales de chimie*. Here Wurtz argued that the newest work on amines—his and others'—demonstrated that the constitutions of these compounds were no longer in doubt. Amines were formed by removing hydrogen atoms from ammonia, and replacing them by the "alcohol radicals"[32] methyl, ethyl, propyl, butyl, and amyl; this was their "mécanisme de la formation." Such a method showed the way of the future, Wurtz averred. It was not admissible to assume, wrote Wurtz, that atoms were randomly arranged in molecules; everything that was known suggested that the arrangements of the atoms were stable, and that they were determinable

30. Ibid., pp. 95–96.

31. Wertheim, "Ueber die Constitution einiger Alkaloide," *Annalen* 73 (1850): 208–212; Anderson, "Action de l'acide nitrique sur les alcalis organiques," *Comptes rendus* 31 (1850): 136–138; Hofmann, "Ueber das Vorkommen des Trimethylamins in der Häringslake," *Annalen* 83 (1852): 116–117; Wurtz to Hofmann, 31 July 1852, Berlin-Brandenburgische Akademie der Wissenschaften. Hofmann's letter from London to Liebig in Giessen announcing the result was dated 20 June; the publication date of the issue of the *Annalen* was 14 July; and Wurtz read the article in Paris on the 30 July. This is an illustration of the efficiencies of scientific journal publication and postal services in nineteenth-century Europe.

32. The term in quotation marks was coined by Kolbe, who would later change the name to "alkyl" radicals.

in principle from all chemical reactions except the most violent ones. The alcohol radical series exhibited incremental addition of the elements of C^2H^2 ($C = 8$), and the alkylated organic bases of Wurtz and Hofmann paralleled perfectly the alkylated organic acids of Kolbe and Frankland. Both kinds of compounds incorporated the homologous series of alcohol radicals.[33]

Wurtz suggested one modification to Kolbe's theory: rather than represent the various acids as higher homologues of oxalic acid (which, he said, was probably dibasic), it would be better if "this distinguished chemist"[34] would depict them as homologues of the monobasic formic acid. The radicals could then be viewed as substituting for the hydrogen of formic acid that was not substitutable by bases (that is, the hydrogen inside the formyl radical), in just the same way that the new amines consisted of radicals substituting for the hydrogen of ammonia. Wurtz also hinted that the monomeric formulas for the free radicals isolated as gases by Kolbe and Frankland were actually dimers (that is, "ethyl" gas was not C_4H_5, but rather $C_4H_5 \cdot C_4H_5$).[35] It is clear, however, that Wurtz was allying himself here to the Kolbe-Frankland program, and not challenging it in any substantial way.

Wurtz's point, driven home repeatedly, was that these investigations showed how to schematically resolve the proximate components of organic compounds. Determination of constitution in this or an analogous manner was always "one of the most important problems to be resolved" by the chemist, for it was necessarily the first step leading to the ability to artificially form such substances in the laboratory. Once developed, such synthetic capability would transform the science, Wurtz affirmed.

Wurtz's work clearly was deeply influenced by the Berzelian-German radical theory (which he thought was placed beyond doubt by recent work), and, although he refuted the copula formulation of organic bases, he by no means rejected it for other kinds of compounds. However, we also see here

33. Wurtz, "Mémoire sur une série d'alcaloides homologues avec l'ammoniaque," *Annales de chimie* 30 (1850): 443–507; see esp. pp. 444–446 and 495–507.

34. Kolbe had published a few superlative papers, but he had not yet begun his academic career, and he was only 32 (a year younger than Wurtz). Wurtz's generous epithet on the occasion of this, his first mention of his German rival, is ironic in view of the later history of these two men.

35. Wurtz, "Mémoire sur une série d'alcaloides homologues avec l'ammoniaque," p. 502.

a much stronger influence on Wurtz of the reformist currents that began to become more powerful just at this time: distinctions between monobasic and dibasic acids, between monomeric and dimeric formulas, consistent molecular magnitudes, hydracid theory, substitution theory, a focus on arrangements, rejection of additive formulas consisting of *isolable* radicals—all these showed the impact of the work of Gerhardt and Laurent. Wurtz mentioned Gerhardt in this paper as the coiner of "homologous series," as a defender of polybasic hydracids and dimeric radical formulas, and as the corrector of Anderson's formula for butylamine.

A great shift in the chemical community began about this time. Many of the older (especially German) chemists, such as Liebig, Wöhler, and Bunsen, regarded the new work on amines and the "isolation" of organic radicals as having definitively validated the Berzelian theory. Liebig wrote to Hofmann at the time of his first paper on substituted anilines, when coincidentally Frankland was working in his laboratory:

Those are just incredibly amazing things that you have found, we are all in ecstasy over them and eager in the highest degree to learn the details of the path that led you to them. With these discoveries it seems to me that the nature of the ammonias has been definitively established, and indeed the constitution of all the organic bases; also relevant is the fact that Frankland has isolated methyl, ethyl, and this week amyl. So what we required as the foundation of the theory is finally from this side here.[36]

Six weeks later Liebig wrote:

I await only a detailed description in order to acquaint myself with these lovely bases. The most remarkable is and remains $3(C_4H_5)$ [i.e., triethylamine]. This was a large leap out in front of the French, although in saying that I do not mean to include our friend Herr Wurtz.[37]

There was no gainsaying Liebig's perceptions. But Liebig had left active research; a measure of his disconnection from the current dialectic was that he had required Hofmann to remind him of his own prediction of Wurtz's new amines.

For those who were strongly engaged in the research dialectic, the situation looked different. Both Hofmann and Wurtz traveled further away from electrochemical-dualist radical theory precisely because of their research that to Liebig looked so confirmatory of that theory. Both accelerated their

36. Liebig to Hofmann, 8 December 1849, in Brock, *Briefe*, p. 88.
37. Liebig to Hofmann, 17 January 1850, ibid., p. 89.

bridge building to the reformist ideas of Gerhardt and Laurent, and also to some of the earlier (now abandoned) notions of Dumas and Liebig. Frankland, too, who had appeared to be firmly in the older school, now began consistently to fit his research into the substitutionist type-theoretical approach.

The one strongly engaged younger chemist who continued to work on salvaging the older ideas was Kolbe, but his case has some unusual and atypical aspects.[38] Coincidentally, in late 1850 Kolbe published a review article simultaneously with Wurtz's, which had the same essential theme: recent investigations on the constitutions of organic molecules. Kolbe was just as convinced as Wurtz that determination of constitutions was central to the science, but Kolbe's intent was in some respects the inverse of Wurtz's. In contrast with Wurtz's commitment to building connections to create a unified science—unified between inorganic and organic, and also between the older and the newer theories—Kolbe's purpose was to build a wall of evidence and argument against the tide of (predominantly French) ideas that he saw threatening to submerge the comfortable electrochemical-dualist radical theory. Kolbe accommodated himself to radicals whose hydrogen was substitutable by chlorine and other elements, and he conceded that there were many unsolved problems. However, he was convinced that the older ideas had plenty of life left in them, and that the recent work on acids and amines, especially his own work but also that of his close friends Frankland and Hofmann, only reconfirmed this.[39]

Kolbe had, and has, a well-justified reputation as one of the founders of synthetic methods in organic chemistry. He was the first to produce what chemists today call a "total synthesis": that of acetic acid, in 1844. The methods which Hofmann and Wurtz (and to a lesser degree Frankland and Kolbe) began to develop in the period 1848–1851 were a different kind of synthetic chemistry: the formation of the greatest possible number of analogous, homologous, and isomeric products through a single reaction route. After this work, the hydrocarbon mantra "methyl, ethyl, propyl, butyl, amyl" became standard in both research and pedagogy; after this work, the study of isomeric relations between similar compounds (such as propyl-

38. Rocke, *Quiet Revolution*, passim.

39. Kolbe, "Ueber die chemische Natur und Constitution der organischen Radicale," *Annalen* 75 (1850): 211–239; 76: 1–73.

versus trimethylamine) grew explosively; after this work, the field of potential research and the number of theoretically possible organic substances became virtually infinite. In fact, Hofmann worried that the assembly-line nature of this sort of procedure "smacks somewhat of dilettantism," but then added that it would have required for him "superhuman strength to leave these magnificent salts unanalyzed."[40]

In 1851, Wurtz, borrowing a page from Hofmann's book, produced a bewildering variety of alkylated ureas: not just ethyl, methyl, diethyl and dimethyl, but methyl ethyl, amyl, ethyl amyl, phenyl ethyl, and so on. At the end of this paper he noted that he had tried without success to prepare triethyl and tetraethyl urea, and that he had just heard from Hofmann that his friend in London had succeeded.[41] This is an indication of the friendly character of the rivalry between these two fine chemists; it is also suggestive of the straightforward nature and prolific power of the multiple-alkylation strategy that both had developed and were avidly pursuing. The strategy could only have been developed by chemists with a high sensitivity to the constitutions of the molecules they were manipulating en masse in their laboratory vessels. As Liebig noted to Hofmann: "One sees in this matter how useful a correct theory is to an investigation, indeed how impossible it is to extract oneself from this complicated process without it."[42]

Professorial Appointment

It is curious that Wurtz's research productivity was high in 1848–1850, when the only Parisian facility at his disposal was his makeshift lab in the Musée Dupuytren, but that there was a significant slowdown during his 3 years in a proper lab on the Rue Garancière (1851–1853). Running the instructional portion of this enterprise may have been time consuming; moreover, this was the period when he courted and wed Mlle. Oppermann. "Ich habe . . . wie man hier sagt, keine grosse Chance gehabt," he told Hofmann in July of 1852, and said he was simply planning to finish up some projects he had already begun. "Auf dem Wege findet man immer

40. Hofmann to Liebig, 29 January 1850, in Brock, *Briefe*, p. 91.

41. Wurtz, "Recherches sur les urées composées," *Comptes rendus* 32 (1851): 414–419.

42. Liebig to Hofmann, 12 April 1850, in Brock, *Briefe*, p. 96.

etwas," he added.[43] Find something he did, to his rival Gerhardt's chagrin. Thirty days after writing this letter he discovered an important and novel substance, butyl alcohol, in potato oil. "Quelle chance il a, ce garçon," Gerhardt remarked enviously—and accurately.[44]

In 1852 Wurtz was also much occupied with acquiring his first professorial appointment. Dumas, who had experienced some queasy moments during the revolution, became ever more politically secure and politically involved in the latter stages of the Second Republic and during the regime of Napoleon III. Professor at the Sorbonne, Minister of Agriculture and Commerce, then senator, member and eventually president of the Paris Municipal Council, member and eventual Permanent Secretary of the Académie des Sciences, vice-president of the Conseil de l'Instruction Publique, one of three Inspecteurs Généraux de l'Enseignement Supérieur pour les Sciences—Dumas had little time or inclination to fulfill his duties at the Faculté de Médecine. Wurtz had been substitute teaching for him there since 1849, and in the spring of 1852 Dumas let it be known that he would retire from his professorship. He made no secret of his expectation that Wurtz would succeed him, and no one seems to have doubted his ability to make that happen.[45]

The open professorship of pharmacy and organic chemistry was announced in the autumn of 1852. In the wake of the coup d'état of December 1851, a decree of March 1852 abolished the elaborate system of juried competitions for professorial chairs in the Université de France, which had placed real power of appointments in the hands of collegial peer committees. The new system called for a "double presentation list," each containing at least two names, to be prepared, one by the relevant Faculté and the other by the Conseil Académique, from which the Minister of Public Instruction (Hippolyte Fortoul) would select one for the approval of the president (after December 1852, of the emperor). The emperor had the power, though he seldom exercised it, of disregarding the lists and selecting anyone he wished. The relatively liberal days of the July Monarchy were

43. Wurtz to Hofmann, 31 July 1852, Chemiker-Briefe, Berlin-Brandenburgische Akademie der Wissenschaften.

44. Wurtz, "Sur l'alcool butylique," *Comptes rendus* 35 (1852): 310–312; Gerhardt to Chancel, 1 October 1852, in Tiffeneau, vol. 2, p. 129.

45. At least, he appears to have told Louis Pasteur this in April of that year.

gone, and the universitaires felt the change in faculty appointments and elsewhere. Many thought that the new system typified the authoritarian and illiberal tone of the early Empire, and regarded it as paralyzing the meritocratic judgment of the faculties, even when the sovereign chose not to exercise his powers.[46]

Wurtz's competitors for the chair were Louis René Le Canu (a pharmacist and a professor at the École de Pharmacie) and Pierre Antoine Favre (a physical chemist and an agrégé at the Faculté since 1844). Each submitted self-nomination documents consisting of a curriculum vitae and a printed annotated list of publications.[47] Wurtz was confident, perhaps almost too much so. He outlined the situation to Hofmann in July, noting that Le Canu was his chief competitor and adding, smugly, "Who would you vote for?"[48]

The candidacies were discussed in a faculty meeting on 3 December. Each candidate had an advocate who had written a supporting brief. J. E. Gavarret's brief favoring Favre was strong but had to work against the presumption that Favre's specialty was inapproporiate for a chair in "pharmacy and organic chemistry." Apollinaire Bouchardat's powerful brief for Le Canu stressed his maturity, his stature, and his emphasis on pharmacy rather than abstruse considerations of chemical theory; the chair, after all, was in a medical faculty, and before Dumas's accession in 1838 it had been devoted exclusively to that discipline.

46. See, e.g., "Prémier Rapport sur l'Organisation de la Faculté de Médecine" (written by Wurtz, Denonvilliers, Tardieu, Béhier, Broca, and Gavarret), in AJ16/6357: "These deplorable and unspeakable doctrines triumphed in the Université, as everywhere else." Behind the right retained by the sovereign to ignore the lists was a "poorly disguised goal of intimidating or at least paralyzing the spirit of independence whose noble tradition had been preserved among the professors of higher education. . . ." Their point was that if the faculty were sufficiently intimidated the emperor would not have to go outside the lists very often, thus preserving the appearance of liberalism in an illiberal regime.

47. Recteur of the Académie de Paris to Doyen of the Faculté de Médecine, 9 November 1852, requesting list, in AJ16/6311; self-nomination documents by Le Canu, Wurtz, and Favre, dated respectively 27 May 1852, 10 November 1852, and 10 December 1852, in AJ16/188; A. Corlieu, *Centenaire de la Faculté de Médecine de Paris (1794–1894)* (Imprimerie Nationale, 1896), pp. 209–212.

48. Wurtz to Hofmann, 31 July 1852, Berlin-Brandenburgische Akademie der Wissenschaften.

The brief for Wurtz was written by M. J. B. Orfila, the distinguished professor of medical chemistry and former long-time dean of the Faculté who a decade earlier had provided the crucial support allowing Wurtz to travel to Giessen. Orfila noted that the successful candidate must know the science absolutely and must teach it with order and method. Wurtz not only satisfied this requirement, Orfila averred; he lectures with "animated oratory [that] captivates the listener to the degree that he listens religiously, with perseverance and without distraction." Wurtz had been teaching for Dumas for 4 years, also for 2 years at the Agronomic Institute in Versailles, and his lectures were always attended by throngs of students. But his even greater strength was in research, "so necessary to increase the luster of our body." Orfila summarized Wurtz's remarkable career, quoting liberally from Dumas's effusive referee report of 1849. Wurtz was a "man of the future," not superannuated (a slap at Le Canu's more advanced years); it would be appalling if the new professor were to fall into inactivity. "Young, industrious, active, ingenious, prodigiously learned, imbued with that honorable ambition which always carries us forward, M. Wurtz, especially if he were encouraged by your votes, would make it his mission to respond appropriately to your good will." As a final note, Orfila added that Wurtz was both competent in and sincerely concerned about pharmacy, and that he had never neglected this subject in the 4 years he had been teaching in the Faculté. Although engaged in chemical theory, he was not a slave to fashionable or abstruse theories: "In my opinion, physiology and pathology, which already so frequently call for help from organic chemistry, may expect to receive new services from this science; and nothing can honor our Faculté in the matter of chemical instruction more than a course in organic chemistry that is disentangled from the (I might almost say) romantic part of the science, and limited to useful and uncontested facts."[49] Wurtz supported this sentiment with a separate document, assuring the

49. Documents by Bouchardat and Gavarret, 2 December 1852; brief by Orfila, 3 December 1852, AJ16/6311: "Jeune, laborieux, actif, ingénieux, prodigieusement instruit, doué de cette ambition honorable qui nous porte toujours en avant, M. Wurtz, surtout s'il était encouragé par vos suffrages, se ferait un devoir de répondre dignement à votre bienveillance. . . . Dans mon opinion la physiologie et la pathologie qui implorent déjà si souvent les secours de la chimie organique, attendent de celle-ci de nouveaux services, et rien ne peut honorer notre Faculté, en fait d'enseignement chimique, autant qu'un cours de chimie organique dégagé de la partie, je dirai presque romantique de la science et limité aux faits utiles et incontestés."

faculty that he was well competent to teach pharmacy and that he had never neglected it. Indeed, that subject had been his principal function for several years as préparateur at Strasbourg.[50]

In the Faculté vote of 9 December, Wurtz received 19 votes to 4 for Le Canu. In the second round, Favre received 14 against Le Canu's 9. The voting in the Conseil Académique a few days later was different: Wurtz prevailed over Le Canu, but by a vote of only 10 to 9; the second round was 14 to 4 for Le Canu, with one abstention.[51] Le Canu was bitterly disappointed and very angry. He printed a letter of protest and sent it to all the professors of the Faculté. Only "blind hostility" had prevented the vote from going the other way. He pointed out that he was the only real pharmacist of the three, and he lost by only a single vote in the Conseil; this was probably only because the dean had used his personal influence against him. Furthermore, the abstention in the second round revealed someone's bad conscience.[52] This protest was to no avail, and by an imperial decree of 2 February 1853 Wurtz was duly appointed Dumas's successor. His salary rose from 1200 to 9000 francs.[53] Fortoul, of course, was under no obligation to accept either the Faculté's or the Conseil's top choice for presentation to the emperor, nor was the emperor required to accept Fortoul's nomination. Whatever the vote, the fix may have been in for Wurtz.

Orfila died a month after Wurtz assumed his new position, and these changes provided an opportunity for a reevaluation of chemical and pharmaceutical instruction at the Faculté de Médecine. Le Canu's complaints may have had a delayed effect, for a consensus developed that pharmacy was indeed being neglected. Wurtz had been devoting ten or twelve lectures to the subject at the end of his course in organic chemistry, but the faculty felt that that was far from sufficient. A blue-ribbon committee was appointed to study the problem. They suggested abolishing Wurtz's and Orfila's

50. Wurtz to Recteur, n.d., AJ16/188.

51. Doyen to Recteur, 9 December 1852, and Recteur to MIP, 24 December 1852, in AJ16/6311; Conseil Académique de la Seine, Ordre du jour for 23 December 1852, in AJ16/184; AJ16/188.

52. Le Canu to Faculté de Médecine, 12 January 1853, AJ16/188.

53. The salary for préparateurs in the Faculté varied somewhat year by year, since their fixed salaries of 1000 francs were supplemented by an "éventuel" calculated on that year's enrollment. In the early 1850s the salary for professors was 6000 francs, plus a constant éventuel of 3000. In 1855 the fixed portion was raised to 7000. See AJ16/6618.

chairs ("pharmacy and organic chemistry" and "medical chemistry" respectively) and replacing them with two new professorships: "organic and mineral chemistry" and "pharmacy" respectively. Dean Paul Dubois reported to the rector in November of 1853 that the faculty had voted by a large majority to accept this recommendation. On 10 December Napoleon signed a decree authorizing the new chairs and appointing Wurtz to that for "organic and mineral chemistry." The pharmacy chair was given not to Le Canu but to a colleague of his in the École de Pharmacie, Eugène Soubeiran.[54]

All previous biographies of Wurtz state that he inherited both Dumas's and Orfila's positions in 1853. In one sense this is true: Wurtz was asked to teach both of his predecessors' subjects, organic chemistry and medical chemistry. However, he did not inherit both salaries, for one of these two lines was given to pharmacy. If cumul often amounted to getting two salaries for one professional activity, Wurtz's case was rather close to the inverse of this: two activities for a single salary.

Wurtz was not one to complain, and it cannot be denied that he finally had a secure position from which to pursue his research. However, his laboratory in the Musée Dupuytren was hardly sufficient for the long term, and much too small for instruction of a group of students in the Liebig pattern. The previous holder of the chair, Dumas, had never tried to create a chemical lab at the Faculté de Médecine, since he had had his private laboratory in the Rue Cuvier at his disposal—which closed in 1848. Moreover, Wurtz was about to lose his own private lab school in the Rue Garancière at the end of 1853. Consequently, he asked the dean of the Faculté for a grant of money with which to create a lab. That summer, he proudly wrote to tell Liebig of his new position and of the "schönes Laboratorium" he hoped to build. Promised 2500 francs, Wurtz was given only 2100; it was not much for the purposes he had in mind, but he made do. Probably in early 1854, he opened the new lab to students and paying customers.[55] It was to be his workplace for the next 24 years.

---

54. Paul Dubois to Recteur, 25 November 1853, and décret, 10 December 1853, in AJ16/6310; Corlieu, *Centenaire*, pp. 279–285; A. Prévost, *La Faculté de Médecine de Paris, ses chaires, ses annexes, et son personnel enseignant de 1794 à 1900* (Maloine, 1900), p. 43.

55. Wurtz to Liebig, n.d. [ca. summer or fall 1853], Liebigiana IIB, Bayerische Staatsbibliothek, Munich; Wurtz to dean, 1 April 1860, F17/4020; Rapport à M. le Ministre, n.d. [ca. 1865], AJ16/6556.

6

Coming to Terms with Types

Wurtz was a winning lecturer. His student and principal biographer Charles Friedel pictured his teacher in the lecture hall of the Faculté de Médecine:

... master of his subject, confident in his audience, walking briskly from the experiment table to the blackboard; expressing himself with an eloquence that was casual, familiar, and lively; speaking with enthusiasm about chemical compounds as if they were an affair of state; sometimes astonishing those who did not know him and who were troubled by this unusual exuberance in a hall of science, but who returned to the subsequent lectures, captivated and charmed; often disconcerting his préparateurs by the unexpected course of his exposition and of his movements—even though his lectures were carefully prepared in advance, and indeed ever more so as his career advanced. This was not a professor calmly relating the results of his preparation of the day before; it was a scholar communicating to his students the science which he had lived, which he had helped to construct himself, and which had been transformed under his eyes and by his work. One felt the heat of battle; not against his scientific adversaries—not a trace of this could be seen in his teaching—but against ignorance and obscurity. And the clear understanding which this exceptional mind had achieved was communicated, warm and lucid, to his audience.[1]

This description was apparently not a bit exaggerated, for one finds many of the same words and expressions in virtually every depiction of Wurtz as lecturer. "There was no need to repeat to him the old precept," Henri Gregor wrote, "that enumerates the three requirements for a good lecturer: 1st, action; 2nd, action; 3rd, action. M. Wurtz was action incarnate. He was passionate about his teaching; he seized his audience powerfully; he communicated to the listener the conviction that animated him. No actor has ever devoted as much ardor in portraying human passions as M. Wurtz did in

1. Friedel, "Notice sur la vie et les travaux de Charles-Adolphe Wurtz," *Bulletin de la Société Chimique* [2] 43 (1885), p. xiii.

describing the combinations of the chemical elements." Émile Picard wrote that he had often witnessed listeners leaving the auditorium "stupefied at seeing . . . how affecting and dramatic a chemistry lecture could be." The average size of his audience in his early years of instruction is not known, but figures from the late 1860s indicate a usual attendance of about 300.[2]

Wurtz's pedagogical excellence was recognized not only by students but also by the authorities of the Faculté and the Université. The earliest official assessment of Wurtz's performance that has survived (ca. summer 1855) concludes: "M. Wurtz has fully justified the choice of the faculty. His course is taught with care, very attractive, very instructive, and attended by a large number of students."[3] Much of Wurtz's instruction, however, was tendered not in the lecture hall but in the laboratory.

Wurtz's Laboratory at the Faculté de Médecine

The École (Faculté) de Médecine was created in the time of Napoleon I, in the buildings associated with the old École de Chirurgie, located between the Rue de l'École de Médecine and the Rue Monsieur-le-Prince.[4] In the mid-century Faculté there were around two dozen professorial chairs and several hundred students. Although laboratories supporting scientific instruction and research were planned and actually ordered in 1796, none was ever built in the first half of the nineteenth century.

2. H. Gregor, "Académie des Sciences, Séance du 12 mai, Présidence de M. Rolland: M. Adolphe Wurtz," unidentified newspaper tearsheet, 15 May 1884, Académie des Sciences, Archives, Dossier Wurtz; Picard, quoted in A. Haller, *Inauguration de la statue de Adolphe Wurtz à Strasbourg, le mardi 5 juillet 1921* (Gauthier-Villars, 1921), p. 9; Vice-recteur Mourier, Renseignements confidentiels [au MIP]: 24 June 1866 (300–350 auditors), 24 June 1867 (233 auditors), and 7 May 1870 (300 auditors), in F17/21890.

3. MIP note, Année scolaire 1854–55 [annual report?], F17/21890: "M. Wurtz a complètement justifié le choix de la faculté. Son cours est fait avec soin, très attrayant très instructif et suivi par un grand nombre d'élèves." Similar compliments are found in the confidential reports from the 1860s cited in the preceding footnote.

4. Today both streets run diagonally in the southwest quadrant formed by the perpendicular Boulevards Saint-Germain and Saint-Michel, on the Left Bank; however, these boulevards were constructed in the period 1855–1866, and when Wurtz first taught in the Faculté it was surrounded by the typical medieval Parisian labyrinth of narrow streets and passages.

Wurtz's lab was created out of a portion of the small lecture theater of the anatomy department of the École, consisting of three rooms on the ground floor and one on the first. Friedel, who arrived there in November of 1854, described the lab as providing sufficient space for Wurtz and twelve students in modest but high-ceilinged vaulted rooms with good natural light. Wurtz's lab bench, scarcely larger than that of his students, was in one of the bays. The balances were kept in the truncated anatomy lecture room, and so were inaccessible when classes were in progress. Smaller spaces accommodated combustion analyses, a storeroom, and a washroom; some of these rooms were later used to accommodate additional students. A small adjoining courtyard was used for experiments involving poisonous gases, explosive materials, or reactions carried out in heated sealed tubes (which sometimes exploded unexpectedly). The laboratory was open daily from 10 until 5; the students took lunch in the small neighborhood restaurants of the Left Bank.[5]

The initial appropriation of the dean sufficed only to create the space within which a laboratory could be developed, and no more funding from the Faculté was forthcoming. From the very beginning Wurtz resolved to "create a modern laboratory sufficient to the demands of the science ... to create in [this] laboratory of the Faculté a true school of chemistry."[6] Since no external money was available, Wurtz generated it internally by charging his students monthly fees. Beginners paid 100 francs a month, advanced researchers 50, but Wurtz sometimes remitted all fees for those without resources who badly wanted to do research. Researchers always predominated over novices, and foreigners were plentiful. A student wrote:

From time to time, sometimes standing, sometimes sitting on an old padded footstool that was losing its stuffing through fifty holes, he assembled his group in an informal colloquium, and there he lavished on all his valuable counsel and fecund opinions. Foreigners lost nothing in this, for M. Wurtz added to his other merits

5. Friedel, "Wurtz," pp. xiii–xiv; F. Papillon, "Les laboratoires en France et à l'étranger," *Revue des deux mondes* [2] 94 (1871), pp. 600–601.

6. Wurtz to Dean, 1 April 1860, F17/4020. The management details of Wurtz's laboratory in the following discussion are taken from this and the following files: "État sommaire de recettes et dépenses du laboratoire de chimie" [1859], ibid.; "Arriéré 1859," ibid.; Wurtz to MIP, 28 November 1865, ibid.; and Wurtz to Dumas, 15 February 1864, Dossier Wurtz, Archives de l'Académie des Sciences.

that of speaking the principal languages of Europe. It was a lovely spectacle to see in this sanctuary the master discussing with abandon, benevolently assenting when the student was correct, or revealing error with an ingenious and keen tactfulness.[7]

By one account, for some time after the lab opened there was only one student.[8] After 2 or 3 years, with about a dozen students paying their way, Wurtz was able to create an annual budget of around 6000 francs. Given one préparateur by the Faculté, Wurtz hired a second with these funds; given one orderly (garçon), Wurtz hired another. ("I attribute the greatest importance to this [janitorial] service," Wurtz averred, "for without cleanliness and order, all goes poorly in a laboratory."[9]) He also supplemented the insufficient salaries of the two state employees from his private accounts.

In the first few years after his appointment in 1853, Wurtz built a proper research lab, spending about 2000 francs for wooden furniture and cabinets, 2000 to install gas burners (an innovation that began to be adopted throughout Europe in the 1850s), 1000 for plumbing and other metalwork, and more for furnaces, locks, instruments, stonework, painting, and so on. To these capital expenses must be added the annual materials budget for glassware and other apparatus, chemicals, and fuel. It was expensive to run a lab, and Wurtz generated essentially all of the necessary funds internally and privately. Certain faculty members were envious of the fact that only one professor had a research laboratory and doubted the propriety of a laboratory for pure chemical research in a school of medicine. Wurtz's procedure was indeed quite contrary to regulations, but the dean and higher authorities closed their eyes to it and did not interfere.

By the beginning of the 1860s, Wurtz had managed to shoehorn even more students into the space. He was regularly accepting 15–18, and his tuition-driven annual budget rose above 8000 francs. In 1862 he extracted from his administration a second small grant, 1500 francs, for stone experiment tables for the courtyard, one of which was furnished with a good hood and chimney. By about 1870 the laboratory's steady-state population

7. Papillon, "Laboratoires," p. 601.

8. That student, Auguste Scheurer-Kestner, described his experiences in *Souvenirs de la jeunesse* (Charpentier, 1905), p. 33. He stayed about a year after Wurtz opened the lab, and so he must have been there when Friedel arrived.

9. Wurtz to Dumas, 15 February 1864, Archives, Académie des Sciences.

was between 20 and 25. Wurtz must have had something like 300 labora-tory students over the course of his career, of which Carneiro located the names of 139 and Pigeard later found an additional 100. Of the students whose nationalities can be determined, only 41 percent were French, and about a third of the French students were from Alsace; stated differently, only 27 percent of his students were non-Alsatian French. Very few were medical students. Foreigners included significant numbers of Austrians, Russians, Germans, Swiss, British, and Americans; also represented were students from the Low Countries and the Ottoman Empire, Italians, Spaniards, Finns, Swedes, Irish, Portuguese, Slovaks, and Romanians. Unfortunately, Wurtz's laboratory registers have been lost or destroyed, so a complete and accurate accounting is impossible.[10]

Wurtz's tactic was precisely the same as Liebig's had been during the first decade of his tenure in Giessen. Denied state support for laboratory edu-cation, Liebig created a private chemical institute, which by 1835 had become so successful that the state was persuaded to take it over. Liebig's model caught on in Germany during the next few years, being adopted for instance by Wöhler at Göttingen and Bunsen at Marburg. By the 1850s and the 1860s German students interested in a practical chemical educa-tion had a number of universities from which to choose, including for the first time many of the great Prussian universities. The Marburg Chemical Institute, under the leadership of Hermann Kolbe from 1851 to 1865, was fairly typical for its day and time. Kolbe's lab was about the same size as Wurtz's, and the budget was likewise comparable; however, Wurtz's stu-dents had to pay more out of their own pockets, since there was no state subsidy: 50 or 100 francs per *month*, rather than the equivalent of 100 or 150 francs per *semester* as in Marburg. Wurtz's salary, on the other hand,

10. A. Carneiro, The Research School of Chemistry of Adolphe Wurtz (Ph.D. dis-sertation, University of Kent/Canterbury, 1992), pp. 114–115 and appendix I; Carneiro, "Adolphe Wurtz and the Atomism Controversy," *Ambix* 40 (1993): 75–95; N. Pigeard, L'Oeuvre du chimiste Charles Adolphe Wurtz (1817–1884) (thèse de maîtrise, Université Paris X Nanterre, 1993), passim; Pigeard, "Un alsa-cien à Paris," *Bulletin de la Société Industrielle de Mulhouse* 833 (1994): 39–43; Carneiro and Piegeard, "Chimistes alsaciens à Paris au 19ème siècle," *Annals of Science* 54 (1997): 533–546. Additional information on numbers and nationalities of students was very kindly provided to me in private communications by Natalie Pigeard in June 1999.

was about twice Kolbe's. However, it must be noted that Kolbe was paid poorly in this period, and Wurtz's salary, converted to thalers or florins, was rather typical for Germany at the time.[11]

Having followed the Liebigian model in its private phase, Wurtz sought the second step, and in 1860 he petitioned his dean to bring the lab under official aegis. Wurtz pointed out that since the lab paid for itself it would not create any financial burden for the state; his students would simply pay their fees into the state coffers, and the time-consuming bookkeeping could be done by accountants rather than by a professional chemist who had better ways to occupy his time. Although this request was approved by Dean Dubois and by the vice-recteur, it was derailed at the ministerial level. He tried again in 1864, and repeatedly in subsequent years, but his requests were always denied. An 1868 memo to a sympathetic Minister of Public Instruction, Victor Duruy, was a real cri de coeur:

I believe I can say, simply and without false modesty, that this laboratory has become a source not just of discoveries, but still more of those ideas that have pushed the boundaries and have contributed effectively to that great movement which has transformed the science and which I have recently attempted to depict [in his history of chemical theories]. Thus, for fifteen years I have been the de facto director of a research laboratory that has never had any official existence. Will I be allowed to claim the title of director. . . ?[12]

I will return to this issue in chapter 9.

11. In the early 1860s, Kolbe had about 20 Praktikanten at a time. They had to pay around 30 thalers per semester for chemicals and honorarium. Adding this income to the lab budget itself (800 thalers per year) yields an annual budget of around 2000 thalers, equivalent to 8000 francs. For these data, and for salary comparisons with Germany, see Rocke, *Quiet Revolution*, pp. 114–118.

12. Wurtz to Duruy, 22 November 1868, F17/4020: "Je crois pouvoir dire, simplement et sans fausse modestie, que ce laboratoire est devenir un foyer non seulement de découvertes, mais encore d'idées qui ont passé la frontière et qui contribué efficacement à ce grand mouvement qui a transformé la science et que j'ai essayé de dépeindre récemment. J'ai donc été depuis quinze ans le directeur de fait d'un laboratoire de recherches qui n'a eu aucune existence officielle. Me sera-t-il permis d'en être le directeur en titre. . . ." On this question, see also Wurtz to Dean, 1 April 1860; Dean to Vice Recteur, 7 May 1860; Vice Recteur to MIP, 18 November 1864; Wurtz to MIP, 10 December 1864; note by ? [Vice Recteur or Dean], 7 July 1865; Wurtz to MIP, 28 November 1865; MIP minute, 6 June 1866; and MIP minute, redaction de M. Dumas, 6 June 1866, all in F17/4020. For a broader discussion, see Pigeard, "Un alsacien à Paris."

Williamson, Gerhardt, and Asymmetric Synthesis

The great chemical discoveries of the late 1840s—synthetic routes to the homologous organic acids and the homologous organic bases, the isolation of organic radicals, and the development of the multiple-alkylation strategy—precipitated a schism in the organic chemistry community between those (mostly older) chemists who saw the dualistic electrochemical-radical theory thereby revalidated or even proved and those (mostly younger) workers who viewed the new substances as providing instantiation, if not verification, of the type-theoretical substitutionist option. The difficulty was that a clever theorist could interpret the new data in either direction. What was still very much needed, therefore, was an experiment that could provide unambiguous evidence one way or the other.

It was the English chemist Alexander Williamson who provided it, beginning with a paper published in the summer of 1850. Williamson had studied with Liebig, but had been greatly influenced by his contacts with Laurent and Gerhardt during his residence in Paris in the years 1846–1849. His synthesis of ethyl methyl ether, a novel asymmetrical substance, appeared to be readily explainable only under Gerhardtian assumptions, not under an electrochemical-dualist interpretation; in fact, he argued that the dualist view excluded the very possibility of such asymmetric compounds. Moreover, he had found a way to create any number of such substances at will.

No longer was the reformed chemistry merely a question of abstract schematology and rationalization, as had been developed by Laurent and Gerhardt; now it had produced actual new substances not envisioned or even envisionable by the older theory. It had demonstrated an impressive instance of scientific "cash value," and promised more to come. These theoretical assets were cashed by several workers in the half-dozen years after Williamson's 1850 paper. Benjamin Brodie, William Odling, August Kekulé, and Adolphe Wurtz all applied the strategy and reasoning of Williamson's synthesis of asymmetric molecules in arguing for Gerhardtian formulas, and Hermann Kolbe did the same in an unsuccessful attempt to turn the tables on his opponents and argue *against* the theory.[13]

13. For which see Rocke, *Quiet Revolution.*

Williamson's work and its continuation by these chemists constituted the essential breakthrough of the reform that had been prepared in the 1840s by Gerhardt and Laurent. However, the course of this breakthrough was not uneventful, and the details are interesting. Williamson's paper appears to have excited very little immediate interest or even attention anywhere in Europe. To be sure, as early as September of 1850 Gerhardt trumpeted it as a triumph of his own ideas, but he published this response only in his own journal, which had poor sales and was little read. A German translation of the ether synthesis paper was published in Liebig's *Annalen*, but a French version appeared only in a minor journal.

The story entered a new phase when, in the spring of 1852, Gerhardt applied Williamson's "asymmetric synthesis" argument to his preparation of various organic acid anhydrides. Williamson had not only provided the model and the logic; he had even predicted the existence of such compounds. As Williamson had produced a novel asymmetric ethyl-methyl ether by double decomposition between an alkali metal compound and an organic halide, so Gerhardt produced a novel asymmetric acetic-benzoic anhydride by double decomposition between an alkali metal compound and an organic halide. As Williamson had argued that the dualistic radical theory would have required his reaction to produce an equimolar mixture of ethyl ether and methyl ether and not a lone asymmetric compound, so Gerhardt argued that the dualistic radical theory would have predicted his reaction to produce an equimolar mixture of acetic anhydride and benzoic anhydride rather than a single asymmetric anhydride. As Williamson therefore pronounced decisively in favor of the type-substitutionist viewpoint, so did Gerhardt.

When Gerhardt failed to mention this precedent in his first paper on this subject to the Académie des Sciences, Williamson wrote a stern but carefully phrased letter to his friend. In response, Gerhardt's second paper to the Académie contained a reference to Williamson's "interesting article, where, adopting my notation, he considers all salts as derivatives of the water type." Gerhardt's dilatoriness in giving proper credit to his friend and Williamson's interest in correcting the record suggest that both men were aware of the stakes in this game. Others agreed that this was a scientific milestone. As one who had been a chemistry student at the time later reminisced, Dumas, "despite his authoritarianism," had "an abundance of goodness and justice" and was "conquered" by the paper.[14] Gerhardt's other former opponents in Paris, including Frémy, Regnault, and Thenard, were converted at a stroke to warm advocates. Hofmann, Liebig, and other major figures in the English and German chemical establishments also concurred.

After his short reports to the Académie, Gerhardt prepared a major article for the *Annales de chimie*, as was the custom in France. It was also customary for such papers to be publicly refereed by a committee of Académie members. The committee in this instance consisted of Dumas (the chairman), Regnault, and Pelouze. Their lengthy report, printed in the 21 March 1853 issue of the *Comptes rendus*, not only proffered extravagant praise of Gerhardt's chemical discoveries, and not only agreed that his reasoning on the basis of those discoveries was sound and his conclusions compelling, but also offered a strong endorsement of the full reform in chemistry that Gerhardt and Laurent had been pursuing for a lonely and penurious 10 years.[15]

Interestingly, Dumas portrayed the reform as having its roots in Humphry Davy's work on hydracids; Gerhardt had successfully championed this viewpoint over the alternative oxyacid theory of Lavoisier himself, which now must be relinquished. It was Lavoisier's theory, Dumas averred, that had failed in "predicting or explaining why alcohol gives four volumes of vapor while ether only gives two; above all why, as the experiments of M. Williamson proved so well, two ethers combine so readily in the nascent state, as an acid and a base do, even though they differ scarcely at all in their properties."[16] It was noted in chapter 3 above that Dumas, in well-known writings of 1828 and 1836, appeared to ally himself (somewhat equivocally, to be sure) to the ideas of Avogadro, Ampère, and Davy; that he collaborated with Liebig on a hydracid theory of organic acids in 1838; and that since 1839 he had been distinctly inimical to electrochemical dualism. Around 1840 he became much more cautious about theorizing in any direction. The situation of the other two members of the refereeing commission was parallel. Regnault, before his turn to physical chemistry (which happened just about the same time as Dumas's turn away from engagement with

14. Scheurer-Kestner, *Souvenirs*, p. 35.

15. Dumas, Pelouze, and Regnault, "Rapport sur un mémoire de M. Gerhardt," *Comptes rendus* 36 (1853): 505–516.

16. Ibid., p. 511.

theories), had done some fine work on chlorine substitution and "mechanical types"; Pelouze had collaborated with Liebig on mellitic acid in 1836, and that investigation had first turned Liebig's attention to the hydracid theory.

Williamson was given credit for his signal contribution both in Dumas's referee report and in Gerhardt's full-length article in the *Annales de chimie*, which appeared that same month (March 1853).[17] The name that was most conspicuously absent from both documents was Laurent's. A third disappointment for Laurent was the fact that Gerhardt had notably ignored him in the manuscript of his major treatise on chemistry. This work began to appear in fascicles in June of that year, but Gerhardt had let friends read the manuscript, and it is possible that Laurent was aware of the omission even before its publication. Laurent had been suffering intensely from tuberculosis for more than 2 years.; discouraged and embittered, he died on 15 April 1853. Laurent's friend Jérome Nicklès, who wrote one of the few contemporary obituaries, declared that Laurent had "met with ingratitude . . . from those on whom he had conferred the greatest benefit, [which] embittered his last moments."[18] Clara de Milt argued persuasively that Nicklès was referring to Gerhardt here, though Dumas could also have been meant.[19]

Wurtz's opinions while this drama was being played out are difficult to discern from surviving documents. At first he appeared oblivious to the coming revolution. In the spring of 1851 he theorized on the formation of carbonic ether (ethyl carbonate) from cyanogen chloride and alcohol. The chloride, he thought, was "decomposed by the water molecule of alcohol"; this resulted in the formation of "ethyl oxide," which proceeded to combine with the carbonic acid released from the chloride.[20] This manner of envisioning the constitution of alcohol and ether, taken straight from the older dualistic radical theories of Liebig and Dumas, directly contradicted the central point of Williamson's first paper on ether synthesis. In August of

17. Gerhardt, "Recherches sur les acides organiques anhydres," *Annales de chimie* [3] 37 (1853): 285–342.

18. Nicklès, *American Journal of Science* [2] 16 (1853): 103.

19. Clara de Milt, "Auguste Laurent—Guide and Inspiration of Gerhardt," *Journal of Chemical Education* 28 (1951): 198–204.

20. Wurtz, "Sur un nouveau mode de formation de l'éther carbonique," *Comptes rendus* 32 (1851): 595–596.

1852 Wurtz published a paper with another explicitly dualistic constitutional formula: that for "sulfobutylate de potasse."[21]

These details in his publications made Wurtz appear quite conventional in 1851 and 1852, not at all an advocate of Gerhardt and Laurent. Of course, one must not presume that there were only binary choices during this theoretically labile period. For the first 16 years of his career, from his first published paper of 1842 until the end of 1858, Wurtz's papers were consistently a blend of old and new. But on the other hand, one might have expected Wurtz to have received Williamson's innovations more warmly, for since the mid 1840s Wurtz had provided multiple indications of his receptivity to certain of the new ideas, including hydracid theory, substitution, and type theory.

Did the political climate influence Wurtz's public face? At the time of his first paper of 1842, French academia was enjoying what seemed (at least in retrospect) to be a golden age of moderate liberalism and meritocratic self-rule, and neither Gerhardt nor Laurent had yet experienced unified opposition from the political or the chemical elite. The atmosphere a decade later was very different: Louis Napoleon was consolidating his power, Dumas and Thenard were firmly in control of academic chemistry, and Laurent and Gerhardt were largely excluded. It may well have appeared to some observers that it was no longer possible to win academic positions by merit at all, only by influence and patronage. In 1850–1852 Wurtz was an agrégé at the Faculté de Médecine, a professor at the National Agronomic Institute, and the director of a private lab school. The first of these was a career-track but not a career-making position, the second was a patronage position that would prove temporary, and the third was labor intensive without being the least bit remunerative. In sum, Wurtz would have had good reason to believe that he still required patronage in order to survive professionally. This could have led him to be cautious about his public theoretical posture.

Two propositions seem probable, at least at first glance. First, Dumas's endorsement of the Williamson-Gerhardt theory must have suddenly altered the politics of the situation. Dumas even tied Hofmann and Wurtz to the new ideas, with regard to their defense of substitutionist type-theoretical views in their development of organic derivatives of ammonia.

---

21. Wurtz, "Sur l'alcool butylique," *Comptes rendus* 35 (1852): 310–312.

Second, 3 months after Dumas's referee report appeared, Wurtz publicly appeared to endorse the Williamson-Gerhardt theory for the first time. Like Dumas, he made no mention of Laurent—who had passed away in the meantime—as a precursor of the theory.

Wurtz's "conversion" to Gerhardtism in July of 1853 was neither unqualified nor unheralded, for his opinions both before and after this point merged elements of the older and the newer theories. In fact, I will argue in the next section that the prima facie case I have just outlined is misleading, and that Wurtz did not convert to Gerhardtism in 1853 for political reasons (or even reveal a prior conversion at that time). His essential change of direction only came at the end of 1854 or the beginning of 1855.

Confrontation with Gerhardt

Auguste Scheurer-Kestner, an eyewitness in Wurtz's laboratory in the Rue Garancière, paints a fine portrait of the relationship between Wurtz and Gerhardt in that important summer of 1853:

> In the evenings he often entertained visits from his friend and compatriot Gerhardt, whose laboratory was close by in the Rue Monsieur-le-Prince. To see these two men discussing with equal passion questions of molecules, atoms, and groupings— what a curious spectacle! One would have thought it was a duel to the death. Both of them violent, especially Gerhardt, they dominated the blackboards of the lecture hall one after the other, each seeking (and never successfully) to convince the other, and departing after this passage of arms, covered with perspiration, the best friends in the world.[22]

The topic of their amicable battles may be inferred from a series of exchanges published that summer in successive weekly issues of the *Comptes rendus*. Wurtz appears to have been following ideas on "mechanical types" derived directly or indirectly from the work of Dumas, Laurent, and Williamson, whereas Gerhardt had preferred a much more conventionalist conception of chemical types and formulas as mere "synoptic" summaries of reactions.

Wurtz expressed structuralist inclinations as early as 1847. In PCl_5, he wrote, two chlorine atoms are held less strongly than the others and "occupy a different position" in the molecule. The molecule bears analogies to a stone building, which is not structurally isotropic and whose parts are

---

22. Scheurer-Kestner, *Souvenirs*, pp. 33–34

not interchangeable; "the stones that are placed at the top or on the sides can be removed more easily than those found at the base." So also may phosphorus pentachloride be constructed in a manner that differentiates the various chlorine atoms, and it may serve as the point of departure for a series of compounds all derived "from the same type."[23] Wurtz was following the lead of his mentor Dumas, who 7 years earlier proposed that certain kinds of reactions do not alter a compound's "mechanical type"; in such cases "the molecule remains unchanged, in that it forms a group, a system, wherein an element has pure and simply taken the place of another. According to this entirely mechanical viewpoint . . . all bodies formed by substitution would possess the same grouping and would belong to the same mechanical type."[24] Here Dumas himself was building on ideas of Regnault and especially Laurent.

This structuralist and realist notion of molecules and molecular formulas was adopted most forcefully and explicitly by Williamson, who treated the chemical formula as "an actual image of what we rationally suppose to be the arrangement of constituent atoms in a compound."[25] Williamson had been a devoted student of Auguste Comte, but here he proved himself an apostate to the positivist creed. In contrast, the faithful positivist acolyte appeared to be none other than Gerhardt, who rejected all attempts, even those of his friend Laurent, to discern the internal details of molecules. Gerhardt developed a new theory of types, which considered them in a purely classificatory sense; formulas were used simply to indicate composition and to summarize chemical reactions and empirical relationships.[26]

On 18 July 1853 Gerhardt published what Wurtz later characterized as "one of his most beautiful papers."[27] Proceeding under the conviction that

23. Wurtz, "Recherches sur l'acide sulfophosphorique et le chloroxyde de phosphore," *Annales de chimie* [3] 20 (1847), pp. 480–481.

24. Dumas, "Mémoire sur la loi des substitutions et la théorie des types," *Comptes rendus* 10 (1840), pp. 163–164.

25. A. Williamson, "On the Constitution of Salts," *Journal of the Chemical Society* 4 (1851), p. 351.

26. Gerhardt, "Recherches sur les acides organiques anhydres," pp. 331–342.

27. Gerhardt and Chiozza, "Recherches sur les amides," *Comptes rendus* 37 (1853): 86–90; Wurtz, "Discours préliminaire," in *Dictionnaire de chimie pure et appliquée* (Hachette, 1868), p. liv.

any of his various types could have virtually any acid-base properties, Gerhardt set out to create an acidic amide; by this means an acid could be classified in the ammonia type. He succeeded in doing this by substituting (for example) benzoyl into salicylamide (in modern terms, this amounts to acyl substitution for the amido hydrogens), creating what he called "secondary amides" and "tertiary amides." Since the number of organic acids was increasing apace, since every acid could be converted into an amide, and since the permutations of these new substances involved as many as three units, the number of new substances of this type could be extended "almost to infinity," Gerhardt noted.

Wurtz's immediate response was to admire the chemical artistry but contest the interpretation. He created the opportunity to offer a different view by devising a closely related reaction, publishing his paper just 2 weeks after Gerhardt's article appeared. Instead of using acid chlorides as reagents, Wurtz used his own cyanic ether (ethyl isocyanate). Wurtz found that this substance reacted readily with glacial acetic acid or with acetic anhydride, producing respectively ethyl acetamide and ethyl acetylacetamide. Gerhardt had synthesized acyl-substituted secondary and tertiary amides, while Wurtz had created the very similar alkyl secondary and alkyl/acyl tertiary amides. Wurtz related these substances not to the type of ammonia, as Gerhardt had, but to the type of water, thus avoiding the perceived anomaly of deriving acidic substances from an intensely basic prototype.[28]

It was in this paper that Wurtz declared that his intention was to extend "the beautiful researches of MM. Williamson and Gerhardt" on the water type; his new reaction provided "indirect confirmation" of the theory. Moreover, "the relations that exist between water and substances derived from the water type are expressed in a nicer and simpler manner with the aid of the equivalents adopted by M. Gerhardt than by employing the notation ordinarily used." Wurtz found that ordinary ether was inert to the action of his cyanic ether, possibly (he speculated) because ether contains two ethyl radicals substituted into the water molecule; they constitute two "large encumbering" pieces that "immobilize" the molecule against reactions. Wurtz was thus closely following Williamson's structuralist ideas—

28. Wurtz, "Sur les dédoublements des éthers cyaniques," *Comptes rendus* 37 (1853): 180–183.

without, it must be noted, actually adopting the new atomic weights which he had taken such care to praise.[29]

The next weekly issue of the *Comptes rendus* contained a second, explicitly theoretical article by Wurtz. Here Wurtz sketched the historical development of the type concept, starting with Dumas, extending through the work of Hofmann and himself, to the "remarkable" work of Gerhardt on acid anhydrides. But he drew a careful line between Gerhardt and himself on the interpretation of the amides. In these articles Wurtz not only retained conventional equivalents but also wrote many formulas in a manner that defied translation into Gerhardtian atomic weights (that is to say, he used odd numbers of equivalents in certain portions of certain formulas). This is one of several indications that Wurtz had not yet fully accepted all of Gerhardt's ideas. In his personal dialogue with Gerhardt, Wurtz was also struggling with the different notions of Laurent, Williamson, and others.[30]

A week later Gerhardt's response appeared. It was completely unimportant, he declared, whether amides were classified according to the water type, as Wurtz insisted, or to the ammonia type, as Gerhardt himself preferred; the crucial issue was how formulas were to be regarded in general. His own conception was that formulas were to be general, flexible, empirical, and synoptic, summarizing reactions and therefore to be formulated after the experiments were concluded. Wurtz, on the other hand, was following Dumas's belief that formulas should reflect actual images of molecules, and thus should be formulated before the experiments and used as a guide to them. According to this (false) idea, Gerhardt averred, compounds belonging to the same chemical type must share fundamental chemical properties. This was exactly the proposition that Gerhardt had intended to explode.[31]

In the fifth and last paper of this polemic, published only 6 weeks after the first, Wurtz had the last word. He declared that his conception of formulas was indeed distinct from Gerhardt's. His formulas had "a true molecular signification," and they were intended to indicate "the arrangement of the simple or compound molecules" in a compound. Types had

29. Ibid.

30. Wurtz, "Sur la théorie des amides," *Comptes rendus* 37 (1853): 246–250.

31. Gerhardt, "Note sur la théorie des amides,"*Comptes rendus* 37 (1853): 281–284.

a "tendency to stability" and were usually conserved in reactions—in particular, in all *substitution* reactions. But the type was not "purely mechanical and inert" with regard to chemical properties; it imprinted "a distinctive cachet" on all its compounds. The converse proposition was also true: compounds from different types differed substantially, even if they had the same numbers and kinds of atoms, such as carbamide and urea.[32] (This proved to be a poor choice of example, as these two substances were soon thereafter shown to be identical.) For Wurtz at this time, types indicate molecular arrangements, and vice versa, and both had an important impact on chemical properties.

In this dialogue Wurtz demonstrated that he had accepted an important part of the Williamson-Gerhardt theory, namely that relating to molecular magnitudes. Ether was a diethyl, alcohol a monoethyl compound; organic acids did not contain water, and salts did not contain oxides; organic anhydrides were like ethers in that the first had two acyl, the second two alkyl components; and so on. Against Gerhardt, Wurtz also agreed with Williamson's realist and structuralist notion of molecules and the formulas chemists used to represent them; he associated that epistemological stance too with Dumas's vision of mechanical versus chemical types. Gerhardt adamantly rejected structuralism, or any deep theory applied to molecular chemistry; this was why he insisted that chemical properties or even chemical compounds themselves did not map unambiguously to specific types. In fact these two propositions are quite separable, though few seemed to realize this then. Nothing in principle prevented one from asserting both that molecules have determinable structures *and* that a given molecule can equally be related to two or more different pattern formulas. Indeed, this was the position that the later advocates of structure theory, including Wurtz, eventually arrived at.

All of this illustrates the disparate character of the reforms associated with the names Laurent, Gerhardt, Williamson, Hofmann, and Wurtz. There could hardly have been a single point of conversion for Wurtz, because there was no party line for him to convert to. From early in his career, Wurtz was persuaded by elements of the new chemistry, including the type-substitutionist viewpoint, and he was receptive to hydracids and

32. Wurtz, "Nouvelles observations sur la théorie des amides," *Comptes rendus* 37 (1853): 357–361.

Avogadrian molecular theory. Sometime between the summer of 1851 and the summer of 1853 he came to accept Williamson's arguments concerning molecular magnitudes derived from the asymmetric synthesis of novel ethers and other organic compounds. By this time he was a fellow traveler, but no blind disciple of Gerhardt.

In the 18 months after his polemic with Gerhardt in the summer of 1853, Wurtz signaled aspects of continued independence from the new chemistry, partly by his language and partly by his formula conventions. Regarding language, he refused to use the Gerhardt-Laurent convention of "atomes" versus "molécules" (later adopted by all chemists), preferring the then-customary denotation of "molécules" as referring to chemical smallest parts (Gerhardt's "atomes"). Regarding the second point, he continued to use odd numbers of conventional equivalents in portions of his formulas, a procedure that emphasized his non-acceptance of Gerhardtian atomic weights. For instance, he commonly wrote "CO" (C = 6, O = 8) as an element of both carbamide and urea formulas, which in terms of the reformed chemistry would symbolize a half-atom of carbon united with a half-atom of oxygen—an impossibility for a Gerhardtian. Similarly, he designated water (again, in equivalents) as "HO," which was half the size of Laurent's and Gerhardt's water molecule.[33] Such a consistent practice could not have been inadvertent; moreover, a "closet" Gerhardtian could easily have sidestepped it by refusing to write resolved formulas and thereby avoiding the incriminating half-atoms in portions of the molecule. In short, Wurtz was no closet Gerhardtian, but an independent theorist working his own way through the multiple puzzles that constituted organic chemistry in the 1850s.

Mixed Radicals and Multiple Types

In two papers published in 1855, Wurtz altered important aspects of his language, his notation, and his chemical concepts. Much evidence suggests that these changes were prompted by Wurtz's reading of recent chemical literature coming out of England and by some remarkable work just published by a junior member of the Parisian chemical community: Marcellin Berthelot.

---

33. In "Sur la théorie des amides" (August 1853) Wurtz wrote no fewer than ten formulas containing moieties with odd numbers of carbon or oxygen equivalents. Such formulas continued to appear in his papers until the end of 1854.

From 1847 on, Wurtz had strong ties to the chemical community in London. His sister was there, as was Hofmann, his old friend from Giessen. Although he had never formally studied English, he soon acquired sufficient command of the language to be able to read (and presumably speak) with facility. By 1850 he had met the dean of English chemists, Thomas Graham, and by 1853 had become acquainted with the new work of Graham's young colleague at University College London, Alexander Williamson. These contacts were facilitated partly by a fortunate professional responsibility. At the beginning of 1852 Wurtz was brought, presumably by Dumas, into a subsidiary position in the editorial collective that published the *Annales de chimie*. Moreover, the *Annales* simultaneously began to include translated abstracts of articles from foreign scientific journals, and Wurtz was assigned to be one of the two editors of this new section (Émile Verdet was the other, responsible for physical chemistry and physics). This innovation may have been due to Dumas, who had long been receptive to the widespread foreign criticism that French science was dangerously insular.

In the summer or the fall of 1853 Wurtz must have written Williamson asking him to summarize in French his three English ether-theory papers of 1850–51, for in the January 1854 issue of the *Annales* such a summary appears in Wurtz's foreign abstracts section, prefaced by a note by Wurtz explaining the circumstances. On 18 April 1854, in the earliest surviving letter of the Wurtz-Williamson correspondence, Wurtz told his English friend that the article had "created a sensation" in Paris.[34] Although the ether-theory work had long been available in principle to Parisian scientists in English and German journals, it appears that few knew of it until this time; even Wurtz probably became aware of the work only late in 1852 or in the first half of 1853.

Wurtz could not have chosen a better time to establish a close connection to English chemistry. Williamson was in his prime, in the midst of a wondrous burst of creative activity that would soon diminish. William Odling and Benjamin Brodie Jr. were also making significant contributions, while Graham was adorning the faculty at University College and Frankland was

34. Williamson, "Sur la théorie de l'étherification," *Annales de chimie* [3] 40 (1854): 98–114; Wurtz to Williamson, 18 April 1854, in Harris Collection, Bloomsbury Science Library, University College London.

pursuing his productive career at Owens College Manchester. There was also a lively community of naturalized and visiting German chemists, including August Wilhelm Hofmann, Hugo Müller, and the enterprising young August Kekulé (who resided in London from December of 1853 until August of 1855).

Williamson, Hofmann, Odling, Brodie, and Kekulé—Londoners all in the period at issue—were simultaneously pursuing lines of work that related to Gerhardt's and Laurent's ideas. The story in London had begun in 1849, with Hofmann's conversion to substitution, the "amide" theory, and the ammonia type, and with Williamson's return from his 3-year postdoctoral period in Paris to take up a professorship at University College London. The next year Williamson's ether series began to appear. About the same time, Hofmann and Brodie began independently to argue for the larger (Gerhardtian) formulas for the hydrocarbon radicals that had been discovered by Kolbe and Frankland in the preceding 3 years; instead of Frankland's and Kolbe's monomeric radicals, Hofmann and Brodie interpreted them as dimers. (See chapter 4 above.) In support of the larger formulas, Hofmann argued that the substances isolated by Frankland and Kolbe had all the properties one would expect of a paraffinic hydrocarbon and none of the properties one would expect to see in a real monomer "radical." Their physical properties, too, fit well with doubled formulas, whereas they would represent anomalous exceptions if the molecules really were monomers.[35]

In December of 1850 Brodie reported attempts by both himself and Hofmann to demonstrate the dimer formulas by synthesizing novel asymmetric or "mixed" radicals, parallel to Williamson's demonstration of the doubled ether formula by synthesizing novel asymmetric or "mixed" ethers. These attempts, however, came to naught.[36] A different approach to the

35. Hofmann, "Note on the Action of Heat upon Valeric Acid," *Journal of the Chemical Society* 3 (1850): 121–134.

36. B. Brodie, "Observations on the Constitution of the Alcohol-Radicals, and on the Formation of Ethyl," *Journal of the Chemical Society* 3 (1850): 405–411. The efforts of Hofmann and Brodie to synthesize "asymmetric radicals" in a Williamsonian fashion, and their arguments contra Kolbe and Frankland on the molecular sizes of these substances, are nicely recapitulated in a contemporary account: *Annual Report of the Progress of Chemistry* (Taylor, Walton, and Maberly, 1850), vol. 4, pp. 234–238.

same complex of issues proved more productive. Early in 1854, Odling, Williamson, and Kekulé independently published papers that reported parallel studies of the "substitution value" or "basicity" of atoms and radicals; in this connection, all urged adoption of the Gerhardtian system of atomic weights.[37]

In effect, what had been in the 1840s an unorthodox French would-be reform was transformed in the 1850s into a respectable Anglo-French alternative to the older German-Berzelian orientation. Wurtz had little directly to do with this transition, but he must have been an interested observer. He was collegially close to Hofmann, and by the end of 1853 to Williamson as well; he had at least met Kekulé.

In London, as in Paris, there were two distinct epistemological approaches to the new chemistry. Just as Wurtz followed an element of Dumas's chemical philosophy in pursuing a realist-mechanist-structuralist approach to molecular chemistry while Gerhardt represented the conventionalist position, so also during his London period Kekulé began to follow a Williamsonian structuralist line while Odling and Brodie preferred to express themselves with more empirical caution. Hofmann's position is more difficult to interpret, but it probably was closer to conventionalism than to realism.

Into this environment stepped a new personality: Marcellin Berthelot, who was 25 years old and Balard's préparateur at the Collège de France when he published his first significant contribution to organic chemistry. Thirty years earlier, Chevreul had shown that fats consist of large organic acids combined with glycerin, in the same way that alcohol combines with acetic acid to form an ester. In short, Chevreul demonstrated that fats are esters consisting of glycerin and fatty acids. In 1853 Berthelot succeeded in showing that they are in fact *tri*-esters, named triglycerides, formed of one molecule of glycerin combined with three molecules of fatty acid. He succeeded in forming many new triglycerides, and for the first time he created monoglycerides and diglycerides too.

37. Odling, "On the Constitution of Acids and Salts," *Journal of the Chemical Society* 7 (1854): 1–22; Williamson, "Note on the Decomposition of Sulphuric Acid by Pentachloride of Phosphorus," *Proceedings of the Royal Society* 7 (1856, read 1854): 11–15; Kekulé, "On a New Series of Sulphuretted Acids," *Proceedings of the Royal Society* 7 (1856, read 1854): pp. 37–40.

In these syntheses Berthelot used naturally occurring fatty acids to regenerate natural fats, but he also employed a variety of other organic and even inorganic acids (such as acetic, butyric, valeric, and hydrochloric acids). Since there were now three independent series of glycerides (mono-, di-, and tri-), since many organic acids now were known, and since for the di- and tri- series the permutation game could be played aggressively, Berthelot remarked (paralleling Gerhardt's recent similar comment) that the number of such derivatives could be extended virtually to infinity.[38]

This stunning contribution to the science helped to make Berthelot's career; he used it as his thèse for the degree of docteur ès sciences physiques. Many took note. Williamson quickly produced two small but interesting papers that must surely have been inspired by Berthelot's work. In one of them (assisted by his student George Kay) he reported the production of what we would call "triethoxymethane," a triple ether prepared from one mole of chloroform and three moles of sodium ethoxide. In another he described the preparation and hydrolysis of trinitroglycerin, a tri-ester of glycerin with three moles of nitric acid—exactly analogous to Berthelot's trichlorohydrin of glycerin, a tri-ester of glycerin with three moles of hydrochloric acid. In this substance, Williamson noted, three hydrogen atoms of the glycerin molecule appear to be replaced by three nitro groups.[39]

Williamson and Wurtz were by this time in continuous communication, and Williamson sent an offprint to his French colleague. Wurtz printed a translated abstract of the paper in the April 1855 issue of the *Annales de chimie*, and followed it immediately with his own commentary.[40] "Glycerin can be considered to be a species of tribasic alcohol," Wurtz wrote; "that is, one that contains three equivalents of hydrogen capable of being replaced by three groups." Glycerin is similar to propyl alcohol, except for the fact that propyl alcohol has only one hydrogen that can be replaced. One could imagine, he

38. Berthelot, "Sur les combinaisons de la glycérine avec les acides," *Comptes rendus* 37 (1853): 398–405; *Comptes rendus* 38 (1854): 668–673; *Annales de chimie* [3] 41 (1854): 216–319.

39. Williamson, "Note on Nitroglycerine," *Proceedings of the Royal Society* 7 (16 June 1854): 130–138; translated abstract in *Annales de chimie* [3] 43 (1855): 492.

40. Wurtz, "Théorie des combinaisons glycériques," *Annales de chimie* [3] 43 (1855): 492–496.

continued, that the hydrocarbon radical of propyl alcohol C^6H^7 could lose two additional hydrogens to form a new radical that "could substitute three equivalents of hydrogen, thus forming a bond [lien] between three linked [conjuguées] water molecules." This paralleled Berthelot's work on polyglycerides, and the interpretation followed Williamson's view of trinitroglycerin. Wurtz provided two ways to visualize the molecule of glycerin:

C^2H^2

    O^2

H

                                            H O^2

C^2H^2

    O^2 H C^6H^5 O^2

H

                                                H O^2

C^2H

    O^2

H

Although Wurtz declared that the second of these was preferable as the simplest representation, he emphasized that the glyceryl radical C^6H^5 could be schematically resolved into the three smaller groups indicated, each of the hydrocarbon groups substituting for one hydrogen of the water molecule (H^2O^2).[41] Here Wurtz was not only using all the formal aspects of Williamsonian water type theory; for the first time, he was also using Williamsonian realist-mechanist epistemology. The propyl radical was "monobasic," the glyceryl radical "tribasic"; the acid moieties in glycerides were held together by a "bond" provided by the glyceryl radical, just as oxygen holds the hydrogen of water into a single compound. The mono- and dichlorohydrins of glycerin represent "mixed types," combining a water type with a hydrogen type, in the same way that Williamson's sulfuryl chloride was a mixture

41. Wurtz later claimed that his resolved formula for glycerin of 1855 represented "the first attempt [to determine] such a distribution of the atoms in a radical," but he quickly added that at the time this was no more than a hypothesis and therefore had no material consequences (Wurtz, *La théorie atomique* (Baillière, 1879), pp. 146–147). He was wrong: there had been many attempts to determine arrangements of atoms in a molecule before 1855, and some of them were more than merely hypothetical.

of types centered on the "dibasic" sulfuryl radical. In all such cases the central "polybasic radical" was the material and mechanical pivot on which the molecular architecture hung. (The concept of basicity was obviously borrowed from the polybasic acids, which exhibit the tendency to unite more than one metal atom into a single salt molecule.)

Wurtz's turn to the letter and the spirit of Williamsonian theory at this time is indicated not only by the preceding considerations but also by much additional evidence. First, his terminology changed: he now began to use Gerhardt's and Williamson's denotation of "molécule" (essentially the same we use today), instead of the older and still customary sense of the word—though he still avoided using "atome" in any context. Second, although he continued to use chemical equivalents, henceforth he never used them in odd numbers. Third, his choice of material for the foreign abstract section of the *Annales* shifted: instead of an even-handed selection of foreign articles from a variety of perspectives, Wurtz's section henceforth became a partisan platform for the reform party in England and Germany. Fourth, ever after and to the end of his life, Wurtz touted Williamson's contribution of the early 1850s to be the crucial turning point for modern chemical theory.[42] And finally, Wurtz's own research program suddenly accelerated. Since the beginning of his research career in 1842 through the end of 1854, Wurtz had published about one or two articles per year. In 1855 his rate of publication suddenly increased to about five per year.

And 1854 was the year Wurtz and Berthelot publicly parted company. Despite the importance of Berthelot's contribution to understanding glycerides, his interpretation of that work was radically different from Wurtz's. Berthelot was contemptuous of all attempts to enter the micro-world of atoms and molecules. He insisted that the only permissible approach to chemical knowledge was that of the thoroughgoing empiricist, and that

42. "M. Williamson, whose work has had such a large part in the development of this theory . . ." ("Mémoire sur les glycols," 1859, p. 474). The newer theory of types "began with the experiments of Dr. Williamson on etherification, and his beautiful discovery of mixed ethers" (*Leçons de philosophie chimique*, 1863, p. 88). The experiments of Wurtz and Hofmann on ammonia types represented "but an isolated point of view" until Williamson's work provided "the most decisive movement" to the science (*Cours de philosophie chimique*, 1864, pp. 30–31). Williamson's etherification papers "mark a new era in the history of chemical theory" (*La théorie atomique*, 1879, p. 144).

formulas could only summarize the gravimetric facts of chemical reactions. Somewhat inconsistently, he also explicitly defended the older dualistic theory that interpreted organic compounds as oxides and salts; ethyl benzoate, he insisted, could only reasonably be interpreted as anhydrous benzoic acid united to ethyl oxide.[43] In respect to epistemology he was allied with Gerhardt, in theoretical terms with Hermann Kolbe and other traditionalists. In both respects he was strongly at odds with Williamson—and with Williamson's promoter, Wurtz. Berthelot, 10 years younger than his Parisian colleague, was not shy about bringing Wurtz up short when he thought that Wurtz had indulged too strongly in hypothesis.[44]

But Wurtz could never let theory rest. In an 1864 lecture at the Collège de France he proclaimed theories "the soul of science."[45] Now that he was more fully committed to the newer ideas, he threw himself into the cause. Three months after his glycerin paper, he published what is arguably the most important article of his life. He was aware of Hofmann's and Brodie's independent attempts to synthesize "mixed" radicals—methyl combined with ethyl, or ethyl with butyl, and so on—by fusing various alkyl iodides together using the agency of zinc. This had in fact been the means by which Frankland had first produced "ethyl," but the reaction is a difficult one, and neither man could get it to go. In July of 1855 Wurtz announced the synthesis of five asymmetric radicals, along with three additional symmetrical dimers, all made by reacting alkyl iodides with sodium.[46]

This was an incontestably brilliant experimental accomplishment (the difficult "Wurtz reaction" is no longer much taught), but in his paper Wurtz focused particularly on the theoretical implications. An example will suffice, Wurtz's novel asymmetric "butyl-amyl." Since it was formed from butyl iodide and amyl iodide, it could only be formulated as consisting of the sum

43. Berthelot, "Recherches sur les éthers," *Annales de chimie* [3] 41 (1854): 432–445.

44. Berthelot, *Annales de chimie* [3] 44 (1855): 350.

45. Wurtz, *Cours de philosophie chimique* (Renou et Maulde, 1864), p. 3. The lectures, printed privately and cheaply, were delivered at the invitation of Balard in May and June 1864. They also appeared contemporaneously in Quesneville's *Moniteur scientifique*.

46. Wurtz, "Sur une nouvelle classe de radicaux organiques," *Annales de chimie* [3] 44 (1855): 275–313.

of the hydrocarbon moieties of the two halides—and so all agreed that it must contain 18 carbon equivalents (or nine carbon atoms). Kolbe and Frankland interpreted the butyl and the amyl radicals themselves to consist of eight and ten carbon equivalents, respectively, while the reformers regarded them as having 16 and 20 carbons, respectively. The physical properties of the new butyl-amyl fit perfectly in series with the larger formulas, but were inexplicably anomalous for Kolbe and Frankland.[47] Kolbe himself was unable to answer Wurtz's new evidence—though he continued, without offering further arguments in support, to advocate the monomeric radical formulas.[48]

Having established what he regarded as the correct formulas for the hydrocarbons (explicitly naming Williamson as his model), Wurtz proceeded to a detailed brief in favor of all essential aspects of the reformed chemistry. Just as the hydrocarbon radicals "methyl," "ethyl," and so on are really dimers, so also are the elementary gases hydrogen, oxygen, nitrogen, and chlorine. Moreover, this idea (Wurtz was careful to stress) was by no means new: it had been advocated by Ampère, Dumas, and Laurent, and it was implied by Wurtz's own work on copper hydride.[49] This is a principal point of the newer chemical theories: elements are *groups* of equivalents, and they combine by substitution, not addition. Wurtz even used the Gerhardt-Laurent-Williamson atomic weights in this theoretical section ("for greater simplicity"), though it would be another 3 years before he adopted them consistently in his papers.

From Type Theory to Structure Theory

Wurtz had struck a rich lode. His glycerin paper marked the beginning point of a research program that occupied him for the rest of his life: exploration of polyfunctional organic compounds. Glycerin was a tri-alcohol; triglycerides were tri-esters; mono- and diglycerides were mixtures of two

---

47. For example, compare the following boiling points (all these compounds were known in the 1850s; the boiling points are given in degrees Celsius): butylene –5, amylene 30, octane ("butyl") 126, nonane (Wurtz's butyl-amyl) 151, decane ("amyl") 174.

48. See Rocke, *Quiet Revolution*, pp. 139–155.

49. Ibid., pp. 301–306.

alcohol functions and one ester, or one alcohol and two esters; glycerin chlorohydrins were mixtures of chlorides and alcohols; and so on. Glycerin was not the first polybasic radical recognized—Williamson's sulfate and sulfuryl are other examples—but it was an early instance, and the first clear case in the organic field. Just as the concept of polybasic radicals was emerging, there were also growing indications that the elements themselves had "saturation capacities," "basicities," or "atomicities"—what today is known as valence. About the same time that Williamson was making his case that the oxygen atom could unite two hydrogen atoms into a molecule of water, Frankland stated a general law of saturation capacities of various inorganic elements. Odling and Kekulé added further ideas in this direction, influenced by Williamson but apparently not at all by Frankland.[50]

The historical accident that ideas of polybasic radicals and "polybasic" elements both emerged relatively simultaneously led to some interesting combinations of the two notions. Ever since the days of John Dalton it had seemed doubtful to most chemical philosophers that the chemical "atom" was also an atom in its literal etymological sense (that is, an irreducible material entity). Many had speculated that the chemists' atoms were nothing but groups of smaller particles; Dumas himself had made this claim in 1836. Adopting such a view would mean that there could be substantial analogies between polybasic radicals (groupings of atoms that could form links to other radicals) and polybasic atoms (groupings of subatomic particles that could form links to other atoms).

Wurtz took up this viewpoint in his "mixed radical" paper of July 1855. Just as glycerin was a tribasic radical consisting of three groups of atoms bound together, each group capable of carrying one alcoholic or ester function, so also might the nitrogen atom and the phosphorus atom each consist of three smaller particles bound together, with each particle capable of forming a bond to an equivalent of hydrogen, say, or chlorine. In other words, just as the molecular structure of glycerin may explain its tribasic character, so also may the atomic structure of nitrogen explain its tribasicity (later called "trivalence").

50. For a discussion of these developments, see C. Russell, *The History of Valency* (Leicester University Press, 1971), pp. 34–61.

Wurtz was well aware that this was no more than speculation. He never developed the idea in detail, but he continued to refer to it favorably in subsequent publications for 14 more years.[51] In 1864 Wurtz opined that atoms may have multiple "centers of attraction" for other atoms, by being formed of a "number of smaller particles, primordial atoms" bound together by an indissoluble force and attracting, in groups, other chemical atoms.[52] In different writings, he referred to these smaller particles as "sous-atomes" or "petits atomes." Wurtz's subatomic speculation is significant not only for indicating Wurtz's ideas on these entities but also for providing a conspicuous manner of visualizing the distinctions between chemical "equivalent" (the weight of a subatom), valence (the number of subatoms in a chemical atom), and atomic weight (the weight of the entire complex of subatoms, that is, the atom itself). The hypothesis was adopted in various forms by a variety of major and minor chemists in the period 1855–1869, including Alexander Crum Brown, Emil Erlenmeyer, Charles Delavaud, Alfred Naquet, and Christian Blomstrand. (It should be noted that Wurtz's idea was completely different from Prout's hypothesis; indeed, it was entirely novel, having no direct analogy to any other of the numerous atomistic speculations of the nineteenth century.)

Elsewhere I have pursued the argument that August Kekulé most likely developed the theories of valence, carbon tetravalence, and structure as a direct response to Wurtz's subatomic speculation of July 1855.[53] Kekulé and Wurtz were acquainted ever since Kekulé's 10-month residency in Paris (1851–52). According to my suggested time line for the development of Kekulé's theory of chemical structure, he devised the basic concepts in London in August or September of 1855, drew up a draft of a paper in Heidelberg in the spring of 1856, then had second thoughts and put it away for more than a year. What is beyond question is that he published

---

51. For a full discussion of this theory, see Rocke, "Subatomic Speculation and the Origin of Structure Theory," *Ambix* 30 (1983): 1–18. Not mentioned in that article is the quotation in the next sentence (or the work from which it comes), or Wurtz's brief discussion of the theory in his *History of Chemical Theory* (Macmillan, 1869), p. 157 (French version, "Discours préliminaire," 1868, p. lxx).

52. Wurtz, *Cours de philosophie chimique*, pp. 56, 74–75.

53. Rocke, "Subatomic Speculation" (recapitulated in Rocke, *Quiet Revolution*, pp. 165–166).

the theory in the form of two related papers, in the November 1857 and May 1858 issues of Liebig's *Annalen*.[54] The first article purported only to be an attempt to destroy the remaining vestiges of the dualistic theory of organic radicals, and to replace it with a fully developed theory of polyatomic radicals. In the process, Kekulé formulated the emergent theory of atomic valence, which he called "Atomigkeit" (atomicity), the valences being "Verwandtschaftseinheiten" (affinity units). Hydrogen and chlorine were "monatomic," oxygen and sulfur were "diatomic," nitrogen and phosphorus were "triatomic," and a footnote asserted that carbon was "tetrabasic or tetratomic."

Kekulé's second paper—really the first full account of structure theory—is a clear and well-developed exploration of the interior architecture of organic molecules, the first such that attempted to achieve a degree of generality while remaining empirically grounded. Proceeding from the standpoint of the newer type theory and the theory of polyatomic radicals (in the realist-mechanist version of Williamson and Wurtz), Kekulé showed that a consistent theory could be constructed by using the accepted atomicities of the various elements to link them together to form atomic-molecular chains—including chains of self-linking carbon atoms. The embryonic theory had only limited explanatory range at this stage (derivatives of simple alkanes), but within these confines Kekulé could reasonably argue that the theory worked well.

In both papers Kekulé portrayed his ideas as nothing more than a generalization and extension of the theory of polyatomic radicals. He was careful to give credit to the initial formulators of that theory, especially to Williamson, Odling, and Gerhardt, but Wurtz came in for particular praise in the introduction to the second paper:

. . . never feeling it necessary to develop his ideas more fully, [Wurtz] nevertheless permitted others of us to infer them by reading between the lines of each of his classic researches, through which the development of my views first became possible.[55]

54. Kekulé, "Ueber die s.g. gepaarten Verbindungen und die Theorie der mehratomigen Radicale," *Annalen* 104 (1857): 129–150; "Ueber die Constitution und die Metamorphosen der chemischen Verbindungen und über die chemische Natur des Kohlenstoffs," *Annalen* 106 (1858): 129–159. For a discussion, see pp. 166–180 of Rocke, *Quiet Revolution*.

55. *Annalen* 106 (1858), p. 136.

In June of 1859, with the publication of the first part of his textbook of organic chemistry (in which he developed the emergent theory in greater detail than in his 1858 paper), Kekulé had an opportunity to stress his respect for and his obligation to his French friend once more. He sent Wurtz a copy, asking his opinion of its contents, "because I believe that the viewpoint that the majority of chemists will have in the future can be little different from the viewpoint that you have now."[56] This may well have been more than a polite formality. It is remarkable that not only Kekulé, but also the other independent simultaneous co-developer of the theory of chemical structure, Archibald Scott Couper, was much influenced by Wurtz. Couper, in fact, was in Paris from August of 1856 until the late fall of 1858, spending most of that time in Wurtz's laboratory; his "new chemical theory" of molecular structure, like Kekulé's, bears significant traces of Wurtz's subatomic speculation.[57] Couper's first publication on this subject appeared in June of 1858,[58] a month after Kekulé's second paper on structure theory, but there is no reason to doubt the independence of the two. Early in 1858 Couper had given his paper to Wurtz for presentation to the Académie, but Wurtz, not yet a member of the Académie, had to obtain leave from Dumas or Balard to present it, and he was slow to do that. According to Albert Ladenburg, who worked in Wurtz's lab a few years later, "Couper was very angry, called Wurtz to account, and became abusive to him. Wurtz was not pleased by this, and expelled Couper from the laboratory. It appears that Couper took this very much to heart, and in Paris it was thought that this was how his illness began. This story is genuine, Wurtz himself told it to me."[59] Couper returned to

56. Kekulé to Wurtz, 1 July 1859, in R. Anschütz, *August Kekulé* (Verlag Chemie, 1929), vol. 1, pp. 157–159.

57. Rocke, "Subatomic Speculations," pp. 9–10.

58. Couper, "Sur une nouvelle théorie chimique," *Comptes rendus* 46 (1858): 1157–1160; "Sur une nouvelle théorie chimique," *Annales de chimie* [3] 53 (August 1858): 469–489.

59. Ladenburg to Anschütz, 12 May 1906, in Anschütz, "Archibald Scott Couper," *Archiv für die Geschichte der Naturwissenschaften und der Technik* 1 (1909), p. 226. Adolf Lieben, another of Wurtz's students during the same 2 years of Couper's residency in Paris, confirmed Couper's independent route to structure theory and the existence of a draft of the article before May 1858; Couper had asked him to read the manuscript of the first article before passing it on to Wurtz (ibid.).

his home in Scotland, became an invalid, and never published another article after 1858.

In the first two issues of his new journal *Répertoire de chimie pure* (October and November 1858), Wurtz reviewed Kekulé's and Couper's new theories. Although he had much favorable to say about both, he gave decided preference to Kekulé's. Couper's theory was broader than Kekulé's, Wurtz wrote, but it suffered from weaknesses that his rival's did not have. Couper's exposition had too many hypothetical elements and too many unclear passages, and several of his formulas were arbitrary (i.e., empirically unmotivated). In addition, Wurtz appears to have been disturbed by Couper's brash, even insolent language. Couper had titled his paper "une nouvelle théorie chimique," had labeled Gerhardt's type theory "pernicieux" and "un erreur," and had neglected to cite a single predecessor or positive influence on his thinking. Kekulé's work was more modest (in both senses of the word); he connected his ideas to those of his predecessors (in particular, he praised Wurtz highly), explicitly downplayed the novelty of the theory, and stayed closely tied to what could reasonably and empirically be argued.[60]

All of this is, in the opinion of the present writer, accurate and perceptive. However, two details of Wurtz's critique ought not stand without comment. For one thing, Wurtz could well have taken the opportunity to affirm that Couper's theory was independent of Kekulé's, and that the manuscript for Couper's first paper was in his hands before the publication of Kekulé's crucial second paper on structure theory. Second, Couper, following the practice of his teacher Wurtz and most other chemists of his day, used the smaller "equivalent" weight of oxygen, and as a consequence his formula for water was HO; Wurtz criticized the weight and the formula as "not in accord with the facts." It is ironic that Wurtz wrote these words late in the summer of 1858, just as he was deciding to adopt the larger "atomic weights" of oxygen and carbon favored by Gerhardt, Williamson, and now Kekulé. It seems likely that Wurtz was persuaded to make this move—the final step in his gradual adoption of the new chemistry—by the advent of Kekulé's structure theory. In any case, it is hard to see Wurtz's criticism of Couper for using $O = 8$ as anything other than

60. Wurtz, *Répertoire de chimie pure* (1858), vol. 1, pp. 20–24, 49–52.

disingenuous and unjust. It certainly seems that Wurtz and Couper had clashed personally, and this probably explains Wurtz's failure to do full justice to his student. After ignoring Couper completely in historical accounts published in 1863 and 1864, he mentioned Couper in a footnote in his 1868 history of chemical theory as an independent co-developer of structure theory.[61]

61. Wurtz, "Discours préliminaire" (1868), note on p. lxxi. However, Couper's name does not appear in connection with structure theory in Wurtz's 1879 historical-philosophical work *La théorie atomique*. When Kekulé's student, successor, and biographer Richard Anschütz began to examine Couper's story, in the first decade of the twentieth century, he could at first learn nothing of the man, not even his nationality. (Anschütz thought at first that the name might have been French.) It was only through Anschütz's persistence that a decent biography of this interesting Scottish chemist could be written.

7

The Campaign

Around the beginning of 1855, Wurtz became fully committed to the reform agenda in chemistry, and, entirely coincidentally, Charles Gerhardt's career finally took off. The power elite in Parisian chemistry, especially Baron Thenard, seized the opportunity to promote Gerhardt's career when a vacancy arose in the Strasbourg Faculté des Sciences. Thenard's sincere desire to see Gerhardt in an appropriate position—and, perhaps equally important, to see him out of Paris—is indicated by his success in arranging also a second position for Gerhardt in Strasbourg, at the École de Pharmacie (an example of provincial cumul). Gerhardt had a wife and three children, and was still in debt from capitalizing his lab school in the Rue Monsieur-le-Prince; Thenard argued that he needed both positions to be able to make the move. The arrangements were concluded by the end of 1854, and in February of 1855 Gerhardt took up his new positions in his home town. His old friend and fellow Strasbourgeois Émile Kopp purchased the Paris lab school from him. Thenard even engineered the election of Gerhardt as Membre Correspondant of the Académie.[1]

Tiffeneau remarked that Gerhardt resented the fact that he had been relegated to the provinces while his younger disciple in the new chemistry,

1. E. Grimaux and C. Gerhardt Jr., *Charles Gerhardt* (Masson, 1900) (hereafter "Grimaux and Gerhardt"), pp. 255–271; M. Tiffeneau, ed., *Correspondance de Charles Gerhardt* (Masson, 1918–1925) (hereafter "Tiffeneau"), vol. 2, pp. 152, 158, 297–299. In handing him the note confirming his election, Thenard said: "When you are old and have power, M. Gerhardt, remember always to protect young people who work. It is one more way of being of service to science." (from a note written by Gerhardt; see Tiffeneau, vol. 2, p. 298)

Wurtz, enjoyed a professorship in Paris.[2] Be that as it may, five letters exchanged between them after Gerhardt's move to Strasbourg indicates that their relationship was good, at least on the surface. Wurtz even interceded twice with Dumas on Gerhardt's behalf—to get a delayed paper printed in the *Annales de chimie* and to ensure Gerhardt's nomination to the Légion d'Honneur—for which Gerhardt was appropriately grateful to Wurtz.[3]

Gerhardt died of a sudden fever in August of 1856, 2 days before his fortieth birthday; he had been professor in Strasbourg just 18 months. By this time, however, Wurtz had already taken on the mission of ensuring the ultimate victory of the reforms in chemistry begun by Gerhardt and his equally star-crossed friend Laurent. In view of the nature of the scientific establishment in France at the time, and of the political conditions, a well-conducted campaign was clearly required. The only question was whether Wurtz would be equal to the task.

The Société Chimique and Its Publications

By the mid 1850s, Wurtz had entered the Parisian chemical establishment, but he was not yet in the inner circle of power. His chair was in the Faculté de Médecine rather than the Faculté des Sciences or the Collège de France, and he was not yet a member of the Académie des Sciences. His official duties were to teach the entire science of chemistry at a rather basic level to medical students, which was distant from his passionate calling—pure scientific research in the specialized field of organic chemistry. His election to the Académie de Médecine in April of 1856 only accentuated this professional dissonance.[4] Moreover, although the ideas of Laurent and Gerhardt had made considerable progress in England and in Germany, it was not yet clear that they would prove acceptable to the majority of French chemists. Indeed, Wurtz was the leading figure inclined to the new

2. Tiffeneau, vol. 2, p. 304. Tiffeneau's statement is supported by two letters from Gerhardt to Louis Orfila, both of which betray a degree of resentment toward Wurtz (ibid., pp. 280–281).

3. Tiffeneau, vol. 2, pp. 308–315; Grimaux and Gerhardt, pp. 280–281, 440–441.

4. Wurtz was elected to fill the vacancy created by François Magendie's death the previous October. See F17/3685.

movement; such establishment figures as Thenard, Dumas, Frémy, Pelouze, and Balard tended to ignore it, as did such younger chemists as Berthelot and Deville.

The first vacancy in the chemistry section of the Académie des Sciences after Wurtz's arrival in Paris in 1844 did not occur until December of 1857, and the result of that election was a serious blow to Wurtz. Edmond Frémy was chosen, which did not particularly surprise Wurtz (for Frémy had "long worked" in the field); but Balard voted for Deville, Dumas voted against Wurtz in the Comité Secret (after favoring him in the open session), and the upstart Berthelot actually received more votes than Wurtz. "In short," Wurtz wrote to Liebig, "this election came out very badly for me, especially considering that nothing is certain for me in the future"—viz., these results suggested that even the next vacancy, whenever that might occur, might well not get him into the Académie.[5] In the event, the next vacancy came 10 years later, and Wurtz was elected; but he could not know this in 1857. (When congratulated in 1867, he replied sadly: "Ten years ago I would have been proud of this election."[6]) Wurtz concluded, reasonably, that his work was more respected in Germany than in France. He had to do something to change the situation.

Wurtz's entry into the editorial collective that ran the *Annales de chimie* had already given him an opportunity to influence the literature read by his countrymen, and from 1854 on he had been making the most of it. The fact that he was responsible for foreign literature was particularly apt; I argued in chapter 6 that this duty may have played a role in directing his own attention to some crucial papers just published in London in the early to mid 1850s. But this was a weak reed to rely on. Fortuitously,

5. Wurtz to Liebig, 3 February 1858, in Liebigiana IIB, Bayerische Staatsbibliothek: "Kurz, diese Wahl ist sehr schlecht für mich ausgefallen, besonders dadurch daß mir für die Zukunft nichts gesichert ist." The *Comptes rendus* for December 1857 (vol. 45, pp. 976 and 994) reveals that Frémy received 45 votes to Berthelot's 7 and Wurtz's 6, even though the Comité Secret had ranked Frémy first, Wurtz and Deville second, and Berthelot and Cahours third. Wurtz ought to have realized that when one candidate so dominates an election the difference between second and third place says next to nothing about how the election would have turned out in the absence of the first-place candidate.

6. A. Gautier, "Ch.-Adolphe Wurtz, sa vie, son oeuvre, sa personnalité," *Revue scientifique* 55 (1917), p. 778.

just 2 months after his lament to Liebig a new opportunity arose, in the form of a new society, recently formed in Paris, that clearly was in need of leadership.

At the end of May 1857, three young préparateurs in chemistry hatched a plan to help them master the burgeoning chemical literature. Giacomo Arnaudon, E. Collinet, and J. Ubaldini agreed to meet every Tuesday night in a room reserved for them in a café in the Cour du Commerce (off the Rue Saint-André-des-Arts in the Latin Quarter), and they were soon joined by a few others. All were young, most were foreign, and none was well or permanently employed in the field. The first meeting of the yet-unnamed group, under the presidency of Arnaudon, was held on 4 June; at the fifth meeting (30 June) they voted to call themselves "la Société Chimique." Meetings consisted of members taking turns in summarizing the recent French and foreign journal articles in chemistry, and in discussing their own research; a principal goal was to prepare themselves for the degree of licencié ès sciences. The first list of members (September 1857?) contains the names of three Italians (Arnaudon, Ubaldini, and A. Pavesi), a Russian (L. Shishkov), a Norwegian (A. Rosing), a Colombian (M. Salazar), a Portuguese (J.-L. Mantas), and five Frenchmen, including Collinet.[7]

By early 1858 the new society had grown to 24 members, but it still retained its character as a foreigner-dominated self-study group of chemists in training. Among the newer members were the Scot A. S. Couper, the German Adolf Lieben, and the Russians A. M. Butlerov and Friedrich Beilstein—all future chemists of note (and all students of Wurtz), but at that

7. For the early history of the Société Chimique, see the first few pages of the first volume of *Bulletin de la Société Chimique* (1858–1862); see also *Mémorial de la Société Chimique de France, 1857–1949*, ed. C Pacquot (Société Chimique, 1950[?]); J. Jacques, "Butlerov, Couper, et la Société Chimique de Paris," *Bulletin de la Société Chimique*, 1953): 528–530; *Centenaire de la Société Chimique de France (1857–1957)* (Masson, 1957); J. Jacques and G. Bykov, "Nouveaux matériaux concernant l'histoire de la Société Chimique de Paris," *Bulletin de la Société Chimique*, 1959: 1205–1210; R. Fox, "The Savant Confronts His Peers," in *The Organization of Science and Technology in France, 1808–1914*, ed. R. Fox and G. Weisz (Cambridge University Press, 1980), esp. pp. 269–272; A. Carneiro, The Research School of Chemistry of Adolphe Wurtz (Ph.D. dissertation, University of Kent/Canterbury, 1992), pp. 88–90, 230–239; U. Fell, "The Chemistry Profession in France," in *The Making of the Chemist*, ed. D. Knight and H. Kragh (Cambridge University Press, 1998); Fell, *Disziplin, Profession, und Nation* (Universitätsverlag Leipzig, 2000).

time entirely unestablished. In March, under its new president, Rosing, the group resolved no longer to limit membership to those holding only the degree of bachelier ès sciences. Gradually the membership had already come to include more Frenchmen and chemists of somewhat greater standing, such as Aimé Girard, Alfred Riche, and Charles Barreswil.

The turning point in the life of the young society came in May of 1858, when two substantial changes occurred: Wurtz, the first chemist of renown, was elected member, and the society resolved to start a new journal.[8] These two events were probably connected, but the details are obscure. What remains unclear is precisely when Wurtz resolved to turn the Société Chimique into a professional society that would, inter alia, act as an agent of reform in chemistry, but it is beyond question that this is exactly what he succeeded in doing during the course of the year 1858. International discourse was at the heart of Wurtz's agenda, and the international character of the new society was striking from the beginning. As I have noted, many of Wurtz's foreign students (Couper, Lieben, Frapolli, Butlerov, and others), and his prize (Alsatian) student Charles Friedel, joined the society ahead of him. Whether he intentionally sent these men as an advance column to soften up resistance to transforming the self-study group into a proper professional society, or whether he first learned of the group from his students, is not clear, but Wurtz's actions after his entry as member seem well focused toward his goals.

The initial publication proposal of May 1858 was for the society to produce a *"Revue chimique"* that "would contain only translations of foreign work."[9] By June the plan had evolved: there were now to be two companion periodicals: a *Répertoire de chimie pure* edited by Wurtz, and a *Répertoire de chimie appliquée* edited by Barreswil, which would summarize and critique both the foreign and the domestic chemical literature. A *Bulletin de la Société Chimique* was also planned, but until the spring of 1859 this periodical was a simple minutes sheet intended to be attached to the volumes of the two *Répertoires*. In June of 1858, Wurtz wrote to Alexander Williamson, among others, to garner correspondents for his *Répertoire de chimie pure*, and told his English friend of his intention to

8. Pacquot, *Mémorial*, p. 62 (extracts from manuscript procès-verbaux of meetings of the society), meetings of 12 and 29 May 1858.
9. Ibid.

use this platform to gain a better hearing for the new chemistry.[10] In July, Wurtz's friend Girard was elected president (the first French president of the society), and a new meeting site was established, in Wurtz's workplace at the Faculté de Médecine.[11]

On 10 November 1858, in the second meeting of the new academic year, meetings were changed from weekly to every second and fourth Tuesday of the month,[12] and atomic weights began to be used in organic chemical formulas in the *Bulletin* (perhaps for the first time in France).[13] At the meeting of 28 December, the appointed time for electing officers for the coming calendar year, there was a confrontation between two factions of the society: those who wished to retain the original limited functions of the group (apparently led by Arnaudon), and those (led by Wurtz and his allies) who wished to make the society, in effect, a national professional body. The vote was 36 to 16 in favor of allowing the society to "expand the circle of its activities." Girard then proposed to make Jean-Baptiste Dumas at once a membre titulaire and the president of the society; he declared that Dumas had given his approval for this in advance. The motion was approved by acclamation. Among the other elected officers for calendar year 1859 were Louis Pasteur and Auguste Cahours as vice-presidents and Wurtz as secretary (in charge of publications).[14]

In effect, the Société Chimique, now boasting 100 members, had just become France's national chemical society, analogous to the Chemical Society of London (founded 17 years earlier). There is no reason to doubt

10. Wurtz to Williamson, 19 June 1858, Harris Collection, Bloomsbury Science Library, University College London. By 1859 Wurtz's *Répertoire* had 500 subscribers, a goodly number for this type of periodical (Wurtz to Butlerov, 15 July 1859, quoted on p. 117 of G. Bykov and J. Jacques, "Deux pionniers de la chimie moderne, Adolphe Wurtz et Alexandre M. Boutlerov, d'après une correspondance inédite," *Revue d'histoire des sciences* 13 (1960)).

11. P. de Clermont to Butlerov, 24 July 1858, and Rosing to Butlerov, 12 July 1858, in Bykov and Jacques,"Deux pionniers," pp. 1206–1208.

12. Three months later, the meeting schedule was changed to every second and fourth Friday evening of the month. Extraordinary sessions could then be held on alternate Fridays.

13. *Bulletin de la Société Chimique* 1, pp. 5, 10, 13, 17, 19–20, etc. (sessions from 20 November 1858 to 25 February 1859).

14. Ibid., pp. 6–7 (meeting of 28 December 1858).

that the maneuverings of Adolphe Wurtz were behind these events. Wurtz wrote in 1858:

Little by little the spirit of the society has changed: originally simply a means of pursuing instruction, it has risen by degrees to a more important rank. As it is constituted today, its goal is to unite into a scientific association chemists not only in France, but also abroad. . . . [The Society] knows but one school, that of Progress.[15]

Further changes were registered in the spring of 1859, when the *Bulletin de la Société Chimique*, which had gradually been increasing in size, was transformed from a record of minutes into a true society "transactions"—that is, an organ for publication of scientific articles. The *Bulletin* continued to grow in the early 1860s, gradually usurping the functions of the *Répertoire de chimie pure et appliquée*; in 1862 Wurtz's journal stopped publication, and 2 years later Barreswil's had also been absorbed into the *Bulletin*, which alone continued in existence. An outlet for longer publications was also soon created, in the form of a series of occasional invited lectures, published in an annual collection under the imprint *Leçons de chimie professées en 1860 [etc.] à la Société Chimique de Paris*. These special lectures began in January of 1860, and continued in this form for 10 years.[16]

A new version of the society's statutes also appeared in the spring of 1859. The maximum membership was now fixed at 500 (set by the government), and the name was thenceforth Société Chimique de Paris. No entrance requirements were set; it was only necessary for a candidate to be proposed by two members and proclaimed by the president at the next session. The society was financed by an entrance fee of 10 francs and annual dues of 36 francs. No distinctions were to be made between members, and foreigners were welcome.[17]

The historical context for the early history of the Société Chimique is nicely framed in an essay by Robert Fox, who argues, inter alia, that growth in the numbers of unestablished scholars and the rapid acceptance of the

15. Wurtz, in first issue of *Répertoire de chimie pure* (October 1858), cited in Bykov and Jacques,"Deux pionniers," p. 1206.

16. The first seven lectures, given during special sessions from January through August of 1860, were presented by Pasteur, Deville, Wurtz, Berthelot, Cahours, and Barral. They were published as *Leçons de chimie professées en 1860* (Hachette, 1861).

17. *Bulletin de la Société Chimique* 1, pp. 67–76 (session of 24 June 1859).

research ideal in France provided favorable conditions for national scientific societies during the 1850s.[18] For present purposes it is also necessary to continue to stress the personal agenda of Adolphe Wurtz. Shut out (at least temporarily) from the Académie, Wurtz created a scholarly society and a journal that would help him exercise a degree of influence in his field.

Wurtz's student Armand Gautier later reminisced how the Société Chimique had run the danger of being perceived as "an entrenched camp of atomists," engendering resentment and rejection by other schools of thought.[19] Certainly Wurtz never tried to exclude the "equivalentist" camp; this would have been counterproductive even if it had been possible, and in any case it would have been very much counter to Wurtz's instincts toward openness and collegiality. Among the younger chemists opposed to atomic weights, Berthelot had become a member of the society in June of 1858, and Deville joined before the end of the year. But the number of Wurtz's allies in the group (his own students; he had little support elsewhere) is remarkable, as is the number of papers using atomic weights and pursuing the reformist agenda. The considerable degree to which Wurtz succeeded in creating a platform for his school while also cultivating openness to alternative views has been well explored by Carneiro and Pigeard.[20]

The scientific material presented on that platform was of course central to Wurtz's campaign, and it is time to turn to this subject.

Glycol, Lactic Acid, and Oxalic Acid

In chapter 6 we followed Wurtz through his two landmark papers of 1855 (one on "mixed" or asymmetric hydrocarbon radicals and one on glycerides as compounds of a "tribasic" glyceryl radical). It is clear from the latter paper (and confirmed by subsequent autobiographical statements) that Wurtz was led to the idea he developed there by Williamson's work on trinitroglycerin and on triethoxymethane. As he later described it, he began almost immediately to wonder about the possibility of an intermediate term between propyl and glyceryl. Both radicals had analogies to water and had

18. Fox, "The Savant Confronts His Peers."

19. Gautier, quoted in Pacquot, *Mémorial*, p. 3.

20. A. Carneiro, "Wurtz," pp. 230–239; A. Carneiro and N. Pigeard, "Chimistes alsaciens à Paris au 19ème siècle," *Annals of Science* 54 (1997), p. 543.

identical carbon content, but the former was monobasic (like propyl alcohol) whereas the latter was tribasic (like glycerin). Might it not be possible to construct a dibasic analogue? On 24 March 1856, after a number of unsuccessful tries, Wurtz treated ethylene iodide (a compound first prepared by Michael Faraday) with silver nitrate, then hydrolyzed the product with potash solution. What he produced was a derivative of ethylene that was analogous to two molecules of water condensed into a single compound—a double alcohol, analogous to the triple alcohol glycerin. He announced this new reaction and new substance to the Académie on 28 July. Because the new compound was intermediate between glycerin and alcohol, he named it glycol.[21]

Just as he had described glycerin as three water molecules tied together by the glyceryl radical, Wurtz viewed ethylene as the material bond holding together the oxygens and hydrogens of the alcoholic functions. Since these functions were not really acidic in character, he decided that "polybasicity" was an inappropriate term. He now began to use a new locution, picked up proximately from Berthelot and ultimately from M. A. Gaudin: "polyatomicity." Ethyl alcohol was a monatomic, glycol a diatomic, and glyceryl a triatomic compound. This was intended to convey the theoretical notion that the three substances were formed schematically from one, two, and three molecules of condensed water, and the experimental notion that they had one, two, and three alcohol functions, respectively.

Over the next months the chemical public was treated to a dozen papers by Wurtz, published at an interval of about one every 6 weeks. Having made glycol from ethylene, Wurtz could then prepare its various esters, ethyl and methyl ethers, and other obvious derivatives. He could also play the homology game: take a substance homologous to ethylene and treat it in the same way—which then provided, moreover, a new substance to derivatize. Wurtz succeeded thereby in synthesizing glycols from propylene, butylene, and amylene, and their derivatives. He showed that ethylene glycol chlorohydrin was identical to the "Dutch oil" (ethylene chloride). He

21. Wurtz, "Sur le glycol ou alcool diatomique," *Comptes rendus* 43 (1856): 199–204; "Mémoire sur les glycols ou alcools diatomiques," *Annales de chimie* [3] 55 (1859), pp. 401, 468; "Histoire générale des glycols," in *Leçons de chimie professées en 1860* (Hachette, 1861), pp. 106–108. Wurtz wrote Liebig on 31 July to tell him of the discovery, being careful to express all formulas in dualistic equivalent formulas (Liebigiana IIB, Bayerische Staatsbibliothek).

also showed that glycerin could be synthesized from propylene bromide (by bromination, acetylation, then hydrolysis). In this way the discovery of glycol proved to be a treasure trove of new reactions (and thus of publications) for Wurtz. Hofmann later called this extended investigation "experimental research that is exemplary for all time."[22] In another memorable phrase, he said that Wurtz "conjures a galaxy of new compounds before the mental eye of chemists."[23]

Similar to the multiple-alkylation strategy pursued by Hofmann and Wurtz around 1850, this sort of follow-through on a major discovery is highly productive of new compounds (and published papers), but often adds little of theoretical interest. However, Wurtz's glycol work also led him to significant new results. He found that glycol could be gently oxidized by platinum black to a substance that had both alcohol and acid character, or more vigorously oxidized to form oxalic acid. The product of gentle oxidation, glycolic acid, was similar to—in fact was a homologue of—the familiar substance lactic acid.[24]

Wurtz immediately realized that this reaction route allowed him to decide a question that had divided chemists for many years: the true formula for oxalic acid. Since glycol was derived from ethylene and ethylene from ordinary alcohol, and since no one had ever doubted that alcohol contained four carbon equivalents (two atoms), Wurtz argued that he now had "certain proof" that oxalic acid was the same carbon-size as alcohol.

22. Wurtz, "Recherches sur l'acétal et sur les glycols," *Comptes rendus* 43 (1856): 478–481; "Recherches sur l'acétal," *Annales de chimie* [3] 48 (1856): 370–376; "Note sur l'aldéhyde et sur le chlorure d'acétyle," *Annales de chimie* [3] 49 (1857): 58–62; "Sur la formation artificielle de la glycérine," *Comptes rendus* 44 (1857): 780–782; "Note sur la liqueur des Hollandais,"*Comptes rendus* 45 (1857): 228–230; "Note sur la formation artificielle de la glycérine," *Comptes rendus* 45 (1857): 248–250; "Sur la propylglycol," *Comptes rendus* 45 (1857): 306–309; "Sur quelques bromures d'hydrogène carbonés," *Annales de chimie* [3] 51 (1857): 84–94; "Sur la formation artificielle de la glycérine," *Annales de chimie* [3] 51 (1857): 94–101; "Sur l'amylglycol," *Comptes rendus* 46 (1858): 244–246. This is just the beginning of Wurtz's work on glycols and their derivatives, and it includes none of the work of his students on the same subject. Hofmann's comment is on p. 346 of his biography of Wurtz.

23. Cited in A. Bauer, "Erinnerungen," *Oesterreichische Chemiker-Zeitung* 22 (1919), p. 117.

24. Wurtz, "Mémoire sur la constitution et sur la vrai formule de l'acide oxalique," *Comptes rendus* 44 (1857): 1306–1310.

This meant, in turn, that it could no longer be doubted that oxalic acid was dibasic. Thus one more of the issues separating the older chemistry from the newer was decided in the latter's favor. Wurtz was directing this argument at Hermann Kolbe, who continued to insist on retaining an updated form of the classic Berzelian acid formulas. According to Kolbe, Wurtz's glycol was not a dialcohol but the hydrate of an oxide; his further oxidation of this oxide, Kolbe averred, resulted in an intramolecular rearrangement to glycolic acid.

I have explored the issues separating Wurtz and Kolbe on the polyfunctional acids and alcohols elsewhere[25]; here I will only summarize their multi-year polemic. Kolbe continued to believe in isolable monomeric hydrocarbon radicals, in water molecules in acids and oxides in salts, in electrochemical-dualist formulas, and in equivalent weights as the true weights of the chemical atoms. He took the writing of chemical formulas with great seriousness of purpose, in the conviction that he could establish and express the true atomic constitutions of molecules. But, by the same token, he adamantly refused to countenance any claim that the spatial orientation or arrangement of the atoms within the molecule could ever be determined, or that atoms could ever form directed bonds to other atoms. What held molecules together, in Kolbe's firm opinion, was coulombic forces exerted isotropically between radicals.

These theoretical ideas led Kolbe to many specific disagreements with Wurtz. Contra Wurtz, the glycols were not dialcohols, but rather oxide hydrates. Contra Wurtz, glycolic and lactic acids were not hydroxy-acids at all, but ordinary acetic and propionic acids (respectively) with an extra hydrated oxide function. All the organic acids could be considered schematic derivatives of oxalic acid, rather than formic acid as Wurtz had been insisting for years. Kolbe was able to score points on Wurtz by effectively demonstrating that glycolic and lactic acids were not dibasic, as Wurtz had initially argued. Thereafter Wurtz stated that these acids were monobasic but diatomic (as opposed to glycol itself, which was diatomic but not acidic at all, or oxalic acid, which was both diatomic and dibasic).

25. Rocke, *Quiet Revolution*, pp. 193–196, 214–230. The events summarized in this and the next two paragraphs took place between 1857 and 1862. The polemic took place entirely in the form of published papers (for which see the above citation); there was essentially no correspondence between Kolbe and Wurtz.

Wurtz was pushing hard in the years around 1860 to obtain a hearing from his countrymen; the newer chemistry, he emphasized time and again, made all these polyfunctional compounds simple to understand. As Wurtz plaintively (and perceptively) noted to Liebig, his arguments were more heeded in England and Germany than in France. In 1861, Wurtz estimated the number of French chemists "who follow the progress of science and who keep current with new ideas" as fewer than 20.[26] Two years later, he was shocked that Dumas opposed the nomination of Williamson as corresponding member of the Académie out of hostility toward reformist chemistry. He wrote to Williamson: "M. Dumas is clearly a man of high ability and he has exercised a great influence on the science. But one cannot remain outside the scientific movement for 15 years with impunity."[27] At the same time as this was playing out in France, Kolbe in Germany was pushing hard to retain a traditional orientation to chemical theory, but his countrymen seemed to heed just as little as Wurtz's did.

Three years after his initial discovery of glycol, Wurtz summarized his work and explored the theoretical consequences in a long "Mémoire sur les glycols" in the *Annales de chimie* and in an invited lecture to the Société Chimique.[28] He depicted his 1856 discovery as a logical outgrowth of the theory of polyatomic radicals, and as an important stage in its development. Williamson, "whose work has had such a large part in the development of this theory," had taken the crucial step in 1851 of depicting radicals as providing a material link between the parts of the molecule; the number of links that the radical provided determined its "atomicity." Kekulé's "extremely

26. "Votre lettre . . . me prouve . . . que vous êtes au nombre des personnes qui suivent les progrès de la science et qui se tiennent au courant des nouvelles idées. Et ce nombre est bien restreint, je ne crois pas qu'il y en est 20 en France." (Wurtz to Scheurer-Kestner, 24 March 1861, Scheurer-Kestner papers, Bibliothèque Nationale et Universitaire de Strasbourg, ms. 5983, ff. 434–435)

27. "Et croiriez-vous qu'un des plus hostiles aux nouvelles idées est M. Dumas? M. Dumas est évidemment un homme d'une haute capacité et qui a exercé une grande influence sur la science. Mais on ne reste pas impunément au dehors du mouvement scientifique pendant 15 ans." (Wurtz to Williamson, 22 [?] May 1863, Harris Collection, Bloomsbury Science Library, University College London, ms. add. 356.)

28. Wurtz, "Mémoire sur les glycols ou alcools diatomiques" (read 3 January 1859, published in April); "Histoire générale des glycols" (read 2 March 1860, published early 1861).

important theoretical article" of 1858 applied the same reasoning to the atoms themselves. But his own discovery of glycol was a crucial stage, Wurtz averred. To be sure, the mere fact of the existence of this compound was not highly significant, but it had fulfilled a theoretical prediction based upon the theory, thereby verifying it. In sum, the "germ" of the theory of polyatomic radicals was established before glycol was known, but the discovery of 1856 made the theory "important and fruitful" by providing a "hitherto vague and unsupported hypothesis" with factual support.[29]

Wurtz also attempted to clarify his views on the relationship between theory and experiment, and his interpretation of chemical formulas. Theory was "the foundation and the goal of science," but it must be fully and carefully founded on experiment. Chemical formulas were a form of theory. A chemical formula did not represent the true spatial arrangements of the atoms; it merely expressed the experimental data provided by a number of different reactions. For example,

$$C^2H^4$$
$$O^2$$
$$H^2$$

(one form of Wurtz's formula for glycol) simply expressed the experimental result that there were two different kinds of hydrogen atoms in the compound: one kind closely associated with the oxygen atoms, the other with the carbon atoms. "It is seen," Wurtz concluded, "that I do not give the formula [for glycol] as the expression of a demonstrated truth; I give it as a hypothesis, but as a hypothesis that is useful and convenient, in the sense that it singularly facilitates the interpretation of the numerous reactions that glycol undergoes."[30]

It was in these papers that Wurtz first consistently adopted Gerhardt's atomic weights. Just weeks before his "Mémoire sur les glycols" was read

29. Wurtz, "Histoire générale," pp. 105, 109, 137–139; "Mémoire sur les glycols," pp. 401, 468–475. See also his similar statement in "Sur l'oxyde d'éthylène," *Comptes rendus* 48 (1859), p. 104, which appeared a week after "Mémoire" was read.

30. "Histoire générale," pp. 110, 139; "Mémoire sur les glycols," pp. 463 and 475. By examining differences between the manuscript and the final version of the latter paper, Natalie Pigeard has demonstrated the powerfully theoretical orientation of Wurtz: see Pigeard, *L'Oeuvre du chimiste Charles Adolphe Wurtz (1817–1884)* (thèse de maîtrise, Université Paris X Nanterre, 1993), pp. 49–54.

to the Académie (3 January 1859), Wurtz had begun to publish, in the milieu of the new Société Chimique, both his *Répertoire de chimie pure* and the first skeleton issues of the *Bulletin* (November 1858), which also used atomic weights. In addition to adopting the new weights, Wurtz also began using the symbols pioneered by Williamson (loosely modeled ultimately after those of Berzelius): barred C and barred O, symbolizing doubled weights for these two central elements in organic formulas. Wurtz also relinquished, finally and for good, references to the letters of his formulas as "equivalents." Each letter was now consistently an "atome," and the entire formula a "molécule."

In short, from the fall of 1858 on, Wurtz was by every measure a fully committed disciple of the Williamsonian (mechanist-realist) sect of atomism. Indeed, all indications suggest that he had adopted all elements of the sect as early as 1854, simply retaining for four more years the more customary semiotic and linguistic conventions. But now he had openly declared himself, and his fate was cast with the reformers.

The Conference at Karlsruhe

Wurtz had waged the first 5 years of his campaign by mounting a vigorous research program, and by creating a platform for himself and his students in the form of the Société Chimique and its journals. What about a more explicit and direct thrust: an international conference to provide focused attention on the issues that still separated chemists? The first international meeting of professional chemists, which took place in the southwest German city of Karlsruhe in 1860, is most closely associated with the name of August Kekulé, the local host Karl Weltzien, and the Italian theorist Stanislao Cannizzaro, but in fact it was a collaborative undertaking by several members of the "modern" chemical school, including Wurtz.[31]

Wurtz and Kekulé had been acquainted ever since the latter's 10-month stay in Paris in 1851–52. Kekulé heard Wurtz's lectures in the Faculté de Médecine (Wurtz was then suppléant for Dumas) as well as Dumas's at the Sorbonne, and he apparently visited all three principal private labs then

---

31. The most recent and most interesting study of this conference, which can also be used as a guide to the earlier secondary literature, is B. Bensaude-Vincent, "Karlsruhe, septembre 1860," *Relations internationales* 62 (1990): 149–169.

operating in Paris, those of Gerhardt, Pelouze, and Wurtz. When he was a candidate for a position at the Zurich Polytechnic in 1854, he procured letters of reference from all three of these men.[32] I have already argued that Wurtz's July 1855 paper had a pivotal influence on the genesis of Kekulé's structure theory of 1857–58.

In November of 1858 Wurtz sent Kekulé the first two issues of his new *Répertoire de chimie pure*, and in a follow-up letter he inquired "whether you still intend to be one of our correspondents."[33] In his reply, Kekulé agreed to be a collaborator, but took the occasion to respond to a passage in Wurtz's review of Kekulé's second paper on structure theory, printed in the first issue of the *Répertoire*. Wurtz had reviewed Kekulé's ideas extremely favorably, but commented that the concept of atomicity of elements was not introduced into science by Kekulé. "I do not think I deceive myself," Wurtz wrote, "that my recent work on the synthesis of polyatomic alcohols gave this idea the experimental confirmation it had lacked. The existence of polyatomic radicals in organic chemistry which I demonstrated gave solid support to the idea of polyatomic elements."[34] In a friendly but carefully worded passage of several paragraphs, Kekulé reviewed the history of the concept of valence, pointing out that his 1854 assertion that the sulfur atom was "bibasic" preceded Wurtz's discussion of the "tribasicity" of nitrogen and phosphorus by at least a year. Wurtz concurred, apologized, and promised to print a retraction—which he did a few months later.[35]

Two months after this matter was privately settled between them, in May of 1859, Kekulé visited Paris for 5 days and was, as he wrote to a friend, "much with Wurtz (terrific fellow!—it isn't possible for two people to agree

32. Kekulé's papers, Wurtz wrote, already "lui assurent une place distinguée parmi les jeunes chimistes" (letter of reference, 15 November 1854, August-Kekulé-Sammlung, Institut für Organische Chemie, Technische Hochschule, Darmstadt).

33. Wurtz to Kekulé, 9 December 1858, August-Kekulé-Sammlung; printed in R. Anschütz, *August Kekulé* (Verlag Chemie, 1929), vol. 1, pp. 145–146. The wording ("si vous êtes toujours dans l'intention d'être un de nos correspondants") suggests lost correspondence in which the question had already once been discussed.

34. Wurtz, *Répertoire de chimie pure* (1858), vol. 1, p. 24n.

35. Kekulé to Wurtz, 15 February 1859; Wurtz to Kekulé, 7 March 1859. Both letters are in the Kekulé-Sammlung, and both are printed in extenso in Anschütz, *Kekulé* (vol. 1, pp. 146–148). Wurtz's correction appeared in his "Mémoire sur les glycols" (p. 470n.).

more closely than we do on the general conception of a science)." It was on this occasion with Wurtz that Kekulé first shared his thoughts on an international meeting. Wurtz was very enthusiastic ("er schwärmte so zu sagen für die Sache"), which pleased Kekulé, for he regarded the Frenchman as "the most important personality" to ensure the success of the project.[36] In June the first installment of Kekulé's soon-to-be-famous textbook appeared, in which the definitive version of the initial stage of structure theory was treated, and he sent Wurtz a copy. Wurtz replied with high compliments, reaffirming his view of the importance of structure theory and of Kekulé's originality in its development, and suggesting that a French translation was called for. He added a poignant personal lament:

> In my own country the true meaning (not to mention importance) of my work on glycols has never been recognized. My work has never even been honored by a Rapport at the Institute. I console myself with the thought that the most competent people in the science (you among them) recognize the part it has played in the development of modern chemistry.[37]

Having broached the idea with Wurtz, in October of 1859 Kekulé met with Karl Weltzien, professor of chemistry at the Technische Hochschule in Karlsruhe, to begin organizing the conference. They planned the meeting for Karlsruhe, a lovely city in a prized natural area (the Black Forest) and, moreover, right on the French border, just a few miles from Strasbourg. Weltzien began writing letters early the next year, and a second meeting of the principals—Kekulé, Wurtz, and Weltzien, along with Adolf Baeyer and Henry Roscoe as well—took place in Paris in late March of 1860. The conspirators had to be careful not to appear to be stacking the deck or preparing to issue binding edicts on world chemistry. Instead, they advertised the upcoming congress as an opportunity for a large number of leading chemists to discuss pressing theoretical ideas and to confer on such conventional questions as terminology, nomenclature, and notation.

As we have seen, by the 1850s a distinct split in the chemical community had developed between older chemists who had taken part in the "theory

---

36. See Kekulé to Erlenmeyer, 16 June 1859, and Kekulé to Weltzien, 14 March 1860, Kekulé-Sammlung (Anschütz, *Kekulé*, vol. 1, pp. 152 and 183).

37. Wurtz to Kekulé, 21 July 1859, Kekulé-Sammlung (Anschütz, *Kekulé*, vol. 1, p. 159). With this letter, his habitual salutation changed from "Mon cher Monsieur Kékulé" to "Mon cher ami et collègue." At the same time, Kekulé's changed from "Werthester Herr Wurtz" to "Werthester Freund."

wars" of the 1830s and the 1840s and then removed themselves from the fray by embracing a variety of positivisms and/or applied fields, and younger chemists who were nearly uniformly taken by the recent Gerhardtian ideas. Kekulé repeatedly referred to this situation in his correspondence. Attendance by the older chemists—Liebig, Dumas, Bunsen, and others—would be necessary to give the congress the required gravitas, but the real work would be done by the younger attendees. Fortunately, enough of the older generation did attend to make the event successful, and Kekulé's prediction was fulfilled.

Nearly 140 people arrived in Karlsruhe for the opening day of the conference on 3 September 1860. The French chemists present included Wurtz, Dumas, Balard, Béchamp, Boussingault, Friedel, Gautier, Le Canu, Nicklès, Oppermann, Persoz, Riche, Scheurer-Kestner, Schützenberger, Paul Thenard (son of Baron L. J. Thenard), Verdet, and several others. Of these, Friedel, Gautier, Scheurer-Kestner, and Schützenberger were students or former students of Wurtz; and of former Wurtz students coming from other countries there were Beilstein, Lieben, Luna, Savich, and Shishkov.[38] One-sixth of the attendees were French (and more than a third of the French came from Alsace and Lorraine). Nearly half were German, which says something about the shifts going on in European chemistry; nonetheless, the oral proceedings were carried on in French.[39] Kekulé and Wurtz had, in fact, plenty of fellow Gerhardtians at the meeting, but this was not always evident from the proceedings.

It appears to have been an exciting but difficult and confusing 3 days. A crucial situation arose when the steering committee came to consider the question of changing notation. Adolf Strecker, a respected mid-career organic chemist just hired at Tübingen—a Gerhardtian, and a friend of Wurtz from his Giessen days—"proposed [that the body] adopt in principle the atomistic notation." Dumas, who was presiding over that meeting of the steering committee, responded by "insisting forcefully on the disadvantages of the current confusion." In seconding (and twisting) Strecker's

38. In addition to these names, Beketov, Bussy, Cahours, Deville, Frémy, Malaguti, Payen, Peligot, Pelouze, and Regnault signed the advance circular, but they failed to attend. Of the very few French chemists who dissociated themselves from the congress, Berthelot was the most significant name.

39. Bensaude-Vincent, "Karlsruhe," pp. 156–159.

proposal, Dumas strongly advised returning to the "universally recognized authority" of the atomic weight system of Berzelius. This put Wurtz in a difficult tactical position. He responded:

M. Wurtz is happy to see that M. Dumas has placed the matter in its true light, and thinks that it is necessary to return to the principle of atomic weights and to the Berzelian notation. Insubstantial modifications in the interpretation of certain facts would suffice, in the speaker's opinion, to place the principles of this notation in harmony with the requirements of modern science. The notation which is appropriate to adopt today is not precisely that of Gerhardt. Gerhardt rendered immense services to science. Today he is dead, and his name, the speaker stated, must be pronounced only with respect. But it seems that this chemist committed two errors. One is concerned only with form, the other is more inherently fundamental.[40]

Gerhardt's first mistake, according to Wurtz, was an error of rhetoric, tact, and strategy: he presented his atomic-weight proposal of 1842 as entirely original to him, rather than as a slight modification of the Berzelian system. His more substantial mistake was to have subsumed virtually all low-oxidation metal oxide formulas to the pattern M_2O.

It is difficult to disagree with Wurtz's analysis. Berzelius's atomic weights of 1826—his final revision of his atomistic system—were identical to Gerhardt's system of 16 years later, with the single exception that Berzelius took the general low-oxidation metal oxide formula to be MO. Neither Berzelius nor Gerhardt felt that the strong chemical analogies between most metal oxides could be ignored, so both men were convinced that all must be subsumed within a single pattern formula; they differed only on what that one formula should be. Both men recognized that the cost of this procedure was the generation of certain anomalies with respect to the weights of certain metals, as determined by physical data such as isomorphism and atomic heats; but both men were chemists rather than physicists, and chemical analogy naturally took pride of place for them over physical measurements or physical laws.

Wurtz, not one to commit any error of form or tact, was making veiled reference here to a modification of the Gerhardtian orthodoxy which he and many other "modernists" had adopted: the assumption of the M_2O formula for alkali metals such as sodium and potassium but the MO formula for the alkaline earth metals such as calcium and magnesium. The loss

40. Anschütz, *Kekulé*, vol. 1, pp. 680–681 (from Wurtz's French-language minutes—see below).

of parallelism in chemical analogy, these men felt, was more than compensated for by the gain in consistency across all methods of determining atomic weight. In particular, it was Mitscherlich's law of isomorphism and Petit and Dulong's law of atomic heats that dictated the diverse oxide formulas; the loss of parallelism was an issue, but once these formulas were posited a beautiful consistency emerged. Such a modification to atomic weights had first been proposed by two men not particularly associated with the reforms: Victor Regnault and Heinrich Rose.[41]

In the final plenary session of the congress Cannizzaro took the floor and spoke for some time in favor of the system of Gerhardt. This speech appears to have impressed all observers with its coherence and profundity. According to the minutes of the speech:

> An ardent defender of the unitary system, [Cannizzaro] would not want to see the notation of Berzelius retained; he would adopt in its entirety that of Gerhardt. A move that would consist in modifying the dualistic system to introduce into it a portion of the unitary system is unacceptable to him; this would be to require chemists to march backwards. . . . The eloquent professor from Genoa then discussed the fundamental ideas of the unitary system; in his remarkable pleading in favor of the theories of Gerhardt, he was at pains to demonstrate that it is impossible, in the current state of the science, to adopt any notation other than that of the unitary school.[42]

All one need do, Cannizzaro affirmed, is double Gerhardt's atomic weights for the alkaline earth metals—as Regnault and Rose had already suggested—to produce a perfectly consistent system. After Cannizzaro's speech, Strecker rose to second Cannizzaro's sentiments and to object to the wording of the resolution as modified by Dumas. The original resolution had contained the name Gerhardt; this had been changed to Berzelius on a majority vote of the steering committee, but inappropriately so, in Strecker's opinion. A number

---

41. Regnault, "Recherches sur la chaleur spécifique des corps simples et composés," *Annales de chimie* [2] 73 (1840), pp. 61–64, 69 (using, in part, H. Rose's work on silver and copper); Regnault, "Note sur la chaleur spécifique du potassium," *Annales de chimie* [3] 26 (1849): 261–267; H. Rose, "Ueber die Atomgewichte der einfachen Körper," *Annalen der Physik* 100 (1857): 270–291, esp. 281, 291 (referring to Regnault's work).

42. "Congrès des chimistes à Carlsruhe," *Moniteur scientifique* 2 (15 October 1860), p. 986. Cannizzaro's speech is reproduced nearly verbatim in Anschütz, *Kekulé* (vol. 1, pp. 682–686). See also "The Congress of Chemists at Carlsruhe," *Chemical News* 2 (6 and 20 October 1860): 203, 226–227. Both of these articles followed the reports of Louis Grandeau of the *Revue germanique*.

of comments on a number of related points followed.[43] In the end, the only resolutions on which the body could decide to vote were to adopt Williamson's barred O, C, and S symbols for the larger (atomic) weights and to declare by acclamation that "the concept of equivalents is empirical and independent of the concept of atom and molecule."[44]

The organizers seem to have been disappointed by the lack of concrete results. After the meeting, Wurtz volunteered to prepare for publication a detailed report of the proceedings. For this purpose he solicited and received retrospective summaries of speeches by the three chemists, besides himself, who had taken the most active part in the discussions: Kekulé, Dumas, and Cannizzaro.[45] Cannizzaro's speech in favor of the system of Gerhardt (and, ultimately, in favor of the ideas of Ampère and of Cannizzaro's countryman Avogadro) occupies no less than a third of the French-language draft report. Wurtz was well familiar with Cannizzaro's views even before the meeting, for he had reviewed the *Sketch of a Course in Chemical Philosophy* in his *Répertoire de chimie pure*, and his treatment of Cannizzaro's speech suggests his high regard for the arguments therein. After the conference ended, Cannizzaro's friend (and one of the founding members of the Société Chimique) Angelo Pavesi passed out copies of this pamphlet, which subsequently made a great impression on more than one former attendee.[46]

But Wurtz was dissatisfied with his draft compte rendu—it is in fact uneven in quality and in level of detail—and the report remained unpublished during Wurtz's lifetime.[47] It is fortunate that the compte rendu, from which much of the preceding descriptions were taken, was saved and eventually published. But the report can be supplemented by reference to eye-

43. Anschütz, *Kekulé*, vol. 1, pp. 686–688.

44. The second of these resolutions is mentioned in only a single source: "Congrès," p. 985.

45. Wurtz to Kekulé, 16 October 1860, August-Kekulé-Sammlung.

46. Cannizzaro's *Sunto di un corso di filosofia chimica* (1858) has often been reprinted and translated. The most recent Italian edition is by L. Cerruti (Sellerio, 1991). The Alembic Club English translation (1910) is reprinted in M. Nye, *The Question of the Atom* (Tomash, 1984). Wurtz's review appeared in his *Répertoire* (vol. 1, pp. 201–205).

47. Wurtz to Kekulé, 6 July 1861, August-Kekulé-Sammlung. There is no further mention of this draft in Wurtz's correspondence.

witness testimony, reports in the *Chemical News* and *Moniteur scientifique*, as well as by published and unpublished correspondence after the meeting between Kekulé, Weltzien, Wurtz, and Lothar Meyer.[48]

Not only was there no published compte rendu from the congress, but there were other planned follow-up events which likewise failed to happen. In closing the final plenary session, Dumas made reference to the idea of a similar meeting in the coming year. This sentiment was paralleled in a curious passage written anonymously. The real purpose of the Karlsruhe Congress, this French writer asserted, was to demonstrate to "Russian, German, English, and even Swedish" chemists that "our French chemists walk with giant steps, which can only be followed." Chemistry was maturing and becoming powerful; thus, it was time

to lay the foundations of a powerful association, to name its leaders, and to compose a charter while waiting for its laws. Three sessions [at Karlsruhe] sufficed, not to establish all of this, but to settle on a watchword for the next meeting. This winter the Société Chimique de Paris will elaborate the plan, and, during the next vacation, the foundations of a true science will be laid.[49]

The meaning of this passage is by no means clear, but in any case there is no evidence that the Société Chimique, or Wurtz as its principal leader and agent, did anything close to what was being suggested here. Although as late as the summer of 1861 Wurtz was expecting both future congresses and a printed compte rendu of the first, neither came to pass. Wurtz and Kekulé, disappointed in the results of Karlsruhe, had both lost interest.[50]

48. Wurtz's compte rendu was first printed in K. Engler, *Festgabe zum Jubiläum der vierzigjährigen Regierung seiner Königlichen Hoheit des Grossherzogs Friedrich von Baden* (Karlsruhe, 1892); a verbatim transcription of French manuscript was republished in Anschütz, *Kekulé* (vol. 1, pp. 671–688). Nye's book *The Question of the Atom* includes both the original French text (from Anschütz) and an English translation (pp. 3–30 and 633–650). Much correspondence about the meeting is reproduced and discussed in Anschütz, *Kekulé* (vol. 1, pp. 183–209), in Engler, *Festgabe*, and in A. Stock, *Der internationale Chemiker-Kongress . . . vor und hinter den Kulissen* (Verlag Chemie, 1933). See also Bensaude-Vincent, "Karlsruhe."

49. "Congrès," p. 984.

50. Wurtz to Butlerov, 14 February 1861, in Bykov and Jacques, "Deux pionniers," p. 120; Wurtz to Kekulé, 6 August 1861, August-Kekulé-Sammlung. Some additional reasons for the sense of failure are given in Bensaude-Vincent, "Karlsruhe."

8

Berthelot

Through most of Wurtz's professional life, his principal Parisian rival was Marcellin Berthelot, a younger man who nonetheless achieved renown about the same time, during the 1850s. There are many interesting contrasts between Wurtz and Berthelot, and any depiction of the former's life and career would be incomplete without a satisfactory understanding of the latter.

Early Career

Berthelot was born in Paris on 25 October 1827.[1] His father, son of a farrier, grew up near Orléans, studied medicine, then became a physician in Paris. Berthelot père, never a wealthy man, continued to support his son financially until the latter was more than 30 years old. Berthelot's mother, Ernestine (née Biard), came from the Parisian bourgeoisie. The family lived on the right bank of the Seine, on a modest street that was later destroyed in the Haussmann projects to make way for the new Hôtel de Ville. "I grew up there," Berthelot later wrote, "surrounded by the love of my family, in the republican tradition, to the roar of the cannon and the staccato of gunfire,

1. Secondary works on Berthelot include the following: E. Jungfleisch, "Notice sur la vie et les travaux de Marcellin Berthelot," *Bulletin de la Société Chimique de France* [4] 13 (1913): i–cclx; C. Graebe, "Marcelin Berthelot," *Berichte der Deutschen Chemischen Gesellschaft* 41 (1908): 4805–4872; L. Velluz, *Vie de Berthelot* (Plon, 1964); R. Virtanen, *Marcelin Berthelot* (University of Nebraska Studies, no. 31, 1965); J. Jacques, *Berthelot, 1827–1907* (Belin, 1987); J. Dhombres and B. Javault, eds., *Marcelin Berthelot* (Société Française d'Histoire des Sciences et des Techniques, 1992). On his birth certificate Berthelot's first name is spelled "Marcelin," but Berthelot preferred and normally used "Marcellin."

in the midst of the barricades, the insurrections of the reign of Louis-Philippe, the Revolution of 1848, and the June days. In my earliest youth, at the most tender age, my oldest memory is that of the bloody men wounded [in the 1834 insurrection and massacre] at the church of Saint-Merri and Rue Transnonain. They had been brought to be saved by my father, who was for thirty years the physician of the charity bureau and friend of the people."[2] Whether because of these tumultuous events or for some more organic reason, Berthelot lived his life accompanied by a vague sense of physical insecurity and unhappiness. Supremely self-confident, he distrusted most of those around him, and he was uneasy about the future. The historian has no need to indulge in post-facto psychology to arrive at these conclusions; Berthelot himself wrote about his feelings, in both public and private contexts. From the age of 10, he later said, he was "tormented by insecurity."[3]

Berthelot's secondary schooling was at the classical Collège Henri VI. In 1846, in a competition among all the lycée students in France, he won the prix d'honneur in philosophy. He had already become close friends with a fellow student lodging next door to him in the same pension, Ernest Renan. Renan came to know Berthelot's parents (Berthelot père was "the first republican I had ever seen," Renan later wrote; "the apparition astonished me"), and introduced his younger friend to Hebrew and theology; Berthelot instructed Renan in elements of science that were beginning to fascinate him. Renan was then going through a crisis of faith, and their intense conversations led to the obliteration of the remaining traces of Christian faith from both men's souls.[4] Their lifelong friendship had a powerful effect on the intellectual development of both Renan and Berthelot. Under the influence of the historian and philologist Renan and of his classical education at the Collège Henri IV, Berthelot became a leading historian and philosopher of chemistry, as well as an outstanding bench and research chemist. His studies were aided by a superb memory and by iron discipline.

2. Berthelot, *Science et libre pensée* (Calmann-Lévy, 1905; written in 1903), p. 60. The Rue Transnonain was eliminated in the Haussmann era in a deliberate attempt to obliterate the memory of this event.

3. See, e.g., the introduction to his published correspondence with his lifelong friend Ernest Renan.

4. Renan, *Souvenirs d'enfance et de jeunesse* (Calmann-Lévy, 1883), p. 333.

Berthelot received his baccalaureate in letters in 1847 and his science baccalaureate in 1848. Even while an adolescent in secondary school, Berthelot began attending lectures at the Collège de France. It was something of a golden age there: in addition to those of the famous philologist Eugène Burnouf, Berthelot heard lectures by Pelouze, Regnault, Biot, Dumas, and Magendie, and came to know Magendie's préparateur Claude Bernard. Although Berthelot had begun to study medicine seriously, by the fall of 1848 he was captured by physical sciences, especially chemistry. In October of that year he had a private conversation at the Collège de France with Regnault, who predicted a golden age dawning for the science of chemistry and suggested that Berthelot enter Pelouze's private laboratory school—which he did in November.[5]

Pelouze's spacious establishment consisted of a large teaching lab for elementary instruction and a number of smaller rooms where research was done by more advanced students. Alphonse Davanne later told of his experiences in "that little world" where "such gaiety reigned." The small lab space Davanne shared with Aimé Girard doubled as a kitchen, whose stovepipe formed the axis of a small spiral staircase; Berthelot's chamber was directly above. According to Davanne, "Berthelot was still young, but I don't know how he did it, that devil of a man, he already knew everything. Whenever we were at a loss, we rapped on the stovepipe. Berthelot came down, picked up a piece of chalk, and used the stovepipe as a blackboard; the formulas twisted around and rose in spirals to the ceiling."[6] In July of 1849, Berthelot became licencié in physical sciences; in 1850, Pelouze made him his préparateur, at a salary of 600 francs.

Pelouze exerted a strong influence on his Berthelot. Pelouze had arrived in Paris from the provinces in 1825, and had happened to meet Gay-Lussac through a fortuitous encounter on the streets. Gay-Lussac offered him a place in his laboratory at the Arsenal, and it was while he was there that he met Liebig, on the latter's visit in 1828. From that time on he was Liebig's

5. Berthelot, *Science et libre pensée*, pp. 50–52; Velluz, *Vie de Berthelot*, pp. 30–31. On Regnault's curious ambivalence between physics and chemistry, see M. Dörries, "Easy Transit," in *Making Space for Science*, ed. C. Smith and J. Agar (Macmillan, 1998).

6. Interview with Davanne, *La presse*, 29 July 1901, cited in Velluz, *Vie de Berthelot*, p. 180.

good friend, correspondent, and sometime collaborator, at the same time that his career was successfully promoted by Gay-Lussac. Pelouze was a republican, taking part in the insurrections of July 1830. Already a member of the Académie des Sciences in 1837 (at the age of 30), and Thenard's successor at the Collège de France in 1846, Pelouze was able to exert considerable influence in the Parisian chemical world, especially after 1848. He was profoundly oriented toward experimental chemistry, and against theory. When Liebig urged him to avoid a speculative approach, Pelouze, then répétiteur to Gay-Lussac at the École Polytechnique, replied: "The illustrious scientist with whom I have had the good fortune to work thinks exactly as you do. I remember often having heard him say the same thing and I will never forget it."[7]

Pelouze was succeeded at the Collège de France by Antoine Balard early in 1851, and Balard quickly named Berthelot his préparateur. Berthelot remained in this position for 8 years, by the end of which time he had become world famous for the scientific work he had published during the 1850s. Balard was always enormously proud of and solicitous toward his student; it was Balard, above all, who helped Berthelot make a career in the difficult Parisian world of the early Second Empire. But this is not to diminish Berthelot's own merits. In addition to true scientific brilliance, marked especially by a self-confident and original cast of mind, Berthelot excelled by his remarkable stamina and appetite for sustained work. His research productivity was phenomenal, and his writing voluminous (and often prolix). In December of 1859 Berthelot was named to a new professorship of organic chemistry at the École de Pharmacie.

Rivalry with Kekulé and Wurtz

During the 1850s, before gaining his first professional position, Berthelot lived the life of a sort of professional postdoctoral student, supported financially by his father. In August of 1858, still unestablished but by then well known through his published research, Berthelot visited Heidelberg and met Kekulé; indications suggest that neither man impressed the other.

---

7. Pelouze to Liebig, 19 June 1832, Liebigiana IIB, Bayerische Staatsbibliothek: "Le savant illustre auprès duquel j'ai eu le bonheur de travailler pense absolument comme vous. Je me rappelle lui avoir entendre dire souvent la même chose et je ne l'oublierai jamais."

Berthelot was not attracted by the ferment in chemical theory that was so prominent especially in Heidelberg; he also did not think much of the German system of higher education. "I see many things [here in Heidelberg]," Berthelot wrote to Renan, "that show me more clearly, in contrast, the advantages of the French system." He continued:

I chat frequently with the Privatdozenten in science: their lives are far more miserable than those of our young instructors, and they are far more completely lacking the instruments of work. The little independence that they have by compensation is more theoretical than real, for a pauper can hardly be regarded as truly free. As for the professors, I don't know if the direction of 50 students obliged to execute a regular course of 150 precisely identical manipulations is not the same, with respect to neutralizing scientific activity, as the exams and double functions [of French professors]. There is one single advantage here: the absence of the ambitious preoccupations that are the undoing of all of our scholars once they attain maturity. But, by the same token, their horizon [here] is narrow and one can scarcely acquire that lively and general feeling for things that one finds in Paris. Everything, I think, must be viewed both from a near and a far perspective; here I am much more optimistic regarding the French system, whose enormous disadvantages we know so well. . . .[8]

Aspects of the German system that Berthelot did *not* see were its diversity and dynamism. Heidelberg was both typical and atypical of Germany. The rich scientific life was evident, especially among the younger chemists: not only Kekulé, but also Emil Erlenmeyer, Adolph Baeyer, Lothar Meyer, Leopold von Pebal, Hans Landolt, Henry Roscoe, Heinrich Carius, Angelo Pavesi, the Wurtzians Friedrich Beilstein and Agostino Frapolli, and others were working there in 1858. But without doubt the star was the Director of the Chemical Institute, Robert Bunsen. Unlike many other chair-holders in Germany, Bunsen withheld all material support from his Privatdozenten; he was a man of great kindness and magnanimity, but he regarded this as an important point of principle. Berthelot must have been rather shocked to see Privatdozent Erlenmeyer's shed-cum-laboratory, or Privatdozent Kekulé's improvised laboratory in his lodgings above the flour merchant (which he facetiously called the "Akademie des Mehlhändlers Goos").

The other point to stress in connection with Berthelot's reaction to conditions in Heidelberg is that Berthelot happened to visit Germany at a crucial transition point in its intellectual history. In the 1850s a kind of competitive entrepreneurial fever began to catch hold in the German states

8. Berthelot to Renan, Heidelberg, August 1858, cited in Velluz, *Vie de Berthelot*, pp. 45–46.

Figure 8.1
Marcellin Berthelot, photographed by Nadar in 1887. Source: Photograph
Collection, Deutsches Museum, Munich.

with regard to academic laboratory sciences, especially chemistry. In 1851
Bunsen had been attracted away from the University of Marburg to Breslau
by a promise of a new laboratory institute; the very next year, the authorities in Heidelberg stole him from Breslau by promising an even more substantial institute, which was completed in 1854. That started a chain of
events. Within 20 years, Königsberg, Greifswald, Berlin, Bonn, Leipzig,
Göttingen, and Munich all had fabulous new chemical institutes, with lavish annual operating budgets and salaries to match. None of this, however,
was yet evident in 1858.

When Kekulé visited Paris in May of 1859, he visited Berthelot twice. He wrote to Erlenmeyer that Berthelot was "scarcely recognizable, [he] now has hair on his head again, but in Paris none on his teeth, [and] was in any case very boring."[9] (In the German idiom, to have hair on one's teeth means to be truculent and argumentative.) This private reference was undoubtedly intended to play upon his and Erlenmeyer's memory of Berthelot's visit to Heidelberg, where he must have been newly shorn and where he may have displayed an overbearing manner toward his youthful peers in an attempt to impress the foreigners. When Kekulé visited Paris once more in March of 1860 and asked Berthelot whether his name could be added to the preliminary circular for the Karlsruhe Congress, Berthelot forcefully declined, adding (in Kekulé's disdainful recounting of the conversation to Weltzien):

This is an undertaking of the Liebig-Dumas school. Among the adherents of this school agreement is possible, but not with others. Of everything that we now believe, in ten years nothing will remain, even if we raise it to a kind of law. If we were to conclude a reasonable sort of agreement, this could only happen by returning to the equivalents of Lavoisier, etc. etc. and more such nonsense. Even with all this he did not say that he would under no circumstances come, only that he could under no conditions participate in the call to such a gathering.[10]

The first line of this quotation is puzzling (if Kekulé was relating the conversation accurately), for in 1860 few would have thought that such a thing as a "Liebig-Dumas school" could be said to exist. The remainder sheds some light on what may have been Berthelot's position. Above all, Berthelot was troubled by the excessively theoretical (read: speculative) character of the science at the time; he wanted to return to a period, before the extravagant burst of ideas in the science in the 1830s, when (as he imagined it) a pristine sort of empiricism had reigned. He could not but have been impressed by the powerful experimenticist and empiricist commitment of such mentors as Pelouze and Regnault, as well as (at one remove)

9. ". . . zweimal bei Berthelot (kaum mehr zu kennen, hat jetzt wieder Haar auf dem Kopf, in Paris aber keine auf den Zähnen, war wenigstens sehr langweilig) . . ." (Kekulé to Erlenmeyer, 16 June 1859, Kekulé-Sammlung, Institut für Organische Chemie, Technische Hochschule, Darmstadt). The letter appears in Anschütz, *August Kekulé* (vol. 1, p. 52) with the text between the parentheses elided.

10. Kekulé to Weltzien, 17 April 1860, Kekulé-Sammlung (Anschütz, *Kekulé*, vol. 1, p. 189).

Gay-Lussac. From the beginning of his career he showed no interest in the views of his fellow republican Gerhardt; the few times he used rational formulas, he showed a decided preference for the dualist style of depicting esters as compounds of acid anhydrides with organic oxides.[11]

By the same token, there was indeed, as we have been at pains to stress at various points in this work, considerable similarity in the theories of Dumas and Liebig, before both men turned to more pragmatic concerns around 1840, and Laurent and Gerhardt quite accurately depicted their own work as an outgrowth of both men's ideas. Pelouze, in concert with his mentor Gay-Lussac, had always been more or less inimical to Dumas, and Berthelot was one of the few in Parisian chemistry at this time to succeed without ever tying himself to the prince of French chemistry. Furthermore, Kekulé and Wurtz, leaders of the new school and organizers of Karlsruhe, could both be said to have been under the strong influence of both Dumas and Liebig. Finally, theories of chemical atoms and molecular constitutions—anathema to Berthelot—had always been particularly salient in the ideas of Dumas and Liebig. The man who was most prominently carrying forth these themes in his own country was Adolphe Wurtz, with whom Berthelot already had a vigorous rivalry.

The course of Berthelot's fundamental work on glycerides in 1853–54, which he used as the basis for his doctoral thesis, was related in chapter 6 above. As Wurtz later stated, it was this work, and that of Alexander Williamson, that led him to his important theoretical considerations on the nature of polyatomic radicals, and to his discovery of the glycols in 1855–56. Clearly there was close interplay between the préparateur at the Collège de France and the professor at the Faculté de Médecine during these years, but it does not appear to have been congenial. Berthelot later emphasized that he had been the first to introduce the concept and the term "polyatomic alcohols," and that Wurtz's work was a simple application of his ideas.[12] Wurtz, for his part, regarded Berthelot as merely a predecessor of

11. E.g., in "Recherches sur les éthers," *Comptes rendus* 37 (1853): 855–858. Here Berthelot was succumbing to the natural instinct to regard older theories as more empirical than newer ones.

12. Berthelot, "Sur les alcools polyatomiques," *Comptes rendus* 45 (1857): 175–178. Many years later, in an otherwise generously worded obituary ("Adolphe Wurtz," *Le temps*, 14 May 1884, p. 1), Berthelot wrote that Wurtz's work on glycols was derived from his prior work on glycerin.

the crucial theoretical breakthrough, the notion of polyatomic radicals, and his successful experimental exemplification of them in the glycol series.[13] During the 1850s Berthelot repeatedly likened his three series of glycerides to salts of ortho-, pyro-, and metaphosphoric acids; Wurtz later pointed out that the correct analogy would be to mono-, di-, and tri-salts of orthophosphoric acid.[14]

In his first paper on glycerin, Wurtz made reference to Berthelot's "lovely researches" on glycerides, which had confirmed and further developed the work of Chevreul 40 years earlier. However, he irritated Berthelot by suggesting an alternative interpretation of a reaction that Berthelot had discovered, the conversion of glycerin into allyl iodide.[15] Berthelot responded that Wurtz's explanation was "more elegant and more direct than that to which we have had recourse in our article." "Nonetheless," he continued, "it does not appear to us to be consonant with the real *experimental* nature of the reaction. . . . We deduced an equation representing the reaction, an equation all but one of whose terms have been *determined by direct weighings*."[16]

Berthelot and Wurtz clashed once more in their simultaneous attempts to unravel the isomers created by different means of tribrominating glycerin. In criticizing Wurtz's interpretation, Berthelot suggested that his rival's approach had led him astray; the difficulty, he wrote, "resides precisely in such subtle isomeric relationships, which often remain unrecognized by theories that only view a compound through its formula and assume that the formula reveals all the secrets of its constitution."[17] Wurtz responded by arguing that Berthelot had confused the compounds being discussed. He

13. Wurtz, "Sur quelques points de philosophie chimique," in Société Chimique de Paris, *Leçons professées en 1863* (Hachette, 1864), p. 121.

14. Berthelot, "Sur les combinaisons de la glycérine avec les acides," *Comptes rendus* 38 (1854), pp. 672–673; *Annales de chimie* [3] 41 (1854), p. 319; "Alcools polyatomiques" (1857); Wurtz, "Discours préliminaire," in *Dictionnaire de chimie pure et appliquée* (Hachette, 1868), pp. lxii–lxiii.

15. M. Berthelot and S. de Luca, "Action de l'iodure de phosphore sur la glycérine," *Annales de chimie* [3] 43 (March 1855): 257–283; Wurtz, "Théorie des combinaisons glycériques," *Annales de chimie* [3] 43 (April 1855), pp. 494–495.

16. Berthelot and de Luca, "Remarques sur la formation du propylène iodé," *Annales de chimie* [3] 44 (July 1855): 350–352.

17. Berthelot and de Luca, "Sur les combinaisons formées entre la glycérine et les acides chlorhydrique, bromhydrique, et acétique," *Comptes rendus* 45 (1857): 178–180.

reiterated his theoretical interpretation, suggesting that the isomers were derived from different "condensed types," and added in conclusion that Berthelot's basic theoretical assumptions were "if not incompatible [with], at least very different" from his.[18]

It was no doubt galling to Berthelot that Wurtz's acknowledgment of his prior work was often scanty and grudging; Wurtz always preferred to emphasize his intellectual debt to Williamson instead. Moreover, it was precisely in connection with the interpretation of the glycerides that Wurtz had entered the most hypothetical of paths: speculations on the presumed internal structure of the chemical atoms themselves, published in the austere and august record of the Académie des Sciences. That these speculations were then at the point of generating a fruitful theory of structure (as I argued in chapter 6 above) could not yet have been known, even by Wurtz; even in retrospect, this remained unacknowledged and unappreciated by Berthelot.

Four months after the last public skirmish recounted above, the Académie voted on its next candidate for the section of chemistry. Wurtz was not only disappointed not to be elected, but stunned to find that Berthelot received one more vote than he. Clearly, he had work to do.

Organic Synthesis: Initial Steps

In chemistry, "synthesis" has multiple meanings, of which probably two are most common.[19] One can take the word as a simple antonym of "analysis," indicating any building up of heavier from lighter substances; in modern organic-chemical parlance this often implies the formation of new carbon-carbon bonds. Alternatively, one may use the word to indicate the preparation of a particular organic compound from simpler materials, especially if the compound is a substance found in organic nature and the simpler compounds are inorganic. In the latter instance chemists are accustomed, following Berthelot's coinage, to refer to a "total" synthesis.

Synthesis in the first sense was common in the first half of the nineteenth century; indeed, the earliest organic synthesis in this sense might be the

18. Wurtz, "Note sur la formation artificielle de glycérine," *Comptes rendus* 45 (1857): 248–250.

19. C. Russell, "The Changing Role of Synthesis in Organic Chemistry," *Ambix* 34 (1987): 169–180.

preparation of ether, which dates to the sixteenth century. By the middle of the nineteenth century, chemists were rapidly building a repertoire of synthetic reactions: Kolbe's electrolysis reaction (which was synthetic as well as analytic); Frankland's organometallic routes; the Williamson ether syntheses; the "Wurtz reaction" of 1855; the multiple-alkylation strategy of Hofmann, Wurtz, and others; the profusion of novel glycerides in the hands of Berthelot—all these and many others accumulated, particularly in the period 1845–1855.

Synthesis in the second sense was also nothing new. The old chestnut that Wöhler's preparation of urea (1828) was the first such reaction, and that it destroyed the mystical "vital force" of immortal memory, was long ago shown to be untrue. On the other hand, Wöhler regarded his deed to have been a significant event in the history of organic synthesis, and most of his colleagues readily agreed; there is no need to react against the "Wöhler myth" to the extent of denying it any significance whatever.[20] Vitalism has a long and complicated history. Suffice it to say that by 1828 a majority of chemists were ready to affirm the possibility in principle of artificial preparation of physiologically familiar substances, and that by the 1840s any meaningful form of vitalism in chemistry was long dead.

After Wöhler, the individual most closely associated with total synthesis at the middle of the century was Wöhler's student Hermann Kolbe. In 1845, while still a postdoctoral student, Kolbe had succeeded in a total synthesis of acetic acid (perhaps using "Synthese" for the first time in a chemical sense), and had predicted that artificial preparation of sugar, starch, and a multitude of other organic products was imminent. By 1858 he expected artificial indigo, alizarin, and quinine to hit the market in the near future. Adolf Strecker synthesized alanine and lactic acid in 1850, and Wurtz and Hofmann reproduced naturally occurring amines about the same time. In 1859 Kolbe prepared salicylic acid from phenol and carbonic acid; 2 years later he made formic acid directly from carbonic acid and water, using potassium.[21] In the 1860s this stream of synthetic novelties became a flood.

Into this context stepped the young and ambitious Berthelot. In August of 1851, just months after being appointed préparateur in the Collège de

20. This, at least, is my argument in *The Quiet Revolution* (pp. 239–242).

21. Rocke, *Quiet Revolution*, passim.

France, he published his first synthesis. After conducting alcohol and acetates through hot porcelain tubes, he had isolated some complex organic substances from the reaction mixture, including benzene, naphthalene, and phenol. This was just Berthelot's third paper, and his first in the field of chemistry. His was not, strictly speaking, a new reaction, for others had observed the pyrogenic production of naphthalene from alcohol; moreover, complex aromatic compounds had often been produced from pyrogenic treatment of essential oils. But Berthelot's work signficantly extended this research, and he was careful to draw out the philosophical implications. He argued that, inasmuch as Kolbe had already formed trichloroacetic acid from inorganic materials such as carbon disulfide and Louis Melsens had shown how to transform Kolbe's product into acetic acid, his own pathway from acetic acid to benzene, phenol, and naphthalene constituted the final link in a synthetic route, in principle, from inorganic materials to these complex organic products.[22]

Beginning in May of 1854, Berthelot, having concluded his initial work on glycerides, pursued a concerted synthetic program that occupied him for the next several years.[23] The beginning of this series of papers was modest enough. It had long been known that sulfuric acid dehydrates many organic substances, including alcohol (to form ethylene) and formic acid (to form carbonic oxide). In two related papers published in 1855 Berthelot revealed that these dehydrations could be reversed: sulfuric acid could be used to hydrate ethylene to ordinary alcohol, and propylene could be hydrated to a new propyl alcohol; furthermore, damp potash transformed carbonic oxide gas to potassium formate.[24] In a referee report on the first of these

22. Berthelot, "Action de la chaleur rouge sur l'alcool et sur l'acide acétique," *Comptes rendus* 33 (1851): 210–211; *Annales de chimie* [3] 33 (1851): 295–301.

23. The date of May 1854 as the start of this program, provided later by Berthelot himself (*Annales de chimie* [3] 43: 500–501), is consistent with other details of the chronology. Berthelot stated that the research had a conscious theme from the beginning (ca. early 1850) (*Annales de chimie* [3] 53 (June 1858): 71).

24. Berthelot, "Sur la reproduction de l'alcool par le bicarbure d'hydrogène," *Comptes rendus* 40 (1855): 101–106; [same title], *Annales de chimie* [3] 43 (1855): 385–405; "Transformation de l'oxyde de carbone en acide formique," *Comptes rendus* 41 (1855): 955; "Nouveau procédé pour préparer l'acide formique," *Comptes rendus* 42 (1856): 447–450; "Recherches sur les relations qui existent entre l'oxyde de carbone et l'acide formique," *Annales de chimie* [3] 46 (1856): 477–491.

papers, Thenard, Dumas, and Balard pointed out that Hennel had rehydrated ethylene a number of years ago, and Chancel had first discovered by another route the new propyl alcohol that Berthelot had produced. However, the rapporteurs praised the contribution highly for the novel aspects and the promise posed by the new technique.[25]

The second of these reactions was clearly synthetic in both senses of the term, for formic acid is undeniably organic in every way (the traditional method of producing it was the destructive distillation of ants), whereas the starting material, carbonic oxide, was arguably inorganic. The hydration of olefins to generate alcohols seems less obviously synthetic, although Berthelot referred to it as such—also incorrectly urging his own originality—in the ringing declaration "Cette synthèse n'a pas encore été réalisée."[26]

By this time Berthelot had achieved a few significant organic syntheses, but his contributions were desultory and unconnected; moreover, his ultimate goal was synthesis directly from the inorganic elements. He had not as yet accomplished any such feat himself, and piggybacking on Kolbe's synthesis of alcohol was less than satisfactory for a man like Berthelot. In the summer of 1856 he published a paper that widened his synthetic range enormously, and in a novel way that led back to the inorganic elements. This was the breakthrough for which he had been searching. In brief, Berthelot found that passage through hot copper could transform carbon disulfide and hydrogen sulfide, or carbonic oxide and marsh gas, into a variety of simple hydrocarbons; also, destructive

25. *Comptes rendus* 40 (1855): 222–224. In subsequent years Berthelot affirmed the novelty and originality of this work relative to that of Hennel. For a vigorous defense of Berthelot, see Jungfleisch, "Notice sur la vie et les travaux de Marcellin Berthelot," pp. lvi–lxi; for a thorough dismemberment of his arguments, see Jacques, *Berthelot*, pp. 85–92.

26. *Annales de chimie* [3] 43 (1855): 385. In the same year, Berthelot found a way to synthesize oil of mustard and oil of garlic from glycerin: hydrogen iodide transforms glycerin into allyl iodide, and the latter treated with potassium sulfocyanide yields oil of mustard. He noted that this series of reactions also provides a direct link between mustard oil and the simple hydrocarbon propylene. (M. Berthelot and S. de Luca, "Production artificielle de l'essence de moutarde," *Comptes rendus* 41 (1855): 21–23; *Annales de chimie* [3] 44 (1855): 495–501.) Nikolai Zinin discovered the same route to oil of mustard simultaneously and independently.

distillation of formates and acetates could yield several low-weight unsat-urated hydrocarbons.[27]

By such means Berthelot had developed synthetic routes that started with inorganics and yielded methane, ethylene, propylene, butylene, and amy-lene, as well as the three aromatics he had already synthesized 5 years ear-lier. From the four olefins named he already had shown how to proceed to the respective alcohols, and the alcohols could be oxidized to acids. Such reaction chains could be extended indefinitely, justifying Berthelot's proud summary: ". . . one can regard as a fait accompli the total synthesis of an immense number of organic compounds." This represented the first appear-ance in print of the phrase "total synthesis." A year later Berthelot added the methyl series to the list of compounds susceptible of total synthesis, by demonstrating that methane could be mono-chlorinated, yielding wood alcohol by hydrolysis.[28] Berthelot's pride in his accomplishments—his oppo-nents regarded it as arrogance—is revealed by his language; the last three paragraphs of this paper contain the personal pronoun "je" no fewer than 21 times. In neither of these two papers did Berthelot mention Kolbe, Wöhler, or any other predecessor in the field of organic synthesis.

Organic Chemistry Founded on Synthesis

In June of 1858 Berthelot published a monograph-length paper on organic synthesis in the *Annales de chimie* that summarized and expanded upon all his work in this field over the preceding several years.[29] Berthelot opened with a quotation from the second (and last) French edition of Berzelius's famous textbook, in which the Swede affirmed the existence of a vital force which "lies entirely outside the inorganic elements." This passage, Berthelot noted, had been "written just nine years ago," in 1849. To be sure, he conceded, "syntheses" were not unprecedented. However, the examples that one might cite were "so rare, so isolated, and so unfruitful,

27. Berthelot, "Synthèse des carbures d'hydrogène," *Comptes rendus* 43 (1856): 236–238.

28. Berthelot, "Synthèse de l'esprit-de-bois," *Comptes rendus* 45 (1857): 916–920; *Annales de chimie* [3] 52 (1858): 97–103.

29. Berthelot, "Sur la synthèse des carbures d'hydrogène," *Annales de chimie* [3] 53 (1858): 69–208.

that the majority of scholars were induced to regard as chimerical the hope of producing organic substances in a general manner from the simple bodies that constitute them." Only general methods of total synthesis, Berthelot averred, could demonstrate the identity of organic and inorganic forces. "This," he wrote, "is the project I have pursued for over eight years, of which the present memoir contains the point of departure." He had, in effect, accomplished his goal: ". . . all the initial and most difficult terms of synthesis have been achieved." The rest of the project was unproblematical, for "accomplishing the total synthesis of hydrocarbons and alcohols [as he had done] is to accomplish the synthesis of a nearly infinite number of organic compounds."[30]

In articulating these views, Berthelot was intent on establishing the originality and high significance of his project, but he was guilty of five serious distortions. First of all, Berzelius could not have written the cited quotation in 1849; he would have had to have done so from the grave. The quoted words, taken from a French translation of Berzelius's textbook published in that year, first appeared in German in 1827 and were written around 1826—not 9, but 32 years before Berthelot was writing.[31] Second, if Berthelot had read two pages further in the 1849 French edition he would have encountered Berzelius stating unequivocally that "vital force" was simply a term used to refer to the complexities of animal chemistry, and that all the phenomena of life are produced by forces which are absolutely identical to those of inorganic nature.[32] Third, throughout his career Berzelius was consistently committed to materialist-mechanist explanations in physiology and chemistry, and to categorical rejection of all mystical

30. Ibid., pp. 69–71, 208.

31. Berzelius, *Lehrbuch der Chemie*, second German edition (Arnold, 1827), 3:i, pp. 135–138 (although there was a Swedish edition of this version of the textbook, this German edition appeared earliest);*Traité de chimie*, second French edition (Didot, 1849), vol. 4, pp. 1–2 (this edition was translated from the third German edition of 1833–1841). Jungfleisch noted that Berzelius was indeed deceased when the cited French edition was published, but he repeated all the other errors committed by Berthelot ("Notice," p. lxxv). Berthelot himself repeated the phrase "ces paroles de Berzelius, écrites il y a dix ans" in his *Chimie organique fondée sur la synthèse* (Mallet-Bachelier, 1860), vol. 1, p. xxi; he repeated the claim once more without essential alteration in *La synthèse chimique* (Baillière, 1876), p. 13.

32. Ibid. (1849 volume), pp. 5–6.

obscurantism. Nor is this understanding of Berzelius's views an instance of historical hindsight: Berzelius's contemporaries were aware of his opinions. In short, Berzelius was never a vitalist of the type Berthelot tried to portray; he always believed in the unity of forces across the organic/inorganic divide, and in the feasibility of organic synthesis.[33] Fourth, as Berzelius went, by and large, so went the entire chemical community. As was noted in the preceding section, virtually no professional chemist defended transcendent vital forces after the early 1830s, and by the middle of the century few in the community doubted the eventual success of wide-scale organic syntheses.[34] In manuscript lecture notes from 1854–55 we know that Frankland taught his students at Owens College, Manchester, that "vital force" was long dead and buried, and that Wöhler had proved the feasibility of organic synthesis as early as 1828—a conviction that can be traced back at least to 1843.[35] Fifth, prior instances of organic synthesis were by no means as "rare, isolated, and unfruitful" as Berthelot wanted to believe.

Berthelot's investigations of the 1850s were novel in one important respect: he was interested primarily in "total" syntheses. This was partly motivated by his desire to create a complete system of synthetic chemistry, for it was this first essential step that was most often missing in the growing synthetic repertoire. With the gap filled, an impressive tableau of artificially manufactured organic substances could then schematically be described. Berthelot's syntheses also plugged a philosophical lacuna, allowing no possible breathing space for vitalism or teleology—which fit well with his positivist and materialist bent.

Such systematic, taxonomic, and philosophical concerns were the guiding ideas behind Berthelot's famous treatise of 1860, *La chimie organique fondée sur la synthèse*. This monumental 1500-page work, composed between June 1859 and August 1860, constituted a full survey of the field from the point of view of synthesis, and could be considered at once as a textbook, a handbook, and a literary and philosophical treatise. Berthelot's own pyrogenic total syntheses occupied pride of place in the discussion,

33. See the various papers in *Enlightenment Science in the Romantic Era*, ed. E. Melhado and T. Frängsmyr (Cambridge University Press, 1992).

34. Recently this point of view has been ably defended by Jacques (*Berthelot*, pp. 80–83).

35. C. Russell, *Edward Frankland* (Cambridge University Press, 1996), pp. 155–157.

and at times he neglected the work of others, especially the Germans—just as the Germans often dismissed Berthelot's syntheses. His argument was that his was the first system of chemistry *founded* on synthesis, and that such a system was first made possible by his own synthetic reactions. Wöhler's and Kolbe's papers were briefly mentioned, but Berthelot argued that these had been "isolated and unfruitful" events. His opinion that urea was just barely definable as an organic compound was not unreasonable; his characterization of acetic acid as likewise isolated and "scarcely more fecund" was more strained. As his vision was focused on total syntheses (for reasons mentioned above) he did not discuss here the myriad recent contributions (mostly German) to synthesis more broadly defined.[36]

The book made Berthelot's reputation; today it is regarded as one of the great classics of nineteenth-century scientific literature. In the introductory matter he disparaged Gerhardt's theory of types and the emerging structuralist school:

Indeed, virtually all the systems constructed in organic chemistry for the last twenty-five years display this common and singular character of being founded nearly exclusively on combinations of signs and formulas. These are theories of language rather than theories of facts. In addition, chemists often take the properties of numbers hidden in their formulas to be the mysterious properties of real entities, an illusion analogous to that of the Pythagoreans. . . .

Berthelot disclaimed any need to treat these theories in his book, for they dealt solely with imaginary bodies:

The pretentions and effects of such theories are by no means without analogy to the syllogistic methods that were invented in the middle ages in order to reduce all questions and problems to a certain number of logical categories determined in advance, from which one may derive in a rigorous fashion their rational solution. In this regard, the symbols of chemistry exhibit a curious seductiveness.[37]

In Berthelot's view, many of his colleagues had been led astray by an overly cerebral, indeed metaphysical approach to science, neglecting the proper grounding in phenomena and in direct experience:

. . . in the pursuit of science all that matters is the discovery of general facts and of the laws that connect each one to all the others. The language in which one expresses them matters little; that is a question of exposition rather than true invention, for signs have no value but for the facts for which they stand.[38]

36. Berthelot, *Chimie organique*, vol. 1, p. cxlviii.

37. Ibid., pp. cxxii–cxxiv.

38. Ibid., p. cxxiv.

As Jean Jacques has remarked, Berthelot was less interested in persuading his colleagues of the field than in making a much broader point:

Berthelot, for his part, never sought through synthesis to verify theories (since he rejected them) or to make an argument for a structure (since he did not believe in them). His goal was essentially philosophical: he wished to demonstrate a certain power of science. . . . He wanted above all to convince writers and journalists, and, through them, the larger public. He synthesized very simple organic compounds—alcohol, formic acid, shortly thereafter acetylene—because, frankly, he would be quite incapable of reproducing more complicated molecules. But the very modesty of his objectives and means valorized his enterprise.[39]

In short, Jacques wrote, Berthelot's "synthesis" was "a kind of intellectual exercise, marvelous and abstract."[40] This is an appropriate perception, and one that carries considerable irony in light of Berthelot's search for an operational phenomenalist science.

Berthelot's enterprise was indeed successful with his targeted audience. The Germans, however, were not so enthralled. Kekulé wrote to his friend Lothar Meyer:

. . . I had gotten Berthelot's book and felt obliged to gulp it down fast, even if unchewed. That brought on more indigestion and now I fear greatly for my mental stomach. If I have to do it, I will now choke it down slowly and in small doses. I think you can tell that I am not exactly crazy about this sorry piece of work. Indeed, I confess that I have hated his bombastic style for some time, especially since his "Synthesis of hydrocarbons" [paper of 1858]. . . . The whole thing is a treatise of vainglory; and even the title is unjustified.[41]

Berthelot's extravagant rhetoric disgusted Kekulé; he recognized high merit in several of the Frenchman's contributions, but failed to recognize "science" in the treatise as a whole. He called it a "chemical-political or political-chemical lampoon." He could scarcely believe Berthelot's attempted appropriation of organic syntheses in general: What about Wöhler's and Kolbe's "*lovely* syntheses"? What about all the new "ethers, anhydrides, amides, and salts" that had appeared recently? Finally, what was all this about finally driving the mystical vital force into oblivion? After all, Kekulé asked rhetorically: "When do educated people ever speak of this any longer, except when drunk?—and the common crowd will certainly never

39. Jacques, *Berthelot*, p. 77.
40. Ibid., p. 78.
41. Kekulé to Meyer, 23 October 1860, August-Kekulé-Sammlung, Darmstadt (Anschütz, *Kekulé*, vol. 1, pp. 204–206).

read Berthelot's massive book, their vital forces would never suffice for that."[42]

Kolbe's reaction was similar. Approached by his friend and publisher Eduard Vieweg for his opinion of the salability of a German translation, Kolbe was decidedly doubtful—despite his receptivity to the anti-structuralist tone of the book. He thought it a curious, old-fashioned, almost ridiculous book, written mainly to showcase Berthelot's own work. He predicted that it would not sell well even in France.[43] Wöhler too was unhappy with Berthelot's apparent attempt to usurp the entire field; the book was so cunningly written that he worried that even some professional chemists might get the impression that Berthelot really had invented scientific organic chemistry, and that total syntheses were novel. He urged Liebig to write one of his famous polemical articles debunking the book and its pretensions— just don't mention my name, he added.[44]

Berthelot treated chemistry above all in the tradition of natural history, in which empirical observation and classification were paramount scientific goals.[45] One of his chief objections to the atomist school was always that its

42. Ibid.

43. Kolbe to Eduard Vieweg, 5 August 1861, Vieweg-Verlag Archiv, Wiesbaden, Kolbe file, letter no. 174. Interestingly, in 1864 Kolbe was asked to judge a manuscript by Albert Ladenburg—then still studying at Heidelberg—titled "Die organische Chemie, gegründet auf die Synthese" (Kolbe to Eduard Vieweg, 26 January 1864, ibid., letter no. 198). Kolbe liked it no better than Berthelot's book; Vieweg declined to publish it, and the manuscript must have been destroyed. In 1879 Kolbe called Berthelot a "very bad chemist" (Kolbe to Heinrich Vieweg, 11 December 1879, ibid., letter no. 447). In this letter he spelled the name both "Berthellot" and "Berthollet."

44. Wöhler to Liebig, 13 October 1863, in *Aus Justus Liebig's und Friedrich Wöhler's Briefwechsel in den Jahren 1829–1873*, ed. A. Hofmann (Vieweg, 1888), vol. 2, pp. 145–146. The paraphrased quotation begins "Das Berthelot'sche Buch ist freilich an sich so jesuitisch abgefaßt, daß. . . ." It is not clear whether Wöhler intended irony in applying "jesuitisch" to the agnostic Berthelot. At this time this was a common German term of contempt applied to scheming Frenchmen.

45. This is a common theme in the Berthelot corpus. Some examples: "Leçon sur l'isomérie," in Société Chimique de Paris, *Leçons de chimie professées en 1864 et 1865* (Hachette, 1866), pp. 207–210, 231; *Leçons sur les méthodes générales de synthèse en chimie organique* (Gauthier-Villars, 1864), pp. 62–66, 521–524; *Die chemische Synthese* (Brockhaus, 1877), pp. 164–168, 182. For a discussion, see M. Nye, "Berthelot's Anti-Atomism," *Annals of Science* 38 (1981): 585–590. Also relevant is Nye, "Explanation and Convention in Nineteenth-Century Chemistry," in *New Trends in the History of Science*, ed. R. Visser et al. (Rodopi, 1989).

adherents had founded their chemistry partly on an uncertain *physical* theory, trusting it even when doing so meant the neglect of clear chemical analogies.[46] Berthelot wanted chemistry to stand on its own, independent of the encroachments of physics and other sciences. It was a proud science, with its own methods, its own characteristic ideas, its own life.

It is therefore ironic that when Berthelot left organic chemistry, in the mid 1860s, he turned to the emergent field of physical chemistry; indeed, today he is considered one of the founders of this discipline. Even in his early, purely organic-chemical work, physical methods and physical analogies often intrude. His favorite synthetic method, electrical pyrogenesis, was physical. Moreover, he often used mathematical language, mathematical analogies, and mathematical calculations in his organic chemistry. It may even be fair to say that Berthelot's affinity for mathematical rigor influenced his philosophy of chemistry, turning him toward the pursuit of scientific laws and away from hypothetico-deduction and transdictive inferences to the submicroscopic realm. In fact, as regards synthetic chemistry in particular, Berthelot explicitly and eloquently rejected the natural-historical model:

Chemistry creates its object. This creative faculty, similar to that of art itself, distinguishes it essentially from the natural and historical sciences. The object of the latter is given in advance, independent of the will and action of the scientist; the general relations which these sciences can foresee or establish are based on more or less probable inductions. . . . They are thus all too often condemned to an eternal impotence in the search for truth. . . . By contrast, the experimental sciences have the power of realizing their conjectures. . . . In this sense the procedure of the experimental sciences is not without analogy to that of the mathematical sciences.[47]

The contradiction is only apparent. Other passages in the Berthelot corpus confirm that he was certain that someday chemistry would become a branch of mathematical physics, but he was just as certain that this day was very far away, and that present attempts to provide mechanistic-

---

46. For instance, Cannizzaro and those who followed his example placed the alkali metals and the alkaline earth metals in different classes (today we call them "monovalent series" and "divalent series," respectively), despite the strong chemical similarities of compounds in the two categories. To say the same thing in the manner of Berthelot: the fact that the "new chemistry" demanded that these two series be distinguished theoretically despite their chemical analogy was compelling evidence that the new chemistry was deeply flawed.

47. Berthelot, *Chimie organique*, vol. 2, p. 811.

materialistic interpretations (such as those of his atomist opponents) were sadly deficient, worse than useless. Thus it was necessary for the foreseeable future to renounce such interpretations and such methods. "We are still today in the era of the founders," he wrote.[48]

In Germany, the trend toward hypothetico-deduction in organic chemistry was accelerated by the advent of structure theory, which led chemists to exactly the sort of active manipulations praised (but not practiced) by Berthelot.[49] The theory was in essence reducible to a series of simple algorithmic and combinatorial rules by which one could imagine all sorts of possible structures and all sorts of ways to examine experimentally which of the possibilities was correct. If one had at least provisional or heuristic faith in the rules, as most active German organic chemists did by the mid 1860s, the theory was extraordinarily powerful.

But another aspect to the German organic-chemical "style" is just as important as hypothetico-deduction. The schematic simplicity of structure theory had to be constantly adjusted upon its encounter with details at the empirically contingent level. All familiar macroscopic analogies— Newtonian mechanics, gravitation, Coulombic attraction—proved heuristically worthless, and even the theoretically inexplicable structuralist algorithms had to be continuously modified at the margin. In effect, chemists had to emulate the naturalists once more, in exploring the mysterious contingent relationships among the atoms in a molecule. In short, it was necessary to maintain an open attitude. The theory was an indispensable guide in a general sense, but flexibility was the most pragmatic course for questions of detail.

48. Ibid., pp. 810–811; Berthelot, *Leçons sur les méthodes générales*, p. 8.

49. The following two paragraphs summarize the discussion on pp. 243–252 of Rocke, *Quiet Revolution* (that discussion itself draws from M. Nye's "Explanation and Convention in Nineteenth-Century Chemistry"). See also Rocke, "Convention versus Ontology in Nineteenth-Century Organic Chemistry," in *Essays on the History of Organic Chemistry*, ed. J. Traynham (Louisiana State University Press, 1987); Rocke, "Kekulé's Benzene Theory and the Appraisal of Scientific Theories," in *Scrutinizing Science*, ed. A. Donovan et al. (Kluwer, 1988); Rocke, "Methodology and its Rhetoric in Nineteenth-Century Chemistry," in *Beyond History of Science*, ed. E. Garber (Lehigh University Press, 1990); and Nye, *From Chemical Philosophy to Theoretical Chemistry* (University of California Press, 1993). On this subject, see also L. Laudan, *Science and Hypothesis* (Reidel, 1981); J. Schuster and R. Yeo, eds., *The Politics and Rhetoric of Scientific Method* (Reidel, 1986).

Here was where the German organic-chemical style proved more fruitful than that of Berthelot and those influenced by him. The state of the field called for open-minded pragmatism and empirical flexibility—qualities for which Berthelot was not noted. Berthelot was too fully imbued with philosophical and methodological convictions which he single-mindedly tried to put into practice. That intensity of purpose provided the engine for his remarkable career, but it simultaneously lamed him in certain respects.

A Chair at the Collège de France

In December of 1859 Berthelot gave up his position as répétiteur at the Collège de France when he accepted a new chair in organic chemistry at the École de Pharmacie, at an annual salary of 3000 francs. This was just when he was engaged in writing his *Chimie organique fondée sur la synthèse*. The work made a splash among the Parisian literati when it appeared the next autumn. The fact that much of the voluminous introductory and concluding matter was accessible to the lay public meant that it had influence outside the science community and outside the academy. The famous historian Jules Michelet was enraptured, writing to Berthelot: "I did not read you, but drank you like a sponge!"[50] Seven years later, Michelet wrote of his century's most notable characteristics, including new inventions, new prime movers, new varieties of plants and animals, all devised by the creative gifts of mankind:

. . . the mineral world resisted; this was our limit, it was said, and never could life be drawn from that source. It was an immutable domain reserved for death. Wrong—the barricade has fallen! . . . See now how minerals and stones are animalized. They have slept enough. It was discovered that these slumberers contain the universal ferment, the intoxication of nature. From the stone springs alcohol. Fermentation, electricity, magnetism: curious powers bringing together the two states of life and of putative death, creating intermediate states, a bridge that goes from one to the other. . . . The flower of life itself, milk flows from the breast of nature which had been thought inorganic. Now one can say: "*All is alive*, or must be. Inertia, the negative, disappears, and is nothing more than a word. . . . O Death, where is thy victory?"[51]

50. "Je vous ai non pas lu, mais bu comme une éponge!" *Revue générale des sciences* 18 (15 May 1907), cited in Virtanen, *Berthelot*, p. 3.

51. Michelet, "Le Collège de France," in *Paris-guide, par les principaux écrivains et artistes de la France* (Lacroix, 1867), vol. 1, p. 142.

This delightful "morsel of uncontrolled lyricism" (in Jacques's pungent phrase) may stand as a symbol of the lay intellegentsia's admiration for Berthelot's public persona.

Berthelot's growing circle was not limited to Renan and Michelet. From 1857 he attended the salons of the mathematician Joseph Bertrand and of Madame Didier, and from 1864 on the biweekly dinners organized by the Goncourt brothers at the Restaurant Magny. Others of the Magny group included Sainte-Beuve, Flaubert, Taine, and George Sand. He also drew useful acquaintanceships from his years as a student at the Collège Henri IV: former classmate and future prime minister of France Émile Ollivier, and former teacher and future Minister of Public Instruction Victor Duruy. Berthelot's public fame was mirrored by more conventional marks of esteem granted by his scientific colleagues: the Jecker Prize of the Académie des Sciences in 1860, and election to the Légion d'Honneur in 1861.

These public appreciations, reflecting a growing sense of Berthelot's stature, culminated in an unusual step taken by his supporters. In a petition addressed to the new minister of public instruction (Duruy) and published in December of 1863, seven members of the Académie des Sciences (Balard, Bernard, Dumas, Bertrand, Pelouze, Regnault, and Deville) urged that a new chair of organic chemistry be founded at the Collège de France, and that Berthelot be installed in it.[52] There can be little doubt that Berthelot had been pulling strings (at least with Balard, Renan, and Duruy); there is also evidence of support for Berthelot in the emperor's circle.[53]

Wurtz, thunderstruck, quickly penned a strong but carefully worded protest to Dumas:

> . . . it seems to me that in proposing M. Berthelot as the candidate naturally designated to fill the chair being created, and in leading the Minister to believe that he is the sole representative of modern organic chemistry in France and the principal author of the new chemical theories, the signatories of the Note have lost view of the rights and legitimate interests of other scientists, among whom I may be permitted to count myself. And I do not refer here to their material interests, speaking for myself at least; for, you know, Monsieur and dear master, I do not aspire to the honor of a double teaching post, and I would not wish to impede in any way

---

52. "Sur la création d'une chaire de chimie organique au Collège de France," *Journal général de l'instruction publique*, 26 December 1863.

53. Velluz, *Vie de Berthelot*, pp. 63–66. This also includes the petition in extenso and some of the associated correspondence.

the generous intentions of Monsieur the Minister with regard to M. Berthelot. I speak of much higher interests, ones that are much more precious to a scholar: considerations that attach to his work and to the honor of his name.

Wurtz then made his point explicit: since he and Berthelot had been pursuing virtually the same subjects, and since the petition implied that no one else could match Berthelot's eminence as a scientist, the clear inference for anyone to draw was that Wurtz's work was judged by all to be inferior to Berthelot's. In this manner, the note "has thrown clear disfavor upon [my] already long career, for those who are able to read between the lines." Wurtz was sure, he wrote, that Dumas had not intended this implication in signing the note, but that had been the effect. He ended with the wish that Dumas should "seize the first possible occasion to tell Monsieur the Minister that among your former students is one who has also made some discoveries and whose work has exercised a certain influence on the development of organic chemistry."[54] Plausibly inferring from a vague reference in this letter, Ana Carneiro suggests that Wurtz had had an understanding with Dumas that he, not Berthelot, would be given a research laboratory at the Collège. She surmises that Dumas's breach of faith could have been due to Balard's insistence, for Dumas's students already occupied nearly all the principal Parisian chemical posts.[55]

54. Wurtz to Dumas, 30 December 1863, Fonds Dumas, Archives de l'Académie des Sciences: "Il me semble, en effet, qu'en proposant M. Berthelot comme le candidat naturellement désigné pour remplir la chaire à créer, et en laissant croire au Ministre qu'il est le seul représentant de la chimie organique moderne en France et le principal auteur des nouvelles théories chimiques, les signataires de la Note ont perdu de vue les droits et les intérêts légitimes d'autres savants parmi lesquels il m'est permis de me compter. Et je ne parle pas ici de leurs intérêts matériels, en ce qui me concerne au moins; car, vous le savez, Monsieur et cher maître, je n'aspire pas à l'honneur d'un double enseignement et je n'aurais voulu entraver en aucune manière les généreuses intentions de Monsieur le Ministre à l'égard de M. Berthelot. Je parle d'intérêts bien plus élevés, de ce qu'un savant a de plus précieux: la considération qui s'attache à ses travaux et l'honneur de son nom. . . . Elle a jeté sur une carrière scientifique déjà longue une défaveur évidente pour tous ceux qui savent lire entre les lignes. . . . J'espère que vous voudrez bien saisir la première occasion pour dire à Monsieur le Ministre que parmi vos anciens élèves il en est un qui, lui aussi, a fait quelques découvertes et dont les travaux ont exercé une certaine influence sur le développement de la chimie organique." What Wurtz did not then know was that the ministerial decision had already been made, on 24 December.

55. A. Carneiro, The Research School of Chemistry of Adolphe Wurtz (Ph.D. dissertation, University of Kent/Canterbury, 1992), pp. 68–69.

Also unhappy, to put it mildly, was Louis Pasteur, director of scientific studies at the École Normale, who had professional roots in organic chemistry and mineralogy. While not gainsaying Berthelot's scientific contributions, Pasteur objected formally, publicly, and forcefully to the procedure of creating a new post for a designated individual, giving no one else an opportunity to apply or compete for the position. Beyond the procedural questions, he had many private reasons to object to Berthelot: Pasteur was a politically conservative Catholic of deep religious convictions, while Berthelot was a highly public atheist, a leftist, a materialist, and a positivist. Writing privately to Dumas, Pasteur was blunt. What were these "new ideas in chemistry that require a new chair" argued in the petition? "A few years ago," Pasteur continued, "we were presented with a scientific revolution by the appearance of these two enormous volumes, *Chimie organique fondée sur la synthèse*. Never has a work been more quickly forgotten! I see in all of this nothing more than manifestations of this [positivist] school, impatient and dangerous, personified by the names of MM. Renan, Taine, Littré, etc. It is Renan who is behind this whole affair and who has had the skill of garnering the signatures of the members of the Académie des Sciences."[56]

Despite the objections and some further questions about the budget and about the appropriateness of splitting Balard's chemical chair, the petition was successful. At first, Berthelot was invited to be simply a "chargé de cours" at the Collège, teaching a "cours complémentaire" on organic chemistry. Accordingly, between February and June of 1864 he offered 32 lectures on "general methods of synthesis," and these were printed at the end of that year.[57] This provisional arrangment was extended in 1865. On 8 August 1865 Berthelot received what he had hoped for initially: the appointment as a professor for organic chemistry at the Collège de France—a chair created essentially by splitting Balard's chair. From that time until his death (more than 40 years later), Berthelot worked in his laboratory at the Collège every day, with great energy and success. Organic synthesis,

---

56. Pasteur to Dumas, 25 December 1863, in *Correspondance de Pasteur*, ed. P. Vallery-Radot (Flammarion, 1951), vol. 2, p. 154. Pasteur wrote on the same theme, but more tactfully, to Chevreul, Balard, Claude Bernard, and Berthelot (ibid., pp. 151–153).

57. Berthelot, *Leçons sur les méthodes générales de synthèse en chimie organique professées en 1864 au Collège de France* (Gauthier-Villars, 1864).

along with some shrewd alliances, had won Berthelot a career—although, ironically, he was already then moving into investigations in physical chemistry, and he would do less and less organic chemistry as time went on.

Wurtz's Countermoves

There were now three newly established professorships for organic chemistry in Paris, but, paradoxically, none could be used to promote the new ideas on molecular magnitudes and chemical structure then emerging in the science. The chairs at the École de Pharmacie and Faculté de Médecine had practice-oriented pedagogical restrictions that excluded any systematic teaching—or even extended discussion—of chemical theory; the Collège de France, on the other hand, was explicitly devoted to pure scientific research, including high theory, but molecular theory was the last thing either Berthelot or Balard wanted to do.

Wurtz naturally found this situation frustrating, particularly since the chemical reforms did not seem to be making much headway in France. His chair at the Faculté de Médecine gave him a secure professional position, and his research laboratory was popular and successful, but he dearly desired a rhetorical platform from which he could make the case for the new chemistry to a wider circle. The Société Chimique and its journals provided him with some leverage, as he had planned; however, since these had become national institutions, it would be inappropriate to guide them with too strong a hand in a single direction, even if that were possible. The Karlsruhe conference had not had the wider ripple effects Wurtz had hoped for. What to do?

In the first half of 1862, Wurtz made two public presentations, in very different contexts, each of which was partly designed to serve his wider goals. On 13 March he presented an "Éloge de Laurent et de Gerhardt" to a special meeting of the Société des Amis de la Science (the fund set up to assist the families of the two early-departed chemists). The occasion, formal and well attended, was presided over by Maréchal Vaillant. Wurtz skillfully blended a subtle advocacy into his eulogy. Three months later, on a visit to the London International Exhibition, he gave a formal lecture to the Chemical Society: "On Oxide of Ethylene, Considered as a Link between Organic and Mineral Chemistry." In that year Hofmann was president and Williamson was vice-president of the Chemical Society, and these close collegial friends of Wurtz no doubt arranged the event. Wurtz depicted his discovery of ethylene oxide as entirely analogous to the oxides of barium, strontium, calcium, zinc, and all other divalent metal oxides, thus demonstrating the close analogies between organic and inorganic chemistry. In the process, he rehearsed all the reasons that had led him and other reformers to adopt the newer atomic weights and molecular magnitudes.[58]

Shortly thereafter, Wurtz became president of the Société Chimique, and in March of 1863 he used the opportunity to give extended lectures on the historical development of the new chemistry that he had been advocating for nearly a decade.[59] This work, published as a monograph titled *Leçons de philosophie chimique* in 1864, is an extraordinary display of historical writing as well as effective advocacy for the newer ideas. Wurtz was first concerned to show that the system of "equivalents," which went back to the earliest years of the chemical atomic theory and even now (1863) was still in nearly universal use in France, had never been—probably could never be—developed in a fully consistent fashion. The proposals of Gerhardt, he related, resolved many of these difficulties by the simple expedient of returning to essential elements of the Berzelian system of atomic weights allied with the concern for combining volumes so characteristic of the work of Gay-Lussac. Gerhardt would have been shrewd, Wurtz declared, to have invoked these two names in urging his reform:

"You see," a diplomat would have said, "I am returning to the notation to which you all are long accustomed. I am only introducing here some alterations made necessary by the progress of science." But the ardor of his convictions and the fire of his character led him to less moderate and less prudent language. He had a militant manner and the tone of a reformer.[60]

58. Wurtz, "Éloge de Laurent et de Gerhardt," *Moniteur scientifique* 4 (1862): 482–513; "On Oxide of Ethylene, Considered as a Link between Organic and Mineral Chemistry," *Journal of the Chemical Society* 15 (1862): 387–406.

59. Wurtz, "Sur quelques points de philosophie chimique," in *Leçons de chimie professées en 1863* (Hachette, 1864); republished from the same plates, with new title page and added preface, as *Leçons de philosophie chimique* (Hachette, 1864), in serialized translation in the *Chemical News*, vols. 10 and 11 (1864–65), and monographically as *An Introduction to Chemical Philosophy, According to the Modern Theories* (Dutton, 1867). Wurtz's letters to Butlerov, Kekulé, and Scheurer-Kestner indicate that the Société Chimique edition appeared in February and the French monograph edition in late June of 1864.

60. Wurtz, "Sur quelques points," p. 33.

Even Gerhardt's improved system was still flawed, Wurtz noted, by an artificial consistency that halved the atomic weights for such metals as calcium and zinc in order to bring the oxide formulas of these metals into the monoxide pattern established by such metals as sodium and silver. In 1849 Regnault had noted that assuming two distinct classes of metals—those whose basic oxides were of the pattern MO and those whose basic oxides were of the pattern M_2O—would bring all atomic weights into consistency with the Dulong-Petit law of atomic heats. This proposal resulted in the modified Gerhardtian system, which resolved virtually all inconsistencies, averred Wurtz, and which was presently being adopted throughout Europe (but notably not in France). Significantly, Wurtz heeded his own rhetorical advice to Gerhardt, and offered this system to his countrymen, not as Gerhardtian, but rather as "almost identical with that of Berzelius."[61]

It was true, Wurtz noted, that the new system required one to assume that a molecule of oxygen consists of two atoms combined together, and similarly for many simple gases such as hydrogen, nitrogen, chlorine, and so on. Many people, he noted (referring to Berthelot, among others) absolutely refused to countenance the idea that atoms of an element may combine with themselves. However, he laid out many arguments, both physical and chemical, that led inexorably to this conclusion.

Wurtz then proceeded to treat the rise of the theory of types and latterly the theory of structure. The original form of the type theory, that of Dumas, was not itself highly fruitful, but it led directly to the newer theory of types, whose fundamental idea, Wurtz averred, he had been the first to enunciate (in his 1849 paper on primary amines). This "idea" was then converted into a successful scientific theory by Alexander Williamson's papers on etherification.[62] Wurtz then described the "theory of polyatomic radicals," to which Berthelot and he himself had contributed essential elements, and the theory of structure, developed principally by Kekulé. (Wurtz's student Couper was largely ignored.) Wurtz concluded his treatise by arguing that the theories that he had just described were no more than a "rejuvenated and developed expression of the law of multiple proportions," hence were far from being mere speculative constructions. The newest theories, he concluded, were nothing more than a "return to the past," and "one can there-

61. Ibid., pp. 45, 82.
62. Ibid., p. 88.

fore say that the ideas that today are becoming established stand midway between the older theories and those that were developed by Laurent and Gerhardt."[63] Once more Wurtz was practicing skillful diplomacy.

These lectures on chemical philosophy were in press when Berthelot, with Balard's essential aid, attained his chair at the Collège de France in the sudden and irregular fashion noted above. Balard also fully recognized Wurtz's merits, having conferred his patronage on the Alsatian chemist virtually since the latter's arrival in Paris 20 years earlier. Wurtz may have spoken with Balard about his disappointment, in terms similar to his letter to Dumas cited above. Whether or not there was an explicit agreement between them, it would seem that Balard and/or Dumas made partial amends to their former student. Balard lent Wurtz his chair at the Collège so that Wurtz could offer a dozen lectures in May and June of 1864, actually overlapping Berthelot's first course of lectures at the Collège.[64] A student by the name of Fernand Papillon transcribed these lectures; they appeared serially in the *Moniteur scientifique*, and were privately printed monographically under the title *Cours de philosophie chimique*, after being "revu par le professeur."[65]

This work is quite rare, and is virtually unknown to historians of science.[66] It has been consistently confounded with the *Leçons de philosophie chimique*, since it appeared in the same year with a nearly identical title; Wurtz himself considered the two works to be the same, the latter only having "a somewhat more elementary form."[67] They are not, however,

63. Ibid., pp. 221–222.

64. Balard to MIP, 11 May 1864, F17/21890. Berthelot's course ended on 21 June, Wurtz's a week later.

65. Wurtz, *Cours de philosophie chimique* (typographie de Renou et Maulde, 1864). The serialized publication appeared in *Moniteur scientifique*, vols. 6 and 7 (1864 and 1865).

66. Hofmann mentioned Wurtz's Collège de France lectures and their publication in the *Moniteur scientifique*, but he erroneously inferred that the resulting monograph was the *Leçons de philosophie chimique*; Friedel stated that he gave lectures for Balard in the Collège and briefly described their content, relying only on the *Moniteur* and not making mention of the monograph. I am not aware of any other reference to these lectures or their publication in the literature.

67. Wurtz to Scheurer-Kestner, 17 June 1864, Scheurer-Kestner papers, Bibliothèque Nationale et Universitaire de Strasbourg, Ms. 5983, ff. 445–446.

identical, and some of the differences are interesting. In his preface, Papillon expressed dismay that chemistry was still being presented to French students as Thenard had taught it half a century ago; no single course in Paris had even so much as taken the newer chemical ideas into consideration. Wurtz, the "independent continuator" of Gerhardt, was not able to deal with theoretical questions in his teaching of medical students, "to his great regret."[68]

In the first lecture, Wurtz declared theories the "soul of science," even if unverifiable, and signaled his intent to treat the newer ideas in chemistry that were "beginning to prevail." These ideas were in a sense revolutionary, but it would be misleading to call them that, since in fact the science had been "in a state of continuous revolution" since Lavoisier. The exposition that followed was similar in outline, form, and content to the treatment in his nearly contemporaneous lecture series before the Société Chimique. Again Wurtz depicted Williamson as marking the first critical turning point, but with some interesting new language. Williamson's was "by no means a purely theoretical view. These substitutions were effectuated in reality [réellement], and they could be followed, so to speak, step by step. . . . Thus, ether quite truly [bien réellement] contains two ethyl molecules."[69] This approach had led Williamson, Kekulé, Odling, Wurtz himself, and others to develop the idea of atomicity of elements. The second critical turn was then due to Kekulé in 1858, who introduced the linking function of carbon atoms. In the lengthy final lecture Wurtz developed his own ideas on atomicity. He provided here his most detailed discussion of his speculation regarding how the presumed subatomic structure of atoms could explain valence bonds in molecules.[70]

Wurtz concluded his lectures with the caution that these and all other theories were intrinsically fallible and hence must not be regarded as eternal verities. However, he believed that they were superior to those older theories that were still being used and believed in France. One would have to be blind, he wrote, to deny the great progress that had been made over the last two decades in chemistry on the basis of these new ideas. He continued:

68. Wurtz, *Cours de philosophie chimique*, p. 3.
69. Ibid., pp. 31–32.
70. Ibid., pp. 74–76.

It is true that on occasion these advances in organic chemistry have been called into question. However, only he who has followed them attentively would have the right to do this, and no such authorized voice has been raised on this subject. Let us therefore maintain such theories, even while regarding them as still perfectible. They will be a sure guide for us as long as they lead us forward.[71]

This, incidentally, is the same argument Galileo used in his *Dialogue on the Two Chief World Systems*. Galileo claimed that all those who defended the Ptolemaic system against the Copernican had seriously studied only the former, whereas all those who had carefully studied both systems had been converted to the latter view. Wurtz was correct that his opponents had not paid close attention to the details of the new chemistry, and his argument, like Galileo's, was effective.

To summarize: In this work Wurtz seized an opportunity to lay his theoretical cards on the table with a frankness that was impossible in his lectures at the medical school, in papers submitted to the Académie des Sciences, or even in his contemporaneous *Leçons de philosophie chimique*. In addition to defending the new ideas, he declared here a strong realist interpretation of scientific theory, while also emphasizing its heuristic and fallibilist character.

Even here he did not rest satisfied, continuing his literary-pedagogical campaign in new directions. Also published by Wurtz in 1864 was the first volume of a *Traité élémentaire de chimie médicale*, for use in his medical school classes. He commented in the preface that he had "not believed [him]self to be authorized" to abandon the use of conventional equivalents that had been used for the past 20 years in the French lycées, but declared that he had long since adopted "the atomic formulas whose usage Berzelius inaugurated" in all his other writings. In the first chapter he distinguished legitimately empirically defined equivalents from those that William Wollaston had specified 50 years ago—the latter being, in effect, nothing more than one version of atomic weights. But he did not further engage theory in this textbook.[72]

71. Ibid., p. 80.

72. Wurtz, *Traité élémentaire de chimie médicale* (Masson, 1864–65), vol. 1, pp. i–ii, 10–16. The preface was dated 1 June 1864, the day he presented the fourth of his twelve lectures to the Collège de France. A second edition was published in 1873. When Wurtz sent Kekulé a copy of this volume, he apologized for his use of equivalents (Wurtz to Kekulé, 8 August 1864, August-Kekulé-Sammlung).

Two years later Wurtz wrote a second textbook, *Leçons élémentaires de chimie moderne*, intended for use by lycée students.[73] Here Wurtz did use the modern atomic notation, and the fact that the work went through seven editions indicates that the textbook had significant popularity. Although the work, intended as it was for beginners, was oriented toward empirical and descriptive chemistry, atoms were introduced early and defended as a *theory*.[74]

In 1863 Wurtz became Officier of the Légion d'Honneur; in 1864 he won a second Jecker Prize (the first was in 1859) and was elected to the Royal Society of London. In 1864 he won the Prix Napoléon III, worth 20,000 francs. In January of 1866 he was appointed Doyen of the Faculté de Médecine of Paris, an indication of the high regard in which he was held by his colleagues and in official circles. After the death of Pelouze, in 1867, Wurtz was elected (with 46 of 51 votes, including Dumas's strong support and way ahead of Berthelot) to the Chemistry Section of the Académie des Sciences. This event was greeted with joy in certain foreign circles. An anonymous British reporter for *The Laboratory* (almost certainly either Maxwell Simpson or George Carey Foster, both former students of Wurtz) wrote:

> This is welcome news indeed. This election has broken the charm by which the Academy has long been the stronghold of conservatism in scientific matters. The prize withheld from Laurent and Gerhardt has fallen to Professor Wurtz, as an acknowledgment of his great scientific attainments, the importance of his original discoveries, his indefatigable ardour and power of work, and his influence over a school of younger chemists. Modern chemistry is no longer a stranger in the council of this great scientific corporation. But beyond this satisfaction of an almost personal character, there is another and a higher reason for contentment.
>
> Tolerance and that spirit of liberalism which accepts views differing from our own is wanting in official France. A fixed programme ties down the professors of all Government schools to a narrowly defined method of teaching. This is the reason, alluded to in the first number of THE LABORATORY, which has kept France from progressing as rapidly in the path of modern science as the neighboring countries. But, in framing these programmes, the Minister of Public Instruction takes counsel from the men officially considered to be the highest in their respective

---

73. Wurtz, *Leçons élémentaires de chimie moderne* (Masson, 1867–68). Despite the imprint, this work began to appear in fascicles before the end of 1866. New editions were published in 1871, 1875, 1879, and 1884, and posthumously in 1892 and 1894, and it was translated into English, Russian, Spanish, Italian, and Arabic.

74. Ibid. (first edition), p. 25.

branches of knowledge, and such are the members of the five academies. An unwholesome and illiberal spirit of restraint will never be defended by M. Wurtz, and his election, therefore, is not only a triumph of [his] theories, but also a warranty for the unfettered development of future doctrines that will rise one after the other.[75]

One would think that all this activity and all these honors for such a well-regarded senior member of the French scientific establishment would have had a strong cumulative effect on Wurtz's colleagues and students. The result was, however, much less than he hoped. Not only did he still not have the best teaching position for a scientific researcher in Paris (Dumas and Balard were still professors at the Sorbonne, Regnault and Frémy at the École Polytechnique, Deville at the École Normale, and Balard, Berthelot, and Regnault at the Collège de France). He was simply unable to make headway in the collegial community against the strong positivist-empiricist thrust of his two chief contemporary Parisian chemical rivals, Berthelot and Deville, especially since older chemists such as Balard, Pelouze, and Dumas also gave his approach no support. Wurtz understood the situation very well. Berthelot and Deville found themselves, by mere happenstance, in league with the older chemists, who had simply failed to keep abreast of the newer research; moreover, their more positivistic philosophical style was more palatable to many outside of the specialist circle. Wurtz's opponents enjoyed, in effect, a significant strategic advantage, and they made the most of it. For Wurtz it was a most challenging and frustrating game to play.

75. *The Laboratory* 1, no. 16 (1867): 283–284.

9

Renovating Laboratory Science in France

Much has been written on the putative decline of science in nineteenth-century France.[1] When Liebig and other foreigners traveled to Paris in the 1820s, they were probably right to view French national science as preeminent in Europe, though, as Robert Fox has argued, at that time there were already cracks in the reigning French scientific paradigms that would soon lead to their dissolution.[2] In addition to such considerations internal to science itself, there were also cultural and political elements at work in Restoration and Orleanist France that proved unhealthy to the national scientific community.

French scientists, who already had a reputation for insularity, began to withdraw from international collaboration as early as the 1820s, and within a few years foreigners began to notice.[3] The trend was not completely invisible to the French, either, nor was it ameliorated by the often risible material resources provided to scientists by the government. Laboratories, whether for research or for teaching, were rare, small, and penurious, often cobbled together privately in a scientist's residence. When

1. For a historiographic discussion, see chapter 12 below.

2. R. Fox, "The Rise and Fall of Laplacian Physics," *British Journal for the History of Science* 4 (1975): 89–136. A well-argued thesis of this article is that the fate of French science in this period was dependent to a surprising degree on the vicissitudes of the personal influence of a small number of crucial figures, especially Laplace and Berthollet. Fox explored this theme for the general case in "Scientific Enterprise and the Patronage of Research in France, 1800–1870," *Minerva* 11 (1973): 442–473.

3. Fox, "Science, the University, and the State in Nineteenth-Century France," in *Professions and the French State, 1700–1900*, ed. G. Geison (University of Pennsylvania Press, 1984), p. 93.

in 1837 Minister of Public Instruction Salvandy commissioned a report from Baron Thenard, Dean of the Paris Faculté des Sciences, on the state of science education in France, his designated rapporteur, Dumas, was not shy about detailing the deficiencies (see chapter 4 above). That same year Jean-Baptiste Biot offered a severe structural critique of the Académie des Sciences. One notable problem, according to Biot, was the public character of the meetings; this created nearly irresistible pressure for members to indulge in oratory rather than serious debate. Even more serious, Biot drew attention to distortions created by the assimilation of the scientific community to the French national bureaucracy.[4]

Such public "insider" French critics of the 1830s (and there were others) were not free from self-serving motives or exaggeration, and there was much reason for pride in French accomplishments in this period, as well. But there is no gainsaying the impression that well-informed observers were genuinely concerned, and for good reason.[5] It was certainly the case with our particular cast of characters; indeed, chemistry was probably the field that provoked the most anxiety, possibly because there were some extraordinarily fine chemists who were then struggling with the deficiencies. From the beginning of the 1830s—just a half-dozen years after Liebig returned from Paris to his new professorship at Giessen—Dumas, Pelouze, and Gay-Lussac in their correspondence with Liebig independently and repeatedly worried about their national disciplinary community. Cumul was "killing" science, and no one in "our poor country" was staying current with important chemical developments—

4. J. Biot, "Remarques sur l'institution récente des Comptes Rendus Hebdomadaires de l'Académie des Sciences," *Journal des savants*, February 1837, reprinted in Biot, *Mélanges scientifiques et littéraires* (Paris, Lévy, 1858), vol. 2. See also an article by Biot published 5 years later (reprinted in Biot, pp. 265–292). For a pertinent discussion, see Fox, "Science, the University, and the State," pp. 81–83 and passim.

5. This despite the evidence cited by Terry Shinn ("The French Science Faculty System, 1808–1914," *Historical Studies in the Physical Sciences* 10 (1979): 271–332) that French scientific productivity continued at an impressive level until the mid 1840s. As Mary Jo Nye has argued ("Scientific Decline," *Isis* 75 (1984): 697–708), there are intrinsic weaknesses to the sort of quantitative measures pursued by Shinn. Moreover, Nye's line of argument, which supports a case for the absence of a scientific decline in France, is troubled by the aggregate character of her empirical base, for the fate of various fields (and even subfields) within French science were sometimes quite variable. I will return to these issues in chapter 12.

especially not with the "prodigious" rate of German contributions, whence "all light is now coming."[6]

No one was more concerned than Dumas. As was noted in chapter 4, in 1838 he was induced by his sometime collaboration, sometime rivalry with Liebig to found his own private teaching-and-research laboratory, which he designed with the Giessen institute expressly in mind. He admitted openly to Liebig that he could not otherwise hope to keep pace with his German friend. Two years later Dumas was placed at the head of a second government commission to examine the health of the Paris Faculté des Sciences. Dumas may seem to have been an unusual choice at this time, for he was then still only professeur adjoint at the Sorbonne whereas he was professor at the Faculté de Médecine; no doubt he was chosen at the instance of his mentor, Baron Thenard, who had powerful influence in the Ministry of Public Instruction. Although pointing out sources of pride, Dumas's commission report struck a note of urgency, and he forebodingly invoked foreign instances of "luxe inaccoutumé" in scientific institutes and materiel that threatened the comparative international position of French science.[7]

Education Policy during the Second Republic and the Early Second Empire

Dumas's four reports on French science education (1837, 1840, 1846, and 1847) resulted in incipient action by the ministry that was only derailed by the Revolution of 1848. The revolutionary conditions also forced Dumas to close his laboratory. This lacuna was quickly filled by the three new similar enterprises during the short lifetime of the Second Republic: the private laboratory schools of Pelouze (begun late 1845), Wurtz (ca. December 1850), and Gerhardt (May 1851). Significantly, and not coincidentally, all three men had been students of Liebig, and it is certain that all three conceived

6. These quotations come from seven letters written by Dumas and Pelouze in the period 1831–1834 (all cited in chapter 2 above). The language of the two Frenchmen on this subject is remarkably similar in both tone and content; since they had neighboring laboratories in the École Polytechnique and were at that time still friendly with each other, it is probable that this subject formed the topic of many conversations between them.

7. Dumas's report is cited and discussed on pp. 5–6 of H. Paul, *The Sorcerer's Apprentice* (University of Florida Press, 1972).

their enterprises as modeled on Giessen, as Dumas himself had done.[8] (There were other commonalities: two were republicans—Wurtz was a center-left liberal—and two were Alsatian reform-minded theorists.)

All of this suggests that the French chemical community was more sensitive to foreign influence and foreign models during the first half of the century than has hitherto been appreciated: Gay-Lussac, Thenard, Biot, Dumas, Pelouze, Gerhardt, and Wurtz constitute quite a chemical honor roll for this period. But the government controlled the scientific establishment (especially the Université de France, for which all these men worked), and even these stars were unable to shake the complacency and indifference of the bureaucrats. One of the leading French historians of education has understandably referred to this era as the "long stagnation."[9] Worse yet, the coup d'état of Louis Napoleon (December 1851) began a decade of authoritarian rule that was unsympathetic to university reform—or even to the universities themselves. The advocates of reform of government-supported, internationally oriented laboratory science would have to wait a bit longer.[10]

Not that they failed to try. It was undoubtedly to reduce the provincialism of his colleagues that, in 1852, Dumas had Wurtz and Émile Verdet begin writing translated summaries of foreign scientific papers in every issue of the *Annales de chimie et de physique*. A few years later, in 1858, Biot republished his critical essays of the 1830s in his *Mélanges scientifiques et*

8. For all of this, see chapter 4 above. Both Wurtz and Gerhardt explicitly mentioned the Giessen institute as a model. The anonymous author of an undated and untitled manuscript history of Pelouze's laboratories (probably Aimé Girard, ca. 1867) likewise averred that Pelouze intended his first teaching lab to be "analoge à ceux dont l'Allemagne s'enorgueillisait déjà [in 1838]" (Dossier Pelouze, Archives de l'Académie des Sciences, p. 5).

9. A. Prost, *Histoire de l'enseignement en France, 1800–1967* (Colin, 1968), p. 224. Prost's work must be supplemented by the following: L. Liard, *L'enseignement supérieur en France* (Colin, 1894); R. Anderson, *Education in France, 1848–1870* (Oxford University Press, 1975); G. Weisz, *The Emergence of Modern Universities in France, 1863–1914* (Princeton University Press, 1983); F. Ringer, *Education and Society in Modern Europe* (Indiana University Press, 1979); and the more recent numerous works of Christophe Charle.

10. It is well known that Napoleon himself had pursued chemical and electrical experiments while imprisoned at the fortress of Ham (1840–1845), and some have concluded from this that he was favorably disposed toward pure science. However, he himself described these experiments as "harmless," pursued for relaxation and distraction, not for scholarly purposes. See F. Simpson, *The Rise of Louis Napoleon* (Putnam, 1909), pp. 216–217 and 347.

littéraires, supplemented by a long scathing footnote lambasting more recent French developments.[11]

More concretely, Dumas lobbied his good friend Henri Milne-Edwards, the dean of the Faculté des Sciences, to carry through on some of the commitments made by the Orleanist government before the revolution. Milne-Edwards petitioned the Minister of Public Instruction Hippolyte Fortoul to provide funds for "practical chemical work" for the students of the Sorbonne. Fortoul's response was a bureaucratic maneuver. In 1855 he issued a decree establishing an "instructional and research laboratory for chemical studies at the Faculté des Sciences" headed by Dumas, but directed that it be located "provisionally" at the École Normale, and he provided no funds.[12]

There were indeed plans drawn up at this time—once more!—to enlarge and renovate the notably decrepit buildings of the Sorbonne, and it was assumed that the new laboratory would eventually move there. The cornerstone of the new edifice was laid in an elaborate public ceremony on 4 August 1855. However, nothing more was done, since the government never provided the necessary money. That abortive cornerstone "became almost legendary," and it eventually literally disappeared. In 1856, rather than approve the removal of several buildings adjacent to the Sorbonne, Dumas, as president of the Paris Conseil Municipal, had them preserved so that the

11. Biot, *Mélanges*, p. 292.

12. Milne-Edwards to Fortoul, 16 January 1854, AJ16/5755; Fortoul to Milne-Edwards, 22 February 1855, AJ16/5126; Procès-Verbaux des Séances de la Faculté des Sciences, 26 February 1855, p. 213, AJ16/5121. The École Normale, which had moved to new facilities in its present location at 45 Rue d'Ulm in 1847, possessed the only state-supported chemical laboratory in Paris; see C. Zwerling, *The Emergence of the École Normale Supérieure as a Center of Scientific Education in Nineteenth-Century France* (Garland, 1990). Even then, Deville was given only 1800 francs per year to run the lab, and when he was appointed (1851) he started with empty rooms. There is no evidence that Dumas or anyone else from the Sorbonne ever conducted laboratory exercises at the École Normale.

13. O. Gréard, *Éducation et instruction*, second edition (Hachette, 1889), pp. 17–25; Gréard, *Nos adieux à la vieille Sorbonne* (Hachette, 1893), pp. 230–244; B. Bergdoll, "Les projets de Léon Vaudoyer pour une reconstruction sous le Second Empire," in *La Sorbonne et sa reconstruction*, ed. P. Rivé (La Manufacture, 1987), pp. 55–56; Liard, *L'enseignement*, vol. 2, pp. 272–273. A total of 800,000 francs was assigned to new construction in the proposed budget of the Université de France for 1855, but operating deficits eliminated this money, and the government refused to provide more (Liard, ibid., pp. 263–264).

space could be rented to the Faculté des Sciences.[13] For the Faculté de Médecine, as well, Dean Paul Dubois was commissioned in 1855 to draw up a plan for a full-scale reconstruction and renovation; his design would have cost no less than 11 million francs. But the plan was never implemented.[14]

Meanwhile, state-sponsored scientific education and research was undergoing a remarkable expansion across the Rhine. The preconditions for this development were laid in the 1830s and the 1840s, by Liebig above all, but the real movement occurred after 1850. Baden, Bavaria, Saxony, Hanover, and especially the six Prussian universities competed in building large, often palatial laboratory institutes for scientific and medical disciplines, then bidding up the prices for the most eminent professors to direct the new labs. In chemistry, the competition began in the early 1850s with new institutes at the Universities of Breslau (in Prussia) and Heidelberg (in Baden), each of which cost the equivalent of around 150,000 francs—an enormous sum never even contemplated by would-be reformers in Paris. But this was just the beginning. By the mid 1860s the University of Leipzig was building a chemical laboratory building costing the equivalent of 400,000 francs, while Bonn had one for 500,000, and Berlin was spending no less than 800,000, exclusive of the high cost of the land. All of this was spent being on chemistry alone; other disciplines benefited similarly from the fever.[15]

It was not only Germany that was renovating its scientific establishment. Looking back to his youth 36 years earlier, John Tyndall wrote:

I had heard of German science, while Carlyle's references to German philosophy and literature caused me to regard them as a kind of revelation from the gods.

14. Wurtz, "Rapport à M. le Ministre de l'Instruction publique sur l'état des bâtiments et des services matériels de la Faculté de Médecine," 1 February 1872, AJ16/6357, published in revised form as "L'état des bâtiments et des services matériels de la Faculté de médecine de Paris," *Revue des cours scientifiques* [2] 1 (2 March 1872): 852–854.

15. J. Johnson, "Academic Chemistry in Imperial Germany," *Isis* 76 (1985): 500–524; R. Turner, "Justus Liebig versus Prussian Chemistry," *Historical Studies in the Physical Sciences* 13 (1982): 129–162; T. Curtius, *Geschichte des chemischen Universitäts-Laboratoriums Heidelberg* (Heidelberg, 1908), p. 14; Universitäts-Archiv Leipzig, Med. Fac., BIII, Nr. 2b, Bd. 2, ff. 37r–38r. All costs in German thaler and gulden have been converted to their equivalents in French francs, using standard currency conversions from the nineteenth century.

Accordingly, in the autumn of 1848, Frankland and I started for the land of universities, as Germany is often called. They are sown broadcast over the country, and can justly claim to be the source of an important portion of Germany's present greatness.[16]

On his return from study with Bunsen in Marburg, Frankland was hired at the new university in Manchester (1850), and he wasted no time in raising (privately) the equivalent of about 300,000 francs for laboratory facilities there.[17] In London, the still relatively young University College had benefited greatly from the presence first of Thomas Graham and then of the Giessen-trained Alexander Williamson, who worked at the new (1846) and well-equipped Birkbeck Laboratory. Most important, the Royal College of Chemistry had been established in London in 1845, where the amazing A. W. Hofmann presided over a substantial number of students in Liebigian fashion. In 1853 the College was taken over by the British Government, forming the nucleus of what would eventually become Imperial College. Generous funding was provided from the proceeds of the highly profitable Great Exhibition of 1851.

By comparison, in 1846 Dumas had proposed that the government devote only 6000 francs capital expense on a new chemical lab for the Sorbonne, a request that was never granted. A few years later, Gerhardt set up what was in his words an "elegant" private chemical teaching laboratory for 15,000 francs. Other than the new facilities for the École Normale (1847), which was a special case, the government of France spent virtually nothing for laboratory facilities during the middle decades of the century.[18] The entire annual non-salary budget of the Faculté des Sciences of Paris in the period 1855–1865 was around 12,000 francs—not sufficient for current expenses, much less for renovations or other improvements.[19] No wonder well-informed critics were unhappy and worried.

And not just critics of French laboratory science. During the Second Empire, public figures began increasingly to look toward Germany for mod-

16. J. Tyndall, *New Fragments* (Appleton, 1892; written in 1884), p. 232.

17. C. Russell, *Edward Frankland* (Cambridge University Press, 1996), p. 149.

18. For a compelling summary of the situation, see Liard, *L'enseignement*, vol. 2, pp. 271–276.

19. Liard, *L'enseignement*, vol. 2, pp. 273–274; *Statistique de l'enseignement supérieur, 1865–1868* (Imprimerie Impériale, 1868), pp. 440–441.

els to renovate higher education and original research.[20] At the beginning of 1858, Wurtz's old associate Charles Dollfuss joined with another Alsatian named Auguste Nefftzer in starting up a *Revue germanique*, whose purpose was to expose their countrymen to invigorating ideas and movements from across the Rhine. In the first issue of the new journal there appeared a long letter in which Berthelot's intimate friend Ernest Renan, a prominent philologist and historian, wrote: "Your plan to publish a *Germanic Review* realizes a desire that I have very often felt. . . . No race has ever possessed a more marvelous aptitude for scholarly research." France has much to learn from her neighbors, Renan assured the editors (and the public). He continued:

You will doubtless have correspondents at the various universities. That is where one must take, as from its source, the rich development of ideas that assures Germany such incontestable superiority in the field of rational speculation. In Germany, education is by no means a narrow and jealous pedagogy of science, as it is in other countries. Their institutions of public instruction are also scientific institutions; they receive stimulus not from a central administration which is usually unacquainted with intellectual matters, hence naturally indifferent or hostile to what is not understood, but rather from scholars and thinkers who take seriously the world of the mind.[21]

Although Renan mentioned physical science, most of his attention here was devoted to the pursuit of philology and history, as was the new journal that he was helping to inaugurate. A few years later, three German-educated Frenchmen began a *Revue critique d'histoire* similarly modeled on the Rankean "scientific" history so cherished by Renan.[22]

Despite such agitation in a narrow and politically not very powerful circle, not much could happen in the repressive atmosphere of the 1850s.

20. On the German influence on French scholarship, see C. Digeon, *La crise allemande de la pensée française* (Paris, 1959); Paul, *Sorcerer's Apprentice*; P. Lundgreen, "The Organization of Science and Technology in France," in *The Organization of Science and Technology in France, 1808–1914*, ed Fox and Weisz (Cambridge University Press, 1980); Fox, "Science, the University, and the State"; Fox, "The View over the Rhine," in *Frankreich und Deutschland*, ed. Y. Cohen and K. Manfrass (Beck, 1990); F. Ringer, *Fields of Knowledge* (Cambridge University Press, 1992).

21. E. Renan, "Les études savantes en Allemagne," letter in first issue of *Revue germanique*, reprinted in *Questions contemporaines*, second edition (Lévy, 1868), pp. 251, 252, 256–257.

22. W. Keylor, *Academy and Community* (Harvard University Press, 1975); A. Rocke, "History and Science, History of Science," *Ambix* 41 (1994): 20–32.

However, the authoritarian phase of the Second Empire ended around 1860, for complex political reasons.[23] In June of 1863, as part of a general liberalizing movement, Napoleon replaced Gustave Rouland, the equally unenlightened successor of Fortoul as Minister of Public Instruction, with a classical historian named Victor Duruy. Duruy's arrival in the ministry altered the politics of the situation considerably.

Making the Argument

Victor Duruy (1811–1894)[24] was well known for his politically liberal and moderately anti-clerical views; his early association with Michelet resulted in his acquaintance with and respect for German scholarship. He had come into repeated contact with Napoleon when the emperor sought his advice in connection with his biography of Julius Caesar. There is some evidence that Napoleon appointed Duruy precisely because he thought Duruy would introduce major reforms, specifically according to a modified German model. When Duruy asked the emperor what he wanted in the new ministry, Napoleon answered only "You'll do fine," and he never provided any more guidance.[25]

23. Napoleon III, almost pathologically sensitive to public opinion, had alienated the Church with his adventure in Italy in 1859 and had disillusioned entrepreneurs and manufacturers by his secretly negotiated tariff agreement with Great Britain. He was thus seeking a new base of political support, and there had always been liberal elements in his own background and in his government that only needed modest encouragement to emerge. Such, at least, is an outline of the conventional explanation for the political shift around 1860.

24. V. Duruy, *Notes et souvenirs* (Hachette, 1901); Liard, *L'enseignement*, vol. 2, pp. 286–295; Fox, "Science, the University, and the State," pp. 93–105; S. Horvath-Peterson, *Victor Duruy and French Education* (Louisiana State University Press, 1984). On French higher education during the Duruy administration, see also Anderson, *Education in France*, pp. 225–239; G. Weisz, The Academic Elite and the Movement to Reform French Higher Education (Ph.D. dissertation, State University of New York, Stony Brook, 1976); Weisz, *Emergence of Modern Universities in France*; Weisz, "Le corps professoral de l'enseignement supérieur et l'idéologies de la réforme universitaire en France, 1860–1885," *Revue française de la sociologie* 18 (1977):201–232; H. Paul, *From Knowledge to Power* (Cambridge University Press, 1985).

25. The evidence for Napoleon's thinking is outlined on p. 176 of Horvath-Peterson, *Duruy*. For Duruy's own description of his association with Napoleon and his appointment, see *Notes et souvenirs*, vol. 2, pp. 171–195, where the quoted phrase of Napoleon is cited on pp. 191–192.

Rouland's dismissal and Duruy's appointment surprised everyone, including Duruy. At this time Dumas was an Inspector General of higher education and vice-president of the Conseil Académique, had served as minister of agriculture, and knew both the emperor and Duruy personally. Since it was clear by his choice that the emperor had wanted a reformer, Dumas was incensed to have been passed over. In his disappointment and anger, he refused to attend the first sessions of the council. Reluctantly, but needing to act in the face of open insubordination, Duruy removed him from the vice presidency. A few months later Dumas came to Duruy to apologize, and from that time on the two men were friends and close collaborators.[26]

Duruy was aware of the less-than-robust state of science education and research in France, especially as regards facilities and funding. However, he was unwilling or unable to address the situation early in his tenure, for another even more unstable predicament was hanging fire, namely Rouland's suspension of Renan from a new chair at the Collège de France for pursuing the controversial and hitherto exclusively German field of higher biblical criticism. (Renan's *Vie de Jésus* was published just 3 weeks before Duruy's appointment.) This sensitive issue absorbed much of Duruy's attention early on, along with larger questions regarding the ideal of academic freedom, to which he was much attached. Duruy's solution, to dismiss Renan from the Collège de France but offer him a position at the Bibliothèque Impériale, failed to placate the strong-minded Renan.[27]

Duruy did take early action in three respects. First, he commissioned an official survey of the current state of French education at all levels. This monumental work, which began to appear in 1868, is an essential source of information on enrollments, budgets, salaries, and much else.[28] Simultaneously, he sent carefully chosen envoys to survey, analyze, and formally report on the educational establishments of foreign countries.[29] Only with such a

26. Duruy, *Notes et souvenirs*, vol. 2, p. 238.

27. For which see Renan, "Destitution d'un professeur au Collège de France," in *Questions contemporaines*.

28. *Statistique de l'enseignement supérieur 1865–1868* (Imprimerie Impériale, 1868).

29. E.g., J. Badoüin, *Rapport sur l'état actuel de l'enseignement spécial et de l'enseignement primaire en Belgique, en Allemagne et en Suisse* (Paris, 1865); J. Minssen, *Étude sur l'instruction secondaire et supérieur en Allemagne* (Paris, 1866); P. Lorain,

meticulous evidentiary base, Duruy thought, could one lay the ground-work—politically as well as empirically—for thoroughgoing reform of, and enhanced support for, higher education. He was undoubtedly correct.[30]

In a third action, Duruy lent support to an initiative by Edmond Frémy to begin chemical practica at the Muséum d'Histoire Naturelle. Frémy, like so many others in his collegial circle, was distressed at the absence of opportunity in his country for laboratory education in chemistry. (Since the closure of Pelouze's laboratory school, in 1857, Wurtz's was the only game in town.) Apparently Duruy could not immediately command any significant funds for new educational initiatives from the budget, but small allocations were provided. Frémy's laboratory was opened to students early in 1864 and therefore can be considered to be the first "laboratoire public français d'enseignement gratuit de chimie." During the ensuing decade, Frémy continued to be active and effective in promoting ideas both for reforms in chemical education, and for expansion of employment possibilities for academically trained French scientists. Until the closure of this laboratory in 1892, Frémy educated a large number of chemistry students, mostly for industrial careers, carefully avoiding theory in his pedagogy.[31]

As a result of such initiatives—and partly independent of them, in view of the liberalized climate—a spate of public laments and pessimistic analyses emerged in the mid 1860s. Just a month before his dismissal from the Collège de France, Renan published a remarkable essay titled "L'instruction

De la réforme des études médicales (Paris, 1868); G. Pouchet, "L'enseignement supérieur des sciences en Allemagne," *Revue des deux mondes* [2] 83 (1869): 430–449; J. Demogeot and H. Montucci, *De l'enseignement supérieur en Angleterre et en Écosse* (1870). Wurtz's participation in this effort will be delineated below.

30. Duruy, *Notes et souvenirs*, vol. 2, pp. 202–203, 301; Horvath-Peterson, *Duruy*, pp. 176–177; Fox, "Science, the University, and the State," p. 97.

31. On Frémy's significant but little-studied role in the teaching reforms of the 1860s and the 1870s, see G. Kersaint, "L'école de chimie de Frémy," *Revue d'histoire de la pharmacie* 52 (1964): 165–172; C. Schnitter, "Le développement du Muséum national d'histoire naturelle de Paris au cours de la seconde moitié du XIXe siècle,'" *Revue d'histoire des sciences* 49 (1996), p. 68; F. Leprieur, Les conditions de la constitution d'une discipline scientifique (thèse de IIIème cycle, Université de Paris I), vol. 1, pp. 132–149; Leprieur, "La formation des chimistes français au XIXe siècle," *La recherche* 10 (1979), pp. 736–737. Danielle Fauque of the Groupe d'Histoire et de Diffusion des Sciences at the Université Paris-Sud XI (Orsay) has also been studying this subject.

Figure 9.1
Edmond Frémy, 1865. Source: E. F. Smith Collection, University of Pennsylvania Library.

supérieure en France, son histoire et son avenir."[32] This essay was full of praise for German scholarship and institutions, like Renan's letter to the *Revue germanique* and his still unpublished *L'avenir de la science*[33]:

It is in Germany that the university regime bears marvelous fruits. . . . Germany has derived from the universities, elsewhere blind and hidebound, the richest, most flexible and varied intellectual movement the history of the human mind has ever known. . . . Germany, with its doctors, created history; not anecdotal, amusing, declamatory, or witty history, whose techniques France knows all too well, but history conceived in parallel with geology, history researching humanity's past, as geology examines the transformations of the planet.

For instance, it was positively shameful that the largest and most precious collection of Latin manuscripts in northern Europe, that of the Bibliothèque Impériale, had been and was being exploited almost entirely by "colonies of German and Dutch" scholars.[34]

A German visiting courses in Paris, Renand notes, is "very surprised." The lack of dignity and respect, the coming and going of the students during the lecture, the inattention of the auditors, the declamatory style of the professor, and above all, the applause at the conclusion, strike the German student as curious. "An attentive listener has no time to clap. This bizarre custom shows him once more that the purpose of the exercise is not to instruct, but to shine." It mattered not to the German visitor that the lecture was free; indeed, that was part of the problem. If one were to pay for the lectures they would have to be more serious, Renan improbably asserted, for "one does not pay to hear a man who has no other goal than to prove to you that he knows how to speak well." France was running the risk of becoming "a nation of orators and editors, without concern for

32. Renan, "L'instruction supérieure en France, son histoire et son avenir," *Revue des deux mondes* [2] 51 (1864): 73–95.

33. Written in 1848–49, *L'avenir de la science* appeared first in 1890 (Calmann-Lévy). The book is permeated by respect for German erudition and concern for "the decay of the scientific spirit" in France. For example: "Face to face with a public the majority of which wishes above all to be *interestingly amused*, the [French] lecturer is bound to enunciate ingenious views, to afford ingenious glimpses, rather than rely upon scientific discussion. . . . Thus by a strange reversal, science with us is only made for the schools, while the school should only be made for science." (Renan, *The Future of Science* (Chapman and Hall, 1891), pp. 100–101)

34. The quotations in this and the next paragraph are from pp. 79, 80, 83, 84, and 87 of Renan, "L'instruction supérieure."

essential matters and for the real progress of knowledge." Furthermore, "the least distinguished of the German universities, such as Giessen or Greifswald, with their small and shabby ways, their impoverished professors with their haggard and awkward appearance, and their pale and starving *Privatdozenten*, do more for the human intellect than the aristocratic University of Oxford, with its millions in revenue, its splendid colleges, its rich salaries, and its lazy *fellows*."

But heaven forbid that he should disparage the English, Renan quickly added. Indeed, there were intelligent men at work there who wished to draw their countrymen from their old ways, and who were turning to Germany for help.

Napoleon's Ministry of Public Instruction did the same, at least in a fashion. As far as the Faculté de Médecine was concerned, just before Duruy's appointment Rouland had asked Dean Ambroise Tardieu to send an agrégé named Sigismond Jaccoud to several German university towns to investigate the organization of medical education there; Jaccoud subsequently submitted a report on his findings, which was then published.[35] Late the next year, Duruy asked Wurtz directly to gather information on foreign support for academic chemistry. Wurtz wrote to colleagues in Germany, Russia, Italy, and even Norway, and he reported to Duruy at length (though not for publication) on the information he had received by return mail. According to Wurtz's report:

Since the time, many years ago, when M. Liebig gathered at Giessen students from the four corners of the globe, and founded a justly celebrated school, the study of chemistry in Germany has simply soared. Immense laboratories have been constructed at Giessen, Heidelberg, Breslau, Göttingen, Karlsruhe, and Greifswald. Numerous works and beautiful discoveries have been the fruit of these useful creations.

There followed a summary, with precise statistics, of the extent to which other states, and especially German ones, were generously building their academic chemical establishments. Even the smallest German university, Greifswald, a town of a mere 15,000 population, had spent no less than 260,000 francs equivalent on a new chemical laboratory. All this information, wrote Wurtz, "furnishes elements of a comparison that is not at

35. S. Jaccoud, *De l'organisation des facultés de médecine en Allemagne* (Paris, 1864).

all favorable to our laboratories." "I am convinced," he continued, "that the organization of these institutions and the state of the study of practical chemistry in the Faculties demand the serious attention of Your Excellency. This is an interest of the first order; it concerns the future of chemistry in France. This science is French, and God forbid that our country should allow itself to be surpassed." But that was exactly what was happening, Wurtz wrote ominously. "The impetus came from our country; but it has been propagated with great power outside of our borders." Wurtz recommended, as he had previously, that the chemical laboratory at the Faculté de Médecine be taken over by the state, and be properly supported.[36]

Duruy and the rector of the Académie de Paris approved a plan drawn up by Tardieu—adapted from Dubois's abortive project of 1855—to renovate and enlarge the entire Faculté de Médecine campus, but further approval was needed both for the funding and by the city of Paris, and once more the plan did not succeed.[37] Meanwhile, the dean of the Faculté des Sciences, Henri Milne-Edwards, was bombarding Duruy with similar memos. "I know of no teaching corps of the same order in Europe," he stated, "whose facilities are as miserable." He wanted at least 10,000 francs yearly for chemistry, and as much for physics—miniscule by the standards of many other European science laboratories, as the statistics

36. Wurtz to Ministre de l'Instruction publique (hereafter "MIP"), 10 December 1864, F17/4020: "Depuis l'époque, déjà éloignée, où M. Liebig rassemblait à Giessen des élèves venus de tous les pays du monde, et fondait une école justement célèbre, les études de chimie ont pris un grand essor en Allemagne. De vastes laboratoires ont été construits à Giessen, à Heidelberg, à Breslau, à Göttingen, à Carlsruhe, à Greifswalde. De nombreux travaux, de belles découvertes ont été le fruit de ces utiles créations. . . . Ils fournissent les éléments d'une comparaison qui n'est point favorable à nos laboratoires. J'ai la conviction que l'organisation de ces établissements et l'état des études de chimie pratique, dans les Facultés, appellent la sérieuse solicitude de Votre Excellence. Il s'agit là d'un intérêt de premier ordre, de l'avenir de la chimie en France. Cette science est française, et Dieu ne plaise que notre pays s'y laisse devancer. . . . L'impulsion est partie de notre pays; mais elle s'est propagée avec une grande puissance au-delà de nos frontières."

37. This according to a memo from Tardieu to the Rector and the Conseil Académique, annual report for 1863–64, November 1864, AJ16/252. The plan is also briefly mentioned in Wurtz, "Rapport à M. le Ministre," 1 February 1872, AJ16/6357.

were proving.[38] "In my opinion," he wrote in 1865, "the state in which these institutions are left is a disgrace for a wealthy and enlightened country like France, and I would like the Conseil [Académique] to endorse the desire I express for the enactment of the improvements that were decreed more than ten years ago."[39]

In January of 1866 Duruy appointed Wurtz dean of the Faculté de Médecine. In a long unsigned document dated January 1866 an analyst (almost certainly Wurtz himself) summarized once more for the Ministry of Public Instruction the comparison between the Paris Faculté de Médecine and German medical schools. The writer described how, early in the century, the French had created "la grande école expérimentale de Paris" while Schelling and Hegel were retailing their "nuageuses et incohérentes" doctrines in Germany; but now the situation was very different, and the Germans were clearly in advance of France. "What France requires is in no sense a reform, but rather a return to the point of departure, and to the *complete* and *effective* application of the priciples established by our forebears." Regarding the German universities themselves, the writer paid tribute to "the marvelous vitality and fecundity of all these little centers of intellectual life, which, though rival and independent bodies, are all animated by the same spirit and have a common source of inspiration—the love of science and the honor of the institution of which they are members." He continued: "Competition at all levels, an effective and constant solidarity among all the members of this corporation that one calls faculty or university, this is the base of the entire mechanism of German higher education. . . ."[40]

38. J. Jamin to Milne-Edwards, 12 October 1864, and Milne-Edwards to Vice-Recteur, 18 October 1864, Procès-Verbaux de la Faculté des Sciences, pièces annexes, pp. 380–384, AJ16/5126: ". . . je ne connais en Europe aucun corps enseignant de même ordre dont l'installation soit aussi misérable."

39. Milne-Edwards to M. le Président [du Conseil Académique], annual report for 1864–65 [ca. November 1865], Procès-Verbaux de la Faculté des Sciences, pièces annexes, pp. 394–396: "A mon avis l'état dans lequel on laisse ces établissements est une chose honteuse pour un pays riche et éclairé comme l'est la France et je voudrais que le Conseil s'associe au voeu que je forme pour la réalisation des améliorations décrétées depuis plus de dix ans."

40. The 22-page manuscript, simply labeled "Ministère de l'Instruction Publique, January 1866," is held in the Fonds Dumas of the Archives of the Académie des

Wurtz's annual reports for 1866, 1867, and 1868 were even more negative in tone, filled with dark warnings about the superiority of German facilities. In response to his complaints, Wurtz was granted funds to renovate the
dissection pavilions and to introduce student practica in the École Pratique
de Médecine, but little more.[41]

It was one thing for the authorities to receive periodic jeremiads from
professors and deans who, having strong self-interest at stake, might be
subject (perhaps even unconsciously) to exaggeration; it was quite another
for them to read detailed publications from Germany on these palatial
new laboratory institutes that were so terrorizing the French scientists.
Probably the best known of these was Hofmann's published narrative of
the construction, not yet completed, of the Bonn and Berlin labs.[42] The
size, expense, and in some measure even the extravagant luxury of these
installations is indicated by Hofmann's offhand comment that at the Bonn
institute the on-site director's residence (of around 6000 square feet)
would be equipped with a magnificent ballroom "amply satisfying the

---

Sciences. Dumas, an inspector general and a vice-president of the Conseil
Académique, may have solicited the report from Wurtz for transmission to the ministry; the near identity of certain phrases in this document with words that were
later used by Wurtz strongly indicates his authorship. "[C]e n'est point une réforme
qu'il faut en France, c'est le retour au point de départ et l'application *complète* et
efficace des principes posés par nos pères. . . . de la vitalité merveilleuse et de la
fécondité de tous ces petits centres de vie intellectuelle qui, indépendants et rivaux,
ont cependant le même souffle de vie, et pour source commune d'inspiration,
l'amour de la science et l'honneur du corps dont on est membre. L'émulation à tous
les degrès, une solidarité efficace et constante entre tous les membres de cette corporation qu'on appelle faculté et université, voilà la base de tout le mécanisme de
l'enseignement supérieur en Allemagne. . . ."

41. Wurtz to Rector and Conseil Académique, annual reports of the Faculté de
Médecine for 1865–66, 1866–67, and 1867–68, AJ16/6566. On this small renovation, see also Wurtz to M. le Recteur, 21 December 1866, AJ16/6348. A few
months later Wurtz widened the scope of his requests, but without success (Wurtz
to MIP, April 1867, AJ16/6360; Procès-Verbaux de l'Assemblée des Professeurs de
la Faculté de Médecine, 31 October 1867, AJ16/6255).

42. Hofmann, *The Chemical Laboratories in Course of Erection in the Universities
of Bonn and Berlin* (Clowes, 1866). The work appeared in English, since much of
the negotiation and preparation occurred before Hofmann's transfer to Berlin in
1865. Even after the move, Hofmann retained his chair at the Royal College of
Chemistry in absentia until 1867.

social requirements of a chemical professor of the second half of the nineteenth century."[43] Hofmann sent a copy of this book to his friend Wurtz, who apparently used it to good effect. It "made a great sensation here," Wurtz wrote back gratefully, "where so little has been done for chemistry, and in general for science."[44] Surviving documents do not specify with whom Wurtz shared this work, but the effects can be inferred from what followed.

The timing was fortunate, for 3 months later, in May of 1867, the Second Empire played host to the civilized world for the Exposition Universelle, and there is no doubt that the German exhibits made a strong impression on all—especially their Parisian hosts. Napoleon invited Liebig to a dinner at the Tuileries Palace, where they conversed at length. Liebig was powerfully impressed by this occasion. During a long private conversation Napoleon spoke to him in fluent German (notwithstanding Liebig's perfect competency in French), and, as Liebig recounted it, knew well how to listen as well as to speak, a rarity for great personages. Duruy came to a dinner given for Liebig by Deville, expressly in order to meet Liebig, where they had a long conversation.[45] Hofmann was in Paris, as well, both at the Palace and at Deville's; he was also invited to Duruy's official residence in the Rue Grenelle, where he briefed Duruy orally and in detail about the two new labs in Bonn and Berlin.[46] According to one report, Hofmann received the grand prize of the entire Exposition Universelle for his work on dye chemistry, amounting to a scarcely credible 100,000 francs.[47] He returned to Paris for about 6 weeks later that year—for what

43. Ibid., p. 36.

44. Wurtz to Hofmann, 2 February 1867, Chemiker-Briefe, Berlin-Brandenburgische Akademie der Wissenschaften: "Ce dernier a fait une grande sensation ici où l'on fait si peu pour la chimie et en générale pour les sciences."

45. J. Volhard, *Justus von Liebig* (Barth, 1909), vol. 2, pp. 404–406; W. Brock, ed., *Justus von Liebig und August Wilhelm Hofmann in ihren Briefen* (Verlag Chemie, 1984), p. 199.

46. A. Hofmann, "Erinnerungen an Adolph Wurtz," in *Zur Erinnerung an vorangegangene Freunde* (Vieweg, 1888), vol. 3, pp. 259–260.

47. J. Volhard, *August Wilhelm von Hofmann* (Friedländer, 1902), p. 110. On this prize see E. Vaupel, "A. W. Hofmann und die Chemie auf den Weltausstellungen," in *Die Allianz von Wissenschaft und Industrie*, ed. C. Meinel and H. Scholz (Verlag Chemie, 1992), pp. 187–189.

purpose is not known, but obviously not on holiday. He may well have been serving as a government consultant.[48]

The Founding of the École Pratique des Hautes Études

Influenced in 1867 by events in Germany and at home, Duruy appears finally to have resolved to push for major reforms in scientific facilities— a thrust that culminated in the establishment of the École Pratique des Hautes Études, an institution modeled on German ideas. On 4 July 1867 Duruy asked the vice-rector of the Académie de Paris to instruct the science and medical faculties to prepare a response to the new information about the "material organization of chemical instruction in Prussia, especially Bonn." The vice-rector turned to Wurtz, who wrote to the Prussian education ministry for further documentation, and personally traveled to Bonn in December.[49] On 9 December, Duruy addressed the Conseil Académique, urging reform by citing the dangers of foreign competition.[50]

Wurtz's report to Duruy was submitted in April of 1868. This was Wurtz's third pass at summarizing German developments for the Ministry of Public Instruction (counting the unsigned January 1866 report that probably was from his pen). The Bonn lab was a "majestic edifice, where all anticipated needs have been abundantly satisfied, not excluding even luxury. But this luxury is worthy of a great people who understand the role of science and the importance of matters of the mind in the march of civ-

48. Hofmann to Liebig, addressed from 66 Rue des Écoles, 9 September 1867, in Brock, *Justus von Liebig und August Wilhelm Hofmann in ihren Briefe*, p. 199 (which contains an unrelated query). "Life in Paris is somewhat more pleasant than it was last spring," he wrote. "There are not so many demands made upon me now, and I can indulge my own interests more. My tasks here will occupy me for about another four to six weeks. Please be so kind therefore as to send your letter here. From my wife I have a good report." It is possible that Hofmann was offering advice regarding the future renovation of French academic science. I have been unable to find any other reference to Hofmann's trip, either in the published literature or in manuscript documents.

49. MIP to Vice-Recteur, 4 July 1867 and 4 March 1868, AJ16/296; Vice-Recteur to MIP, 16 December 1867, F17/21890. The dates of Wurtz's trip were 17–25 December 1867.

50. F17/12958, cited in Paul, *Knowledge*, p. 44.

ilization."[51] Wurtz confessed to a depressed feeling while touring the lab with its new director (Professor Kekulé) and the architect, considering by comparison the miserable state of French facilities. Many authoritative voices had recently been raised to address the situation, which had as much political as it had purely scientific implications. Wurtz wrote:

> To these voices may I be permitted to add my own? I believe that I have some right to do so: for fifteen years, I have founded and directed at the Faculté de Médecine a research laboratory with resources that I have created myself, without encouragement from the administrations that preceded yours, but not without some success, I may say without false modesty. For my part, I believe myself authorized to declare that the current situation in France regarding scientific instruction cannot be prolonged without placing in peril the supremacy of French science and the intellectual future of our country.

Wurtz concluded by offering to go to Berlin and to Leipzig, so that he might personally inspect those new chemical institutes, as well.[52] This wish was soon to be granted.

Between Wurtz's December 1867 trip and the submission of his report there occurred another event that would help to break the logjam preventing reform. From the start, Duruy had wished to help the universities, including science facilities, but his fellow ministers and the emperor himself were in general uninterested and unconcerned. His ministry was starved for funds. In such a political and budgetary climate there was little of importance that Duruy could accomplish.

In early September of 1867, Louis Pasteur took steps to alleviate his own penurious situation. His colleague at the École Normale, Henri Sainte-Claire Deville, had managed to create what was probably the only

---

51. Wurtz to MIP, 8 April 1868, F17/21890. "C'est une institution grandiose où tous les besoins prévus ont été largement satisfaits, d'où le luxe même n'est point exclu. Mais ce luxe est digne d'un grand peuple qui comprend le rôle de la science et l'importance des choses de l'esprit dans la marche de la civilisation."

52. Wurtz to MIP, 8 and 16 April 1868, F17/21890; Wurtz to Recteur, 8 and 11 April 1868, AJ16/296. "A ces voix me sera-t-il permis de joindre la mienne? Je crois y avoir quelque droit: depuis quinze ans, j'ai fondé et dirigé à la Faculté de Médecine un laboratoire de recherches avec des ressources que je me suis créées moi-même, sans encouragement de la part des administrations qui ont précédé la vôtre, mais non sans quelque succès, je puis le dire sans fausse modestie. Je me crois donc autorisé à déclarer, à mon tour, que la situation qui est faite actuellement en France à l'enseignement scientifique ne peut se prolonger sous peine de mettre en péril la suprématie de la science française et l'avenir intellectuel de notre Pays."

state-supported research laboratory in France, by the politically direct method of taking his case directly to the emperor and touting the future technological utility of his research, especially that on aluminum; the resulting annual grants were outside of any budget.[53] Pasteur tried precisely the same tactic, and at first it seemed to work just as effectively. He had had some contact with Napoleon, who clearly recognized Pasteur's merits, and so he dared to write the emperor to request a new laboratory, citing promised fruits of his research. The day after Napoleon read Pasteur's letter, he wrote to Duruy, directing that his minister take care of the matter. Duruy consulted the ecstatic Pasteur, and plans were drawn up. However, about the time Wurtz returned from Germany, Pasteur learned that the money for his new laboratory had not in fact been approved. He was consumed both by disappointment and by rage. An expense of perhaps 80,000 francs had been shelved, a tenth of the cost of the Berlin chemical institute, at the same time that the new Paris Opéra was being built at a cost of more than 50 million. And think of all the new science and industry that was thereby being sacrificed![54]

Not attempting to disguise his outrage, Pasteur wrote a masterly diatribe entitled "Le budget de science" about the scandalous condition of academic science in France, and in early January he sent it to the Empire's official journal, *Le Moniteur universel*. Pasteur began with a (now customary) summary of the huge and richly endowed new German institutes. And what about France? "France has not yet begun. . . . She has slumbered in the shadow of her old glories." Pasteur related that a few days ago two members of the Académie des Sciences spoke with "one of our finest chemists" who was suffering from pneumonia from working in one of

53. R. Vallery-Radot, *La Vie de Pasteur*, sixth edition (Hachette, 1915), p. 204. Zwerling (*Emergence*, pp. 13–14) comments that the MIP was providing more than 20,000 francs per year in the mid 1850s for Deville's research on aluminum at the École Normale, outside of budget. He does not give a source for this information. Liard (*L'enseignement*, vol. 2, pp. 264 and 273) mentions that Deville's lab was unique in being properly supported, and that this was due solely to the personal favor of the emperor; but he cites no figures. Prosper Moissonnier states on p. 3 of *L'Aluminium* (Gauthier-Villars, 1903) that in 1854 Napoleon provided Deville no less that 170,000 francs to create an experimental site to develop aluminum preparation on an industrial scale.

54. Vallery-Radot, *Vie de Pasteur*, pp. 160–163, 204–215; *Correspondance de Pasteur*, vol. 2, pp. 345–351.

Paris's "unhealthy, damp, dark, unventilated" laboratories. And where was this disreputable place? The Collège de France![55] Claude Bernard called it "the tomb of scholars." Is the Sorbonne's chemical laboratory superior? Not at all; the small and dismal lab there is no better. "And it is named—oh, what a mockery!—the 'laboratory for advanced education and research.'" He added that the young and very distinguished man who works there has asthma.[56] "Who would believe me," Pasteur thundered, "when I say that there is not a penny in the budget of the Ministry of Public Instruction devoted to laboratories for physical science?" A few savants had been managing by creative accounting to divert small sums of money from instruction to personal research. Most others—Dumas, Fizeau, Foucault, Boussingault—simply spent their own money. The situation was simply a scandal.[57]

The editor of the *Moniteur universel* did not feel he could publish such a fiery attack, but was not unsympathetic. He suggested that Pasteur approach M. Conti, the emperor's secretary. Conti showed the manuscript to Napoleon, who was alarmed by what Pasteur had revealed. Conti then suggested to Pasteur that he publish the article in another journal, or as an offprint (Pasteur did both); meanwhile, on the next day (9 January), Napoleon consulted with Duruy. "Pasteur is right to expose these matters," said Duruy; "it is the best way to correct them." Napoleon asked Duruy to prepare for him a short summary of the Parisian institutions of higher education and research, so that he could understand the nature of the problem more clearly. Duruy ended that report by stating that all the institutions named needed major renovations, and that he was currently preparing a detailed proposal to rectify the situation. He implied that it would be expensive; but Duruy, like Pasteur and Wurtz, was now fully

55. The unnamed scientist was presumably Berthelot. It is unlikely that Pasteur would have referred to Balard as "one of our finest chemists," and by this time he probably would have referred to Regnault as a physicist.

56. Possibly Louis Troost, as suppléant for Deville, who had recently been named professor there. But neither Troost nor Deville (42 and 49 respectively) was really young at this time.

57. *Vie de Pasteur*, pp. 215–216; Pasteur, "Le budget de science," *Revue des cours scientifiques* 5 (1 February 1868): 137–139; reprinted as *Le budget de science* (Gauthier-Villars, 1868); second report, with additional material, in *Quelques réflexions sur la science en France* (Gauthier-Villars, 1871); modern reprint: *Louis Pasteur* (Éditions Raisons d'Être, 1947).

engaged in the battle, and would no longer be satisfied with half measures. Now, thanks to Pasteur, he had Napoleon's sympathetic ear, and victory appeared to be within reach. At the end of that month, the imperial couple personally inspected the laboratories of the École Normale and the Sorbonne.[58]

Two months later (16 March 1868), Napoleon summoned Claude Bernard, Pasteur, Deville, and Milne-Edwards to his private study to discuss the problem. Bernard argued for an invigoration of provincial faculties, for the institution of elective courses outside the required examination syllabus that would emphasize cutting-edge science ("cours d'instruction libre," as opposed to the more customary "cours d'instruction réglementaire"), for the creation of university laboratories that would support students preparing science doctorates, and for the integration of disparate faculties into true "university centers." Duruy's annotations on the handwritten summary subsequently submitted by Bernard include the comments that the relatively poor provincial university towns could never engage in the kind of bidding wars for professors that had so invigorated German science; and his own budget was sufficiently strapped that he could do little.[59]

Pasteur also made the most of the opportunity with the emperor. He declared clearly and forcefully that pure science was the source of applied science and technology, hence of industrial and national strength. French academic laboratories were "palpably inferior," and cumul was another "plague." He explained what was happening in Germany: intense competition meant that top scientists had to be wooed by the various independent states, which were being forced by upward bidding pressure to use expensive new laboratory institutes as prizes. These laboratories were not only large and well equipped, but often also beautiful (indicating national pride

58. *Vie de Pasteur*, pp. 216–218; *Correspondance*, vol. 2, pp. 360–361; Duruy, report to Napoleon, January 1868, report in *Notes et souvenirs*, vol. 1, pp. 302–305.

59. "Les villes [de province]," wrote the annotator (almost certainly Duruy himself), "font déjà beaucoup de sacrifices pour leurs Facultés; il sera bien difficile de leur faire prendre l'habitude allemande de surenchérir les professeurs. . . . J'y réussirais mieux avec un budget moins pauvre." (A. Miles, "Reports by Louis Pasteur and Claude Bernard on the Organization of Scientific Teaching and Research," *Notes and Records of the Royal Society* 37 (1982), p. 106)

in scientific achievement), and usually included an on-site residence (which makes the working life of a scientist far easier and hence more productive). Duruy's annotations indicate that he agreed that the state of French labs was dreadful; that "it is unfortunately true" that French scientists need to be either independently wealthy or extraordinarily dedicated to succeed; and that he was already on record against the practice of cumul.[60] After this meeting, Napoleon assured Duruy that he could count on dramatic increases to help rescue French academic science. Money for new teaching labs for Berthelot, Pasteur, Bernard, and Wurtz was immediately put into the proposed budget.[61]

This was the situation when Wurtz's latest report landed on Duruy's desk. Four days later, on 12 April, Duruy sent it to Napoleon, with a cover letter. The developments in Germany were "distressing for your Minister of Public Instruction and a threat to French science. However, if the works approved by Your Majesty for the Sorbonne and for the École de Médecine are executed, all will be corrected." Duruy was concerned, he wrote, for he had been told that the head of the Commission on the Budget (and Napoleon's chief adviser), Eugène Rouher, was seized in a fit of false economy. The budget was also being stretched out of all proportion by the Haussmann urban renewal projects, then reaching a climax of activity. "I am told," Duruy wrote to the emperor, "that [Rouher] will refuse me my poor fifty thousand francs which I have requested for Claude Bernard, Berthelot, Pasteur and Wurtz, etc. We will only get out of our misery if, when returning this note to me on Wednesday, the Emperor would kindly say loudly enough for M. Rouher to hear, that these works must be finished, even were millions to be added to the loan. I have been put into such poverty that I have become inventive; but if faith can move mountains, the most absolute devotion does not suffice to build schools."[62] (This incident is indicative of the extent to which political power in the late empire had fragmented; it was no secret,

---

60. Ibid., pp. 106–112.

61. *Vie de Pasteur*, pp. 218–221; Duruy, *Notes et souvenirs*, vol. 1, pp. 313–314; Wurtz to MIP, July 1868, F17/21890; Pasteur, *Pour l'avenir*, pp. 35–37.

62. Duruy, *Notes et souvenirs*, vol. 1, p. 312. In a subsequent letter to Napoleon (11 June, in ibid., pp. 313–314), Duruy mentioned that Berthelot's teaching lab in the Collège de France would be ready by the end of the year, and Pasteur would have one, as well, but that Claude Bernard's planned facility, costing 40,000 francs, could not be built for that amount.

even to the emperor, that the real budgetary authority had been taken over by Rouher. Parisian wits at this time referred not to the "gouvernement" but to the "Rouhernement."[63])

Simultaneously, Duruy went public, giving a speech and having it published in the *Moniteur* and in the *Revue des cours scientifiques*. Last year at the Exposition Universelle, he said, "foreigners were astonished at the defective installation of our scientific establishments." What was necessary for France, Duruy affirmed, was what existed already for many years in Germany: research labs that serve as nuclei for research schools, where a master can surround himself with student-collaborators like a "famille scientifique." The proposed budget had a line to help create such research laboratories, he wrote; it was only a small amount, but he hoped it would grow. In addition, the new budget, if approved, would create teaching laboratories as well, similar to the new physiological laboratory just completed in St. Petersburg, at a cost equivalent to 3 million francs.[64]

One would have thought that by now Duruy would have had enough data, and analysis of that data, to choke a legislative body. However, in a ministerial decree of 5 June 1868 he directed his best emissary, Dean Wurtz, to go one more time across the Rhine on the fullest, longest, and finest fact-finding mission yet. Wurtz spent the months of June and July[65] visiting academic chemical, physiological, and anatomical laboratories in Bonn, Berlin, Greifswald, Leipzig, Heidelberg, Göttingen, Munich, Karlsruhe, Zurich, and Vienna. The resulting report became justly well known and influential. In addition to full reportage, including architectural drawings of the various labs and detailed information on costs and budgets, Wurtz also wished to instruct and enlighten—not so much Duruy, to whom the report was formally addressed, but other influential readers. He described the ideals of group research, concluding that "a laboratory is . . . not simply a temple of science, but a school as well."[66]

63. A. Cobban, *A History of Modern France* (Brazillier, 1965), vol. 2, p. 191.

64. Duruy, speech of 18 April, *Revue des cours scientifiques* 5 (1868): 343–344.

65. In Wurtz's letter to Duruy of 8 June 1868 (F17/21890) he speaks of his "très-prochain départ." Wurtz's letter 30 July to Kekulé from Bad Griesbach (August-Kekulé-Sammlung, Institut für Organische Chemie, Technische Hochschule, Darmstadt) specifies his return date to Paris as "next Sunday" (i.e., 2 August).

66. Wurtz, *Les hautes études pratiques dans les universités Allemands* (Imprimerie Impériale, 1870), p. 11. The report was dated March 1869.

Wurtz included the now expected description and praise of the German universities. The great expansion of their science departments Wurtz dated (reasonably) to Bunsen's departure from Marburg to Breslau in 1851 and then to Heidelberg in 1852; then the movement experienced a great acceleration with the building of Hofmann's two palatial labs in the period 1863–1868. "Public opinion supports it, and the governments make laudable efforts to promote it." In contrast, Wurtz lamented, such a movement had scarcely begun in France. France possessed the men, the ideas, the genius, and the will, but not the material resources. The expansion of the Sorbonne, promised for 14 years, had never happened; the Faculté de Médecine was squeezed in a space a third of what was needed, with deplorable facilities. The situation was desperate. "Make no mistake," Wurtz warned; "this is an interest of the first order, for the intellectual life of a people feeds the sources of its material power." An "army of workers" was laboring productively in the German university laboratories, producing far more science than in France; on the other hand, an expenditure of merely a few million francs would rectify the situation.[67]

Even Wurtz was astounded at what he saw on this trip, especially the new science campus being built in grand style at the University of Leipzig. The chemical institute was even larger (in student capacity) than those of Berlin and Bonn, and there were equally impressive institutes for physiology and for pathology under construction. Wurtz wrote to Duruy from Leipzig: Saxony "seems intent on conquering, in scientific and intellectual authority, what it has just lost in political autonomy" after being gathered to the Prussian orbit as a result of the Austro-Prussian war. He noted that the proposal for enlarging the Faculté de Médecine which he had recently made on Duruy's invitation would have to be expanded as a result of this new information.[68]

67. Ibid., pp. 12–20, 82.

68. Wurtz to MIP, 5 July 1868, F17/21890: "ce petit pays de Saxe qui semble vouloir conquérir en autorité scientifique et intellectuelle ce qu'il vient de perdre en autonomie politique." For a summary of the development of science education in Saxony and the University of Leipzig during the years 1860–1875, see Rocke, *Quiet Revolution*, pp. 265–278. Saxony had been an ally of Austria in the Austro-Prussian War, and after Königgrätz was forced by Prussia to join the North German Confederation. Thereafter the country was a loyal Prussian ally; later, it was a member state in the German Empire.

Between January and July, Duruy developed a plan to tie all existing Parisian scholarly institutions into a new bureaucratic entity which he named the École Pratique des Hautes Études (EPHE), which was also designed to form a leverage point for expanded state support of higher education. In a word, just as the École Normale was intended to educate future teachers, the EPHE was intended to educate future savants. The 50,000 francs that Duruy had requested (and did indeed receive) on the 1868 budget was to constitute only a down payment on this ambitious scheme. Even before Wurtz returned from his last tour of German labs, Duruy submitted a draft of the proposed legislation before the emperor. "I have the honor," Duruy wrote, "to propose for Your Majesty's signature, two draft decrees that MM. Dumas, Claude Bernard, Pasteur, Wurtz, Jamin, Milne Edwards, Balard, Frémy, Alfred Maury, etc., consider excellent. They are the result of much hard work and have been fully discussed in many meetings. The Conseil Impérial [Académique] has adopted them unanimously. . . . In case that some doubts may be raised in the mind of the Emperor, I pray that His Majesty will kindly ask me to call on him. It seems to me moreover that I have done nothing in all this but to realize the thought that the Emperor deigned to express to me six months ago. . . ."[69]

Along with the draft decrees was a long memo of justification. Cleverly playing on the anxiety of the day, Duruy did not hesitate to use military metaphors to gain support for his proposals. Academic laboratories, the "arsenals of science," were being built at great expense abroad—even in Russia and the United States—and constitute "a serious menace against one of our most legitimate ambitions. . . . Our masters . . . regard themselves as disarmed in the face of their rivals." Duruy's prescription was modeled explicitly on "institutions of the type by which Germany has found the means amply to develop experimental science, which we study with an anxious sympathy." Young Frenchmen hear "witty, elegant, sometimes eloquent lectures, often even applauded (a custom that I would gladly see disappear) . . . but they are only auditors, not students." Oral instruction is insufficient; it is necessary that "eyes see and hands touch," in the midst of a group of devoted workers surrounding the master. Instruction must be no longer be solely oral and didactic, but "pratique," by which he meant not oriented to applications, but rather conducted in the manner of a German university

69. Duruy to Napoleon, 15 July 1868, in *Notes et souvenirs*, vol. 1, p. 315.

Praktikum—learning by doing, and mixing pedagogy with research. Up until now, France had had only one such school, that of chemistry—by which he meant Dumas's Rue Cuvier lab (1838–1848) and Wurtz's lab at the Faculté de Médecine (1853–1868); if we were to create similar practical training laboratories in other disciplines, think of how scientifically eminent the country would become.[70]

The EPHE was to comprise a coordination of teaching labs, research labs, and humanistic seminars spread over a variety of Parisian institutions, and divided into four sections: mathematics, physical sciences, natural history and physiology, and history and philology. Where labs did not exist, they would be built; a director would be appointed for each lab, with an appropriate salary; staffing with répétiteurs would be increased to a proper level; each section would be run by an appointed commission; and the whole organization would be directed by a Conseil Supérieur. The student body would consist of those studying for the licence, agrégation, or doctorat—though no degrees would be awarded by the EPHE itself—and even foreigners could enroll. Duruy averred that one could create all this with little real structural change, for a firm foundation already existed; furthermore, the costs would be very modest, for he was not proposing to construct any new large buildings. He asked for a mere 100,000 francs for 1869 to pay for new research labs, with an additional 80,000 francs for teaching labs. This would suffice to provide for the Muséum d'Histoire Naturelle, the Collège de France, and the École Normale. The Faculté de Médecine would have to wait until next year. The Sorbonne was another matter, which would take longer to rescue; the EPHE, in Duruy's view, was a "young plant, which, pushing between the stones of the old Sorbonne, would one day bring it down."[71] The modesty of this initiative is indicated by the fact that a full reconstruction of Parisian facilities of higher education would have cost perhaps a thousand times the amount of these new appropriations.

Napoleon signed the decrees on 31 July 1868, and the EPHE was born. That day Dumas called on Duruy, and learned the details. "Ah," he cried,

70. Duruy, "Rapport à l'Empereur . . ." of 15 July 1868, F17/13614; printed with insubstantial changes in Duruy, *L'administration de l'instruction publique de 1863 à 1869* (Delalain, 1870), pp. 644–658.

71. Ibid.; *Notes et souvenirs*, vol. 1, p. 316.

"I came to the world thirty years too late!" To Duruy's protests, Dumas responded: "This decree reminds me of the lost time, the futile efforts, the difficulties that I have had to surmount in order to found a laboratory, instruments, financial means; in a word, all that you have just given to my successors. Think of what I could have done with such resources."[72] In view of Dumas's great name, and for political reasons, Duruy designated the École Centrale, which had been nationalized a decade earlier, a "laboratoire libre" of the EPHE, and he named Dumas as titular director (even though he had not been scientifically active for many years).[73]

Dumas and Wurtz were both named to the Conseil Supérieur of the new organization. The commission for the physical sciences section consisted of Wurtz, Balard, Frémy, Jamin, and Paul Desains, and eight labs were included in the section, including those of Wurtz, Deville, Balard, Berthelot, and Frémy. Registration was opened to prospective students in October, and it proved popular. Hoping for at least 40 students and optimistically expecting about 80, the EPHE quickly garnered around 400 registrants, of which only 264 could be accommodated. Most of the labs that were able to accept students began doing so, but often only with difficulty, by the official start of the academic year in November.[74]

Wurtz made modest requests of Duruy after the creation of the EPHE: 2000 francs for repairs and 3000 annual budget for his research lab, and nothing more for the instructional lab, which (as we have seen) had been established with a small budget 2 years earlier. He did note, however, that the city of Paris had recently abandoned rooms in the Musée Dupuytren, which could be converted to much more appropriate labs, if the opportunity were seized. (This building had been the site of Wurtz's first laboratory of his own in Paris, during the years 1847–1850.) At the Muséum d'Histoire Naturelle, Frémy (who, as we have seen, had had a head start since 1863) planned for 50 students in his teaching lab, while urging a further expansion. "Let us not forget," he noted, "that the despicable

72. *Notes et souvenirs*, vol. 2, pp. 238–239.

73. MIP note, 16 April 1884, F17/4010.

74. Duruy, *Notes et souvenirs*, vol. 1, p. 317; EPHE, printed notice, January 1869, F17/13614; MIP arrêté, 2 January 1869, ibid.; MIP report for EPHE, n.d., ca. 1870, ibid.; proof sheets for enrollment report, n.d. [ca. 1872], ibid.; *Rapport sur l'École Pratique des Hautes Études, 1871–72* (Delalain, 1872).

Prussians are constructing *Palaces* for chemistry." A year later Frémy had no fewer than 75 students. At last an instructional chemical laboratory began to be built at the Sorbonne; the director (Deville) was able to accept students beginning in January of 1869. Milne-Edwards reported this to the Conseil Académique with pride and gratitude, while still pleading, with one eye on the Germans, for much more ("... the Conseil knows that the Université de France is poor, very poor ...").[75]

Substantial growth in enrollment in the EPHE continued after 1868; the amount added to the budget was on the order of 100,000 francs per year. Although not a "bricks-and-mortar" initiative, and although the great influx of funding and building fervently desired by the scientists did not occur, the EPHE was a significant success in several respects: a few new labs were built, equipped with proper annual budgets; new salary lines were added; enrollments burgeoned; the German practicum model gained an institutional toehold in France; a new emphasis on original research in French institutions of higher education was established. Duruy was well aware of the modesty of the EPHE compared to the aspirations of the French academic elite. The initiative represented his best attempt to get the best deal he felt was achievable, in view of the many obstacles in the way of more substantial reform, and his assessment was probably accurate. Although Duruy was forced out of the ministry only a year after he invented the École Pratique des Hautes Études, he was immoderately proud of what he considered his "quiet revolution" in French education.[76]

However, for Wurtz little had changed. As dean of his Faculté, he continued to plead with Duruy's ministry for the desperately needed complete

75. Wurtz to MIP, 23 November 1868; Schützenberger to MIP, 17 March 1869; Frémy to Cher Monsieur et ami, 4 November 1868; all in F17/13614 ("[N]'oublions pas que ces affreux Prussiens construisent des *Palais* pour la chimie ..."). Frémy to MIP, 28 October 1869, ibid. Milne-Edwards, *Exposé fait au Conseil académique de Paris sur l'enseignement supérieur des sciences* (Imprimerie Impériale, 1868). Frémy claimed (annual report, 9 November 1871, F17/13614) that "all the German laboratories" have a maximum of ten students per instructor. This was true nowhere in Germany, and it could only have been invented (by Frémy or someone from whom he gained this impression, probably orally). It illustrates that the German case could also be used by the French for false or exaggerated claims.

76. Duruy to Napoleon, 17 January 1869, in *Notes et souvenirs*, vol. 1, p. 319. On Duruy's strategy with the EPHE, see Horvath-Peterson, *Duruy*, p. 195.

reconstruction of the medical campus.[77] When not occupied by his duties as dean, he continued to pour heart and soul into his research laboratory, crammed into the small rooms he had carved out of the Faculté de Médecine more than 15 years earlier. Moreover, the drama described in the last two sections of this chapter was played out while the Second Empire was careening through a minefield of domestic and international crises. Napoleon was ill, distracted, and erratic. War with Prussia soon would break put an end to his dreams. After the national catastrophe, in the new environment of the Third Republic, academic scientists would finally see some of their wishes fulfilled.

77. E.g., Wurtz to MIP, 12 March 1870, AJ16/6348; *Rapport sur l'École Pratique des Hautes Études, 1873–74* (Delalain, 1874), p. 33.

10

The Atomic War

The last three decades of Wurtz's life were dominated by one great professional goal: to win a victory in France for the forces of atomism and theory over opponents who advocated the purportedly more empirical and more certain equivalents. I will begin my analysis of this conflict by outlining the brief for the use of equivalent weights—a position that was represented with particularly energy by Berthelot.

The Case for Equivalents

The use of equivalent weights in chemical formulas goes back to the origins of the chemical atomic theory.[1] Although Dalton's first version of this theory was formulated as early as 1803, a more pragmatic birth date for the theory might be taken as 1814, when central contributions simultaneously emerged from Wollaston and Thomson in England, Ampère and Gay-Lussac in France, Schweigger and Kastner in Germany, and Berzelius in Sweden. A serious problem emerged immediately. Each of these men had different notions regarding how best to determine (that is, to guess at) the true formulas of the simple molecules—water, marsh gas, ammonia, nitric acid, carbonic acid—that had to be used to create the theory. This speculative leap could not be sidestepped, for the empirical combining weights for

1. For documentation of the following discussion, see A. Rocke, *Chemical Atomism in the Nineteenth Century* (Ohio State University Press, 1984), passim. In this chapter, and at many points throughout this book, I use the terms "atomism" and "atomist" as they were conventionally used in late nineteenth-century France (i.e., to designate the reformist chemistry developed in the 1850s), and not as I defined "chemical atomism" in 1984.

any given element only provides a group of weights related by small integral multiples; the assumption of a molecular formula is necessary to provide a basis on which to select one of the various multiples for each element as its unique atomic (or equivalent) weight.

Since different authors made diverse choices for these formulas, different versions of atomic weight tables competed from the start. William Hyde Wollaston captured the rhetorical high ground. In an influential article published in 1814 in the *Philosophical Transactions of the Royal Society*, he proclaimed that his own preferred relative weights for the elements, which he called "equivalents," were derived without any reference to microphysical hypothesis. The contrast between (what appeared to be) the hyper-hypothetical Dalton and the sober empiricist Wollaston could not have seemed greater, and Wollaston's equivalent weights immediately became more or less standard throughout Europe. The only significant dissenters in these early years were Berzelius (who advocated a more theoretically developed system of "atomic weights" but whose influence was not yet strong) and Gay-Lussac (who, for complicated reasons, differed with Wollaston on the assumed formulas for carbonic oxide and carbonic acid gas, and who therefore used different respective relative weights for carbon and oxygen).

The situation became considerably more complex after 1830. Berzelius's stature had risen dramatically in the intervening years, and his atomistic system, in its revised form of 1826, had acquired greater authority. Simultaneously, atomic weights and molecular formulas began to be used in a new, confident, and successful heuristic fashion to guide chemical investigations.[2] The influence of Berzelius's official and unofficial students, including Liebig, Wöhler, and (slightly later) Bunsen, led to a strong resurgence of Berzelian atomism in Germany. Great Britain was more or less immune to this movement, the British continuing to prefer the "equivalents" of their countryman Wollaston. In France, Berzelius had some real influence, but preference there went to the modified system of Gay-Lussac, which was being strongly championed by Dumas. For about 15 years after

2. U. Klein, Experimente, Modelle, Paper-Tools (Habilitationsschrift, Universität Konstanz, 1999); "Paper-Tools and Techniques of Modelling in Nineteenth-Century Chemistry," in *Models as Mediators*, ed. M. Morgan and M. Morrison (Cambridge University Press, 2000).

1830, then, there was a different set of elemental weights in use in each of the three principal European countries. Each different weight system required an entirely distinct set of formulas for all known chemical compounds (the numbers of which were growing explosively), and these differences had ripple effects that cascaded through pedagogy, texts, reference works, and research programs.

In the 1840s, this confusing multiplicity of systems, which at least had had a degree of stability, became even more chaotic, to the point of bewilderment. After 1842 in France, Gerhardt and Laurent advocated a fourth atomic-weight system, a modification of the Berzelian system that Berzelius himself found intolerable. Their early campaign did not prosper, partly because they were arguing against a new concerted campaign, initiated by Leopold Gmelin, to replace the Berzelian system by the (apparently) certifiably empirical Wollastonian equivalents.[3] At first Gmelin's success in this endeavor was at best mixed; however, Baconian empiricism and "textbook positivism" were by then in ascendance, and Berzelius was increasingly viewed as superannuated and overly theoretical.[4] By about 1845, chemists of various nationalities were coming over en masse to the Wollaston-Gmelin flag. At last, it seemed, all European chemists were beginning to agree on a single system, moreover one that was putatively non-theoretical and hence could be adopted permanently.

Around the middle of the century this enviable goal appeared to be within reach of the international community. Ironically, this was just when a coterie of leading theorists began touting the least popular of the four systems, the Gerhardt-Laurent reformed version, not merely as simpler or more convenient but as true—the one system that had to be adopted because of its unique claim to veridical character. There was much implicit and explicit conflict in the 1850s over these issues, all over Europe, until the Karlsruhe Congress was devised to try to create concord. The Congress was, of course, a brainchild of the reform camp, and the reformers did have much impact in Karlsruhe. However, as we have seen, the advocates of equivalents also achieved a major victory by having the following proposition approved in plenary session: "The concept of equivalents is empirical, and independent of the concepts of atom and molecule." The fact that this

3. I described this episode in the fourth section of chapter 3.

proposition prevailed by acclamation raises the suspicion that the motion was the result of parliamentary maneuvering by the equivalentists, as a counterattack against the reformist tone of the entire gathering. In any cases, the reformers went home vaguely dissatisfied with the results of the meeting.

One can heartily sympathize with those young chemists, including Berthelot, who were educated in the peak years of discord, the late 1840s and the early 1850s. Every authority seemed to have his own preferred set of weights, based on his own preferred theories; and the referents of these theories (namely the atoms being specified in the formulas) were utterly beyond direct human perception. "These continual variations in signs," Berthelot wrote in 1860, "are more destructive than useful for true progress in organic chemistry. They destroy the ties connecting its ideas to the most general laws of mineral chemistry; they obscure the regular filiation of ideas and the progressive connections of discoveries; finally, they tend to strip chemistry of its true character."[5] The unabashedly conjectural theories of the microworld of Ampère, Laurent, Gaudin, and others tended, in the eyes of empirically inclined chemists such as Berthelot, to discredit even the more cautiously phrased atomistic notions of Berzelius, Dumas, Liebig, Gerhardt, and Kekulé. This quagmire of uncertainty was profoundly unsettling, and logistically difficult to deal with. Moreover, it reached a crescendo during one of the most unsettling and violent periods of French history. Finally, Berthelot's personal life was filled with a sense of insecurity, despite his own massive (perhaps overcompensating) self-confidence. There should be little wonder, then, at the appeal of the apparent empiricism and permanence of the system of chemical "equivalents."

Over and above these underlying cultural and psychological factors, Berthelot had many substantive complaints against atomism.[6] In chapters 7 and 8 I treated some of the arguments he used desultorily, principally against Wurtz, in the earliest stages of this conflict. One of these complaints

4. For an effective critique of the view that Comtean positivism directly influenced European chemistry of the 1840s, see B. Bensaude-Vincent, "Atomism and Positivism," *Annals of Science* 56 (1999): 81–94.

5. Berthelot, *La chimie organique fondée sur la synthèse* (Mallet-Bachelier, 1860), vol. 1, pp. xxiii–iv.

6. See M. Nye, "Berthelot's Anti-Atomism," *Annals of Science* 38 (1981): 585–590.

was essentially semiotic, and appears to have been prompted by the publication of Gerhardt's *Traité de chimie organique* (1853–1856). In that work, Gerhardt professed an empiricist commitment, but one that was not to Berthelot's liking. In the latter's view, Gerhardt's system led to an empty game of formulas, numbers, and language, rather than to concrete experimental investigations of real substances. Gerhardt was intent on defending one unique symbolic system, with one set of numbers and one set of words, and Berthelot found his sanctimonious ontological claims for that one version repellent and fatuous. What difference does the symbolic language make, Berthelot insisted, if the facts remain unchanged in all the various alternative systems of representation?

Perhaps the principal success claim made by advocates of atomic theory was to be able to explain the internal constitutions (later called "structures") of molecules, and to do so with greater cogency than equivalentists were able to do. In the early 1860s Berthelot took on the atomists directly. "The constitution of bodies," he wrote, "can be envisaged in two ways: 1. From the point of view of positive science, i.e., precise relations that exist between facts; 2. From the point of view of speculative science, i.e., constructions imagined by the human mind to represent things." For positive science, chemical constitutions could only express "the totality of the physical and chemical properties of the body, reduced to the simplest and most general relations that experiment establishes."[7] Accordingly, Berthelot insisted on using what he called "équations génératrices," as contrasted to "formules rationnelles"—that is, he used formulas that indicate how the substance was formed, rather than what its internal construction was thought to look like. He took it as a point of principle that all non-derivativized organic compounds (called "composés unitaires") be written integrally, without indicating supposed submolecular groupings.

The essential problem, Berthelot wrote, was that rational or constitutional formulas depicted putative static arrangements of hypothetical atoms—a compounding of conjectures. In common with atomists, Berthelot affirmed that molecules of chemical substances must contain

7. Berthelot, *Leçons sur les méthodes générales de synthèse en chimie organique, professées en 1864 au Collège de France* (Gauthier-Villars, 1864), pp. 41–42.

smaller particles; however, science simply had no epistemic access to that level of the world. In Berthelot's words:

A body, simple or compound, can be considered as constituting a certain system of material particles, exhibiting a definite mass, all maintained at certain distances, and each animated by its own speed and movements of vibration, rotation, translation, etc.; certain of the movements change with temperature and various other circumstances. It is the totality of these movements which characterize the body, and which determine the effects it produces on other bodies, on our senses in particular. . . . [Thus,] in a compound body the elements no longer exist, properly speaking.[8]

To specify more than we really know about a molecule is a violation of the dictates of proper ("positive") science.

Indeed, what must above all be sought in the representation of an idea is not to particularize it by individual symbols, to make of it a kind of property; but rather it is necessary to give it the expression that is most general and abstract as possible, stripped of all hypotheses, so that its consequences and its relationships of analogy with the totality of known phenomena appear in all their simplicity. Thus it is that in physics the discussions about the general properties of matter, like divisibility, porosity, impenetrability, etc., and about the application of these properties to the explanation of calorific, electric, magnetic, luminous, etc. phenomena, after having long been in dispute, have finally disappeared; the accumulation of discoveries has obliged scholars to exclude all vague explanation, to always report the facts in their simple, clear, and truly determinate relations.[9]

Semiotic analysis and a physicalist-reductionist philosophy of matter provided the context for Berthelot's animus against contemporary atomistic accounts of chemistry, but there was much more to it than that. Atomists themselves admitted inconsistencies in their accounts, only arguing that in a broader vision of the theory these problems receded in importance. Not so for Berthelot, as he continued to argue (somewhat out of place) in a monograph on isomerism published about the same time as his lectures to the Collège de France.[10] One issue was methodological in character. Berthelot was committed to constructing the science of chemistry from chemical data alone, whenever possible, for doing so (he was convinced)

8. Ibid., pp. 63–64.

9. Berthelot, *Chimie organique*, p. cxxv.

10. The following arguments are taken from Berthelot, "Leçon sur l'isomérie," in Société Chimique de Paris, *Leçons de chimie professées en 1864 et 1865* (Hachette, 1866) ("Application of the Study of Isomeric Bodies to the Discussion of Atomic Theories").

increased the empiricism, the rigor, and the internal consistency of the resulting system.[11] In contrast, atomists had combined data from gravimetric analysis of chemical reactions—stoichiometry and "equivalents"—with conclusions drawn from physical measurements of vapor densities, specific heats, and crystallography.

Not only was the atomists' way of proceeding methodologically suspect, in Berthelot's view; it was also rife with inconsistencies. For instance, the equivalentists' weights for oxygen, sulfur, magnesium, zinc, tin, and calcium violated the Petit-Dulong law (the inverse proportionality of elemental specific heats to atomic weights), so the atomists had proposed to double these elemental weights to bring generality to the law. This was ludicrously ad hoc, Berthelot thought. By the same procedure, every specious scientific law could be saved by appropriate juggling of data. Instead of doing the sensible thing and discarding this failed "law," atomists had resorted to shameless trickery. To the atomists' argument that a beautiful consistency resulted from this adjustment to elemental weights, thus justifying the sleight of hand, Berthelot pointed out that even in the atomistically "repaired" system there were other problems. For one thing, not every elemental atomic weight fit the Petit-Dulong law, even after the revision. For another, the proportionality constant at the heart of the "law" was only an approximate number, not a real constant at all. Finally, this whole process involved using suspect physical data to violate, in several notable instances, strong chemical sensibilities. For example, the alkali metals and the alkaline earth metals, so closely analogous in their chemistry, would have to be classed in quite different formula patterns (e.g., caustic potash as K_2O, caustic lime as CaO).

So much for the Petit-Dulong "law." What about use of vapor densities to judge molecular sizes? Ever since Dumas had attempted, in the late 1820s and the early 1830s, to save the generality of the hypothesis that equal volumes of gaseous elements under similar conditions contain equal numbers of their molecules, chemists had strained to perceive a consistent pattern—quite without success, in Berthelot's view. Simply stated, the exceptions to Avogadro's hypothesis were notorious. Some had suggested that these anomalies could be removed by assuming variable submolecularity—i.e., that mercury vapor consisted of single atoms of mercury, while oxygen gas

11. Berthelot, "Leçon sur l'isomérie," p. 231.

consisted of molecules containing two atoms each, and arsenic and sulfur had even higher "atomicities." To Berthelot, this represented another instance of regrettable ad hoc trickery to save an eminently discardable hypothesis. Moreover, Berthelot simply could not accept, at this time or later, the notion that two atoms of oxygen, or two of sulfur (etc.), could possibly combine.[12] What would be the basis for such combination, Berthelot wondered? According to the model of coulombic attraction, only different atoms could ever chemically combine.

Finally, Berthelot demonstrated that the atomic theory was internally inconsistent in a dramatically obvious fashion. On the basis of vapor densities, atomists had urged that the equivalent (or atomic) weight of oxygen should be 16 rather than 8. These numbers, of course, express hypothesized ratios, the numerator being the weight of an oxygen atom and the denominator the weight of a hydrogen atom. However, also in the atomist system, the equivalent of hydrogen would itself have to be doubled to match the measured ratios of vapor densities between hydrogen gas and many compound gases, such as nitric oxide. But in order to continue to maintain the oxygen/hydrogen ratio of 16, doubling the hydrogen weight would then require quadrupling the equivalent weight for oxygen. There was, in other words, Berthelot thought, no way to create an internally consistent system, even on the atomists' own terms.[13]

Blatant internal inconsistencies, constitutive use of deeply flawed generalizations that had no right to be called laws and of hypotheses that were merely conjectures, methodological errors, and philosophical solecisms— atomic theory was not only wrong but embarrassing, not only unhelpful but positively damaging, especially since the atomists had not even been able to agree on what version of the theory to use. The solution was clear: return to empirical, gravimetrically determined equivalents. If there is heuristic value in atomistic formulas, as the atomists insisted, then the same value will also inhere in the equivalentists' formulas, for they represent precisely the same thoughts, just expressed in a different language. But the latter language was philosophically preferable, for it represented, in sharp

---

12. ". . . for there is no chemical attraction—i.e., affinity—whatever, unless one opposes two molecules of different natures" (Berthelot, *La synthèse chimique* (Baillière, 1876), p. 167.

13. Berthelot, "Leçon sur l'isomérie," pp. 224–229.

contrast to atomistic formulas, a permanent encapsulation of positive scientific knowledge.

It could hardly be denied that Berthelot had built a compelling case for equivalentism and against atomism. In the anti-theoretical climate of mid-nineteenth-century Paris, many rallied around Berthelot's flag, including especially Henri Sainte-Claire Deville (1818–1881). Deville had been born just 3½ months after Wurtz, and had graduated M.D. in the same year (1843). Patronized by Thenard and Dumas, Deville spent a few years at the Faculté des Sciences in Besançon before succeeding Balard at the École Normale in 1851. Two years later he also began serving as suppléant for Dumas at the Sorbonne, an activity that led to his full appointment there upon Dumas's retirement in 1867. His reputation at both institutions was justly high, and he built an extraordinary reputation in experimental chemistry, first in the organic and then in the inorganic field. But theory, particularly atomic theory, was not within his range of interests.[14]

The "Debate" at the Académie

Berthelot left his research program in organic chemistry shortly after publishing the monographic versions of the lectures he had given at the Collège de France and the Société Chimique, turning to the study of chemical dynamics and thermodynamics. However, about 10 years after leaving the field, he published a revised and updated monograph on chemical synthesis. It is not obvious why he did so, in view of his inactivity in this field; indeed, some long passages in the new book were simply lifted from his massive 1860 treatise.[15]

One possible motive was his old rivalry with Wurtz. In September of 1874, Wurtz delivered the plenary opening address to the Association Française pour l'Avancement des Sciences in Lille, on the subject "La théorie des atomes dans la conception générale du monde." This was an eloquent, passionate review of the history of atomic theory, in both chem-

14. On Deville, see J. Dumas, *Discours et éloges académiques* (Gauthier-Villars, 1885), vol. 2, pp. 302–328; J. Gay, *Henri Sainte-Claire Deville* (Gauthier-Villars, 1889); D. Gernez, *Notice sur Henri Sainte-Claire Deville* (Gauthier-Villars [1894]).

15. Berthelot, *Synthèse chimique*. Pages 166–171 of this book are virtually identical to pp. cxxi–cxxvi in volume 1 of *La chimie organique fondée sur la synthèse*.

Figure 10.1
Henri Étienne Sainte-Claire Deville. Source: Photograph Collection, Deutsches Museum, Munich.

istry and physics, from the Greeks to the present. The speech ended with a remarkable religious flourish. The human mind, he said, will never rest satisfied with even such a complete and satisfying theory; it will instinctively seek the very origins of things, and " il est conduit à les subordonner à une cause première, unique, universelle: Dieu." That very summer Wurtz had been invited to offer a "special course" at the Sorbonne, on atomic theory and organic chemistry. Wurtz's rhetoric there, too, was powerful, and the course was extremely successful; it led to his appointment to a new chair there in 1875. Neither of these episodes would have been welcome to Berthelot.[16]

All Berthelot's arguments against atomism were reprised in this new monograph on synthesis. Berthelot proclaimed that "all chemists" agreed that matter consists of atoms, for there seemed to be no other way to account for the existence of the laws of stoichiometry. However, what now went by the name of "atomic theory" was different, he wrote; it was, in essence, a deeply flawed theory of the vapor state, a set of speculations constructed upon Avogadro's hypothesis. Moreover, the theory, bad to begin with, had become interlarded with numerous ad hoc auxiliary hypotheses, which had been designed to patch holes in this leaky intellectual vessel but which actually contradicted its axiomatic base. Hence, "there remains nothing more here than an ingenious and subtle work of fiction, and new linguistic conventions."[17]

The ad hoc assumptions Berthelot was referring to here related to various problems regarding the vapor densities of elements and compounds. For atomists, the only truly correct formulas were those that used the reformed atomic weights for the various atoms in the molecule and the two-volume convention to fix the molecular magnitude. However, there were anomalies (discussed in the preceding section). One of these was chloral hydrate, whose atomistic formula matched the amount of the substance that occupied four volumes rather than two. Atomists had suggested that the compound dissociated at high temperature into two discrete molecules, chloral and water vapor, thus occupying twice the

---

16. Wurtz, *La théorie des atomes dans la conception générale du monde* (Masson, 1875), p. 60; Berthelot, *Synthèse chimique*. On Wurtz's special course of 1874, see AN F17/21890.

17. Berthelot, *Synthèse chimique*, p. 164.

space of the unified compound; for atomists, the substance would occupy two volumes if it could be vaporized without dissociation. There was experimental evidence to support dissociation of hot chloral hydrate vapor, but both Berthelot and Deville found the evidence unconvincing. Ironically, in the late 1850s and the early 1860s Deville himself had done much to establish experimentally the fact that many substances dissociate thermally,[18] but he did not believe that the dissociation hypothesis could explain anomalous vapor pressures. For Deville (and Berthelot), chloral hydrate vapor occupied four volumes in an undissociated state. For them, it was one of many instances that disproved Avogadro's hypothesis of "equal volumes—equal numbers," one of the foundations of the atomic theory.

Soon after Berthelot's new book appeared, Deville's former student Louis Joseph Troost (1825–1911) read a paper at the Académie des Sciences supporting Deville's view of the matter. (In 1874 Troost had succeeded Pasteur at the Sorbonne; he was therefore an obvious rival of the somewhat older and scientifically more prominent Wurtz, who was given his new chair for organic chemistry at the Sorbonne in 1875.) Troost claimed to provide a new experimental demonstration that chloral hydrate was not dissociated in the vapor state, and hence that it represented a clear violation of Avogadro's hypothesis.[19] This paper initiated a dramatic 3-year controversy that involved the publication of more than 40 papers in the *Comptes rendus* and sparked the passionate participation of Wurtz, Deville, and Berthelot. The drama was accentuated by the eminent cast of characters:

18. For a summary of early studies of dissociation, including Deville's, see J. Partington, *A History of Chemistry* (Macmillan, 1962–1971), vol. 4, pp. 494–499.

19. Troost, "Nouvelle méthode pour établir l'équivalent en volume des substances vaporisables," *Comptes rendus* 84 (1877): 708–711. Troost introduced a sample of hydrated potassium oxalate into hot chloral hydrate vapor; the measured density of the gas then increased, which indicated to Troost that water vapor must have been given off by the salt. But if the salt emitted water vapor, there could have been little or no water vapor present originally, otherwise mass effect would have inhibited the release of water vapor from the salt. But if water vapor was not present initially, it could only have been because the chloral hydrate was not dissociated. A second experiment involved exposing *anhydrous* potassium oxalate to chloral hydrate vapor, whereupon the salt failed to hydrate. This could only mean that no water vapor was present, hence that chloral hydrate was undissociated in the vapor state.

the three professors of chemistry at the Sorbonne, plus the professor of organic chemistry at the Collège de France.[20]

Wurtz's first response to Troost's article pointed to methodological weaknesses that had invalidated the results; Wurtz's own modified repetition led to the opposite conclusion, that chloral hydrate was dissociated in the vapor state. Deville then came to the defense of his protégé: Wurtz was wrong, Troost's work was correct, chloral hydrate invalidated Avogadro, and there was no need for "recourse to the hypothesis of atoms and molecules."[21] Wurtz responded by providing a broader defense of chemical atomism and arguing that equivalents were neither more empirical nor more consistent than atomic weights. In his historical references he was careful to choose French champions of atoms, such as Gay-Lussac, Dumas, Ampère, Dulong, and Petit. "In summary," he wrote, "the preceding discussion shows that the system of chemical equivalents, which won out ca. 1840 over the atomic notation of Berzelius, did not take any account of the discoveries of Gay-Lussac on the combinations of gases with each other, and that maintaining the principle of equivalence in chemical notation would return the science to the time of Dalton, Wollaston, and Richter. This would be an anachronism, more precisely a retreat, and science does not retreat."[22]

That Berthelot was angered by this pronouncement is clear from the tone of his entry into this controversy, published in the *Comptes rendus*

20. The volumes and initial page numbers of these articles are as follows: vol. 84, pp. 708, 711, 977, 1108, 1183, 1189, 1256, 1262, 1264, 1269, 1274, 1347, 1349, 1407, 1472; vol. 85, pp. 8, 32, 49, 144, 400; vol. 86, pp. 31, 971, 1021, 1170, 1394; vol. 89, pp. 190, 271, 306, 337, 429, 803, 1062, 1099; vol. 90, pp. 24, 56, 112, 118, 337, 341, 491, 572. The first article (Troost's) appeared on 9 April 1877; the date of the last one (by Wurtz) was 15 March 1880. Sixteen articles were by Wurtz, ten by Berthelot, eight by Troost, six by Deville, and one by Moitessier and Engel. For discussions of this episode, see P. Colmant, "Querelle à l'Institut entre équivalentistes et atomistes," *Revue des questions scientifiques* [5] 33 (1972): 493–519; N. Pigeard and A. Carneiro, "Atomes et équivalents devant l'Académie des Sciences," *Comptes rendus* 323 (1996): 421–424.

21. Wurtz, "Recherches sur la loi d'Avogadro et d'Ampère," *Comptes rendus* 84: 977–983; Deville, "Sur la loi des volumes de Gay-Lussac," *Comptes rendus* 84: 1108–1112.

22. Wurtz, "Sur la loi des volumes de Gay-Lussac," *Comptes rendus* 84: 1183–1189.

immediately after Wurtz's article. He repeated many of the criticisms of the atomic system that had appeared in his recent book. "In truth," he wrote, "we do not see molecules, and we have no way to count them." Atomic theory was a ramshackle structure of ad hoc adjustments to a weak hypothesis. "How does this constitute above all a new theory, a modern chemistry, which is appropriate to oppose to the chemistry of Lavoisier and Gay-Lussac, which we all teach? . . . Who has ever seen, I ask again, a molecule of a gas or an atom?" Finally, he invoked the rigor of the physicalist model:

Today, a certain number of chemists, less accustomed to the precision of physical ideas, pretend to replace the rigorous definition of laws themselves by images that represent those laws, that is, by hypotheses, variable with each generation, each sect, each personality. To depict the true character of those hypotheses is by no means to make science retreat, i.e. to abandon acquired truths; rather, it is to permit scientists, relieved of superfluous baggage, to advance with greater certitude in the research of the real laws of molecular mechanics.[23]

Wurtz's rebuttal a week later provided not only new experiments to support his position regarding chloral hydrate but also new arguments for atoms and against equivalents. If atomic theory could be faulted for "inconsistencies," equivalents were worse. Why, Wurtz asked, did Berthelot use true stoichiometric equivalents for some compounds (HO or H_2O_2 for water), and yet write ethyl alcohol, glycol, and glycerin as mono-, di-, and triatomic alcohols?[24] It was, after all, Berthelot's own work that had helped establish glycerin as triatomic and thus helped gain support for a central thesis of the atomistic system: the idea of polyatomic radicals led to polyatomic atoms, the crucial concept now called "valence." Moreover, Berthelot's own thermochemical data supported the notion that certain elementary gas molecules contain two atoms each. Certainly the Petit-Dulong "constant" varied, but it was stable to within 20 percent; and who

23. Berthelot, "Réponse à la note de M. Wurtz, relative à la loi d'Avogadro et à la théorie atomique," *Comptes rendus* 84: 1189–1195. Bensaude-Vincent rightly comments ("Atomism and Positivism," p. 90) that Berthelot here defends a crude version of "polemical positivism" for purely tactical reasons against Wurtz; his hyper-empiricism here is inconsistent with more sophisticated positions he had taken in earlier writings.

24. In other words, Wurtz was suggesting that Berthelot's formula for glycerin, $C_6H_8O_6$, should by rights (in Berthelot's terms) be reduced to $C_3H_4O_3$, and that his formula for glycol, $C_4H_6O_4$, should be reduced to $C_2H_3O_2$.

was not struck by the fact that the two numbers whose product was this constant—elemental specific heats and atomic weights—individually varied by sevenfold and thirtyfold, respectively? In regard to the formula for water as H_2O: "Is it really necessary to repeat here what Gerhardt successfully established thirty years ago?" Let Berthelot find a compound with an odd number of equivalents of oxygen, and his system would be proved; he would never succeed in this. And finally, Wurtz concluded, the empiricist-physicalist appeal was likewise misplaced. All scientists, especially physicists, knew that the luminiferous ether existed, but no one had ever seen that hypothetical entity. Pace Berthelot, hypotheses were crucial to the life of science.[25]

Such arguments meant little to Berthelot. Wurtz had failed to understand, he wrote, that the two systems under discussion were only two different languages, two different conventions. What Wurtz called "valence" Berthelot preferred to call "multiple equivalence." "Valence or equivalence, it is the same idea and the same word; I am pleased to confirm our perfect agreement in this regard." He strongly disclaimed any desire to see hypotheses driven from science, insisting that he wanted only to recognize them as intellectual scaffolding. To be "faithful to the traditions of the French school," it was necessary to seek "true scientific laws."[26] Aha, Wurtz rejoined, his opponent had now yielded on the crucial point. By conceding the existence of "multiple equivalents," Berthelot had shown that not all substances could or should be expressed as composed of simple stoichiometric equivalents—and that was what valence affirmed, as well. But equivalence and valence were by no means the same thing; Berthelot's claim of their identity simply demonstrated that he had never really learned the system he was opposing—a system, moreover, that was in general use "in all European countries."[27]

25. Wurtz, "Sur la notation atomique," *Comptes rendus* 84: 1264–1268.

26. Berthelot, "Atomes et équivalents. Réponse à M. Wurtz," *Comptes rendus* 84: 1269–1274.

27. Wurtz, "Sur la notation atomique; réponse à M. Berthelot," *Comptes rendus* 84: 1349–1352. In the simplest and most common definition, valence equals atomic weight divided by equivalent weight; thus the valence of oxygen is two because that is the ratio of its atomic weight to its equivalent weight. Wurtz was correct in saying that under no legitimate definition could they be considered identical.

After a 6-month intermission, Act II began. Troost published new defenses of his original article. Wurtz performed additional experiments of his own modified design, offering results by two different methods that suggested Troost was in error and his conclusions were false.[28] Berthelot was now really irritated. Certain chemists who had studied this issue, he wrote, had allowed themselves to be captured by systematic bias ("quelque parti pris systematique"). Wurtz's experiments were flawed not only by this prejudice but also by his incaution and inexperience with precise physical measurements. Wurtz responded cooly to these charges.[29]

It was time now for Deville to weigh in again, on the side of Troost and Berthelot and against Wurtz. Troost's work had provided a "crucial" experiment that had "cut all the partisans of atomistic chemistry to the quick." For more than 20 years, Deville wrote, he had done his best to exclude "absolutely gratuitous hypotheses" such as atoms and molecules from his teaching. "I am of the opinion," he concluded, "along with my wise friend M. Berthelot, that a dangerous path is being followed in chemistry, one which for some years has been resolutely rejected by the great minds that have founded the mechanics of heat, thermochemistry, and modern physiology." Wurtz responded once more, systematically dismembering Troost's putatively crucial experiment. He then summarized five different experimental arguments taken from work by six prominent scientists that had conclusively demonstrated the presence of water vapor in hot chloral hydrate vapor, hence the fact of dissociation. But he did not expect to be able to convince M. Deville, in the obvious presence of "what appears to be a systematic bias [un parti pris]."[30]

28. Troost, "Sur la vapeur de l'hydrate de chloral," *Comptes rendus* 85: 32–34; "Sur les vapeurs de l'alcoolates de chloral," *Comptes rendus* 85: 144; "Sur la vapeur de l'hydrate de chloral," *Comptes rendus* 85: 400–402; "Sur les densités de vapeur,"*Comptes rendus* 86: 331–832; Wurtz, "Recherches sur la loi d'Avogadro et d'Ampère," *Comptes rendus* 86: 1170–1175; "Note sur l'hydrate de chloral," *Comptes rendus* 89: 190–192.

29. Berthelot, "Remarques sur la note de M. Wurtz relative à l'hydrate de chloral,"*Comptes rendus* 89: 271–273; Wurtz, "Réponse aux remarques de M. Berthelot sur ma note concernant l'hydrate de chloral," *Comptes rendus* 89: 337–338; Wurtz, "Réplique aux observations de M. Berthelot,"*Comptes rendus* 89: 429–430.

30. Deville, "De la température de décomposition des vapeurs," *Comptes rendus* 89: 803–806; Wurtz, "Réponse aux remarques de M. H. Sainte-Claire Deville sur la température de décomposition des vapeurs," *Comptes rendus* 89: 1062–1065.

Deville was now cut to the quick. He felt personally attacked, and he deplored the discourtesy and intolerance demonstrated by Wurtz's statements—not realizing, or at least not acknowledging, that Berthelot had been the first to accuse Wurtz of bias. "I assume," Deville wrote, "neither the law of Avogadro, nor atoms, nor molecules, nor forces, nor particular states of matter, absolutely refusing to believe in anything that I cannot see or even imagine. . . ."[31] Even if it were proved that complex compounds dissociate in the vapor state, Deville averred, it would not matter to him. All depends solely on experiment, and "I accept what I see." Berthelot, in his final statement on the subject, said that he regarded the matter as settled in his favor and he was content to leave the ultimate judgment to "competent men." Wurtz offered a similar final comment.[32]

The Case against Equivalents

A small but important side plot was played out during the intermission between the two acts of this little drama. Charles Marignac (1817–1894), a Swiss who had studied under Dumas and Liebig, was a professor of chemistry and mineralogy at the Académie de Genève, and a thoroughgoing modernist in chemistry. In September of 1877 he published a commentary on the 20 papers that had already appeared, focusing on the crucial points at issue between Berthelot and Wurtz and clearly favoring the latter's position.[33] Marignac pointed out that Berthelot had failed to provide an empirical definition of "equivalent," or indeed any definition at all. He then proceeded to show that two different meanings of the word were implicitly

31. Deville, "Quelques observations sur une note de M. Wurtz," *Comptes rendus* 89: 90, 56–57; Deville, untitled remarks, *Comptes rendus* 89: 341–342.

32. Berthelot, "Nouvelles remarques sur la chaleur de formation de l'hydrate de chloral," *Comptes rendus* 89: 491–492; Wurtz, "Réponse aux observations de M. Berthelot," *Comptes rendus* 89: 572.

33. Marignac's article and Berthelot's response, originally published in *Moniteur scientifique* (19 (1877): 920–926 and 1254–1257), also appeared in *Bibliothèque universelle* (59 (1877): 233–249; 60 (1877): 343–350). They were republished in Marignac's *Oeuvres complètes* (Masson, 1902–03, vol. 2, pp. 649–660 and 661–667), in *American Journal of Science* ([3] 115 (1878): 89–98 and 184–189), and (most accessibly and with commentary) in M. Nye's book *The Question of the Atom* (Tomash, 1984) (pp. 230–249). The following citations are of the latter source.

being used: what I shall here call "chemical equivalent" and "conventional equivalent."[34]

The chemical equivalent was simply the amount of one element that replaces another chemically. For instance, the relative proportion of potassium and sodium in the two respective chloride salts could provide chemical equivalents for the two metals; chlorine's chemical equivalent could then be determined by examining hydrogen chloride, all these elemental weights being related to the lightest element, hydrogen. Hydrogen was arbitrarily assigned the dimensionless weight of one. Chlorine then became 35.5 (for 35.5 grams of chlorine combine with one of hydrogen), sodium 23 (23 grams of sodium combine with 35.5 of chlorine), potassium 39, and so on. These were indeed empirical quantities determined by an empirical definition—the data came directly from gravimetric combining proportions. However, Marignac averred, equivalentists wrote as if every use of "equivalent" denoted exactly this, whereas in fact a different implicit meaning was usually being (unconsciously) indicated.

The reason a second meaning was necessary was that in a multitude of cases chemical equivalence provided no guidance for determination of elemental weights. One problem was that chains of chemical replacements extended only so far. In Marignac's words: "When we deal with bodies which have not a similar analogy, and particularly if they do not perform the same functions, the idea of equivalence has no meaning."[35] An even more intractable problem was the prevalence of multiple proportions. The chemical equivalent of oxygen was 8 from the case of water, but 16 from the case of hydrogen peroxide. The problem was then further compounded. If the chemical equivalent of oxygen was 8, then the chemical equivalent of carbon was 6 from carbonic oxide or 3 from carbonic acid; if oxygen was 16, the choices for carbon were 12 and 6. Could one decide for carbon by using hydrides of the element? Not at all. If benzene was chosen as the test compound, carbon was 12; if ethylene was chosen, carbon was 6; if methane was chosen, carbon was 3. In fact, nearly every different hydrocarbon (scores of which were known by 1877) dictated a different chemical equivalent for carbon.

34. These are my coinages, not Marignac's, but I believe that they summarize Marignac's argument faithfully. In *Chemical Atomism in the Nineteenth Century*, I define "chemical equivalent" and "conventional equivalent" as I do here.

35. Nye, *Question of the Atom*, pp. 232–233.

Let me put Marignac's argument differently: If chemical equivalents per se were to be used to notate chemical formulas, every formula would be of the form XY, which was tautologically unhelpful. Practically speaking, a single number must be chosen for each element as a precondition for writing formulas. Chemical equivalents, though truly empirical, do not yield single numbers at all: for each element we obtain a series of numbers, all related by being integral multiples of the smallest of the series. We are free, if we like, to choose one of these numbers for each element by convention, and to use that number in calculating formulas for all the compounds in which that element appears. It is then no longer an empirical "chemical equivalent" but rather a "conventional equivalent."[36] In Marignac's words (slightly paraphrased):

It is proved by experience that we may assign to each element various weights, all multiples of the same number, and that, collectively speaking, these weights express the weight proportions according to which the elements combine with one another. For each element, we may choose *one* of these weights to express the equivalent of the body. . . . In principle, it matters little which of the weights is chosen. Practically, however, one of the weights is chosen, according to reasonable guidelines, by borrowing from the atomic theory.[37]

A notorious case, used repeatedly by Marignac, was the elemental weight used for nitrogen by every chemist in the world. $N = 14$ was derived by no single chemical replacement value or gravimetric experiment. If nitrogen had a true "equivalent," it would probably have to be 4.67, for that is the chemical equivalent for the only hydride of nitrogen, ammonia; instead, that number had been arbitrarily tripled so that ammonia could be written as NH_3. "Why then should this number [14] exist?" asked Marignac. Though he did not answer his own rhetorical question, the answer was clearly that this number uncomplicated the formulas for other compounds of nitrogen, especially the oxides. But to decide on a single unchanging equivalent weight violated all the sense of the empirical definition of the term. Doing so—adopting unique conventional equivalent weights for each of the elements—signified, willy-nilly, an entry into the theoretical realm.

36. With "conventional" I intend to invoke the concepts of scientific *convention* and heuristic *convenience*, not to signify the sense of *customary* or *ordinary*. My claim is that conventional equivalents are used instrumentally in theory building as conventions and not as empirical entities.

37. Nye, *Question of the Atom*, pp. 233–234. I emphasize that this is partially a direct quote and partially my paraphrase.

The process was precisely the same as what the atomists did. At the point when a single weight was chosen for each element in order to calculate formulas, there was no longer any procedural distinction between equivalentists and atomists. Wollaston's "high ground" was on the same plane as Dalton's, after all.

There was, however, an important ontological distinction to be made: equivalents were explicitly conventional, whereas atomic weights purported to be true. Equivalents were difficult to define but easy to determine; one simply made a more or less arbitrary choice between alternatives. Atomic weights, on the other hand, were easy to define but difficult to determine. Marignac wrote: "I consider atomic weights, and I believe that many chemists agree in this, as being only equivalents, in the determination of which arbitrary conventions have been replaced by scientific considerations, based on the study of physical properties."[38] Marignac not only had no qualms about introducing physical considerations into the choice of weight multiple to be accepted as the "atomic weight" of an element; he believed that to be required, since the goal here was not convenience but truth. Berthelot, on the other hand, was not consistent in his preferences. He had attacked certain physical data, but not others. In Marignac's words: "If we only admitted physical laws that are absolute, we should have to reject them all." Moreover: "If he does not say so expressly, his whole argument proves that, in his opinion, no account is to be taken of these physical properties, when they disturb the usage established for [equivalent] weights that have been adopted for a long time in chemical notations." Finally, it simply was not true that atomists had never been able to agree. The very fact that for at least 20 years a single system of atomic weights had been gaining strength, and that the vast majority of chemists across Europe (outside France) now accepted it, suggested the strength of the evidence underlying the choices made to establish the system.[39]

Berthelot's response to Marignac was somewhat supercilious and not very much to the point: "The new atomic school has not, it appears to me, justified the pretension of changing the very base of chemical doctrines, and of founding a new chemistry, essentially different from the old. The only

38. Ibid., pp. 234–235, 237.
39. Ibid., pp. 235–237.

thing it has done is to intermix the meshes of its hypotheses with our demonstrated laws, and this much to the detriment of the teaching of positive science." He suggested that molecular mechanics was the field that truly offered new scientific harvests.[40] Marignac's rejoinder, confected with exquisite tact, contained one sharp reproof that repeated a point he had made earlier, perhaps too subtly. Berthelot, he suggested, was suffering "a state of mind in which things seem very natural because we are accustomed to accept them."[41]

Marignac did not oppose Berthelot's anti-atomism in one important respect: "I am nearly ready to agree with M. Berthelot in his opposition, and I have certainly no idea of defending the atomic theory, but merely the chemical notations founded on the atomic weights." This was a crucial point, an apparent paradox that went to the heart of the matter. "I know of no case," he added by way of clarification, "in which an atomic weight has been determined by a method founded on the indivisibility of atoms."[42] In other words, atomist notation still allowed perfect freedom to speculate on the possible divisibility of the referents of the Cs and Hs in an organic formula. To put it in the inverse fashion (as Marignac did not): Operationally considered, equivalentists as much as atomists treated their Cs and Hs as inviolable units. Atomic or equivalent weights? Chemists had been led astray for decades by the names applied to these entities, for "atomic weights" did not necessarily imply metaphysical indivisibility in the chemical units being specified, nor did "equivalent weights" signify true chemical equivalence. Marignac was turning one of Berthelot's arguments around, suggesting that it was the equivalentists, not their opponents, who were being naive and unsophisticated about the language they were using.

Many of Berthelot's arguments in favor of equivalents and against atomic weights can be seen as fallacies in light of Marignac's critique. For instance, in the matter of the Petit-Dulong law or Avogadro's hypothesis, atomists had no qualms about using these generalizations as guides to pick the one multiple per element that worked best. Why not? This was nothing more

40. Ibid., pp. 243–246.
41. Ibid., pp. 247–249.
42. Ibid., p. 236.

than using physical properties to make a choice among specific alternatives offered by chemistry, but between which chemistry was powerless to decide. To be sure, there were theoretical and experimental weaknesses in the physical laws, but no physical law is absolutely precise, and in any case the rectitude of the method was certified by the consistency and heurism of the results. Berthelot regarded this as shameless trickery to alter what he thought were time-tested, empirically certified weights. In fact, we can now see that equivalents were neither time-tested nor empirical, and we can understand why Marignac concluded that Berthelot was simply defending old habits. As for Berthelot's crucial argument for the inconsistency of the atomist system (outlined on p. 308 above), the argument holds only if one assumes that all elemental gases have molecules consisting of one atom each. Just as when he attempted to claim identity of the denotations of "equivalence" and "valence," this example only revealed that he had never succeeded in truly understanding the other system.

Nowhere in the nineteenth-century scientific literature have I found these issues portrayed with such clarity as here. Marignac was simply more effective than Wurtz at pithily laying out the central issues, and one can only wonder if the issue might not have come out differently had Marignac been the point man at the Académie all during the debate. There may simply have been too much bad blood between Berthelot and Wurtz.

It would appear that Marignac's analysis, which I believe is worth adopting, was never generally comprehended in his day. Others besides Marignac attempted during his century to show that equivalents were not empirical and that atomic weights were not metaphysical quantities, but examples are rare, and in few cases was this message heeded.[43] Even atomists tended to accept the equivalentists' terms of the debate, and these mis-

---

43. Wurtz's student Paul Schutzenberger, at least, appears to have heeded. In his *Traité de chimie générale* (Dunod, 1880, vol. 1, p. 252n.) he wrote: "The system of equivalents adopted in classical instruction in France is a mixed and bastard system, sometimes based on substitution values, and sometimes, by contrast, proceeding in an inverse fashion from considerations of this type. . . . For carbon [for instance] a number was chosen positioned between 3 (the substitution equivalent) and 12 (the atomic weight). The rules that determined the choices were arbitrary and multiple; they were not able to lead to a well-ordered system; it is therefore regrettable to see certain influential scientists refuse to reject them." (cited in J. Jacques, *Berthelot 1827–1907* (Belin, 1987), p. 206)

understandings—as Marignac and I both conceive them to be—have persisted to the present day.[44]

The Aftermath

Beginning in the early 1860s and increasingly thereafter, Berthelot had a problem, one that reached a critical point in the aftermath of the events just recounted. Structure theorists were having significant success explaining the increasingly important phenomenon of isomerism in organic chemistry, and any system of formula notation that hesitated to posit intramolecular atomic groupings, as his did, had little hope of matching that achievement. Berthelot recognized and discussed this problem in the important monograph on isomerism (1866) from which I quoted above.[45] One tactic he used there was to try to reduce the territory to which isomerism "in the strict sense" applied. So, for instance, he distinguished the latter from "metamerism" and "kenomerism," which he defined as isomerisms that were created by the formation of new compounds through addition, substitution, or elimination reactions. The unstated implication was that different reaction routes to an apparently identical substance produced isomers of that substance, because different routes led to products with differing internal arrangements of their components.[46]

44. In 1984 I published a book whose major thesis was Marignac's argument. When I first formulated the argument, I was not yet aware that anyone in the nineteenth century (or the twentieth) had defended this thesis; however, while researching and writing that book I discovered several who had, including Laurent and Cannizzaro. I became aware of Marignac's contribution only through Nye's book *The Question of the Atom*, which was published in the same year as mine (see Rocke, *Chemical Atomism*, pp. 18–19, nn. 50 and 57). Analogous to Marignac's experience, I am not at all certain that this viewpoint has yet prevailed in the peer community. However, see the important recent work of Ursula Klein, cited in note 2 above.

45. Berthelot, "Leçon sur l'isomérie," in Société Chimique de Paris, *Leçons de chimie professées en 1864 et 1865* (Hachette, 1866). The lecture was given on 17 April 1863 (the date given in this volume, 27 April, is erroneous); however, the printed version of this "lecture" is the size of a substantial book, and the references indicate that the text contined to be edited, revised, and expanded until January 1866. See G. Ciancia, "Marcelin Berthelot et le concept d'isomérie (1860–65)," *Archives internationales d'histoire des sciences* 36 (1986): 54–83.

46. Berthelot, "Leçon sur l'isomérie," pp. 65–125.

However, for "isomers in the strict sense"—such as benzyl versus cresyl alcohol, or benzoic acid versus salicylaldehyde—Berthelot simply cited the same explanation, qualitatively speaking, that had proved so useful to structuralists. Such pairs of compounds, he wrote, prove that "there can exist several different arrangements in the interior of each of the elementary groups that constitute a definite compound."[47] Arrangements of what? Interior of what? Groups of what? Here Berthelot was compelled to posit a materialistic and qualitatively structuralist explanation, but his pen stuck on the words "atom" and "molecule," and he was unwilling to hazard guesses about what precise submolecular arrangements might explain particular cases of isomerism. The reason for this hesitancy appears to arise, as we have seen before, from assuming a physicalist analogy. Perhaps, Berthelot mused, the fundamental matter out of which substances are formed represents a mathematical function, and simple bodies its determined values, or perhaps the former represents an equation and the latter its various solutions. "The various simple bodies," wrote Berthelot, "could indeed be constituted by a single matter, distinguished only by the nature of the movements which animate it. The transmutation of an element would thus be nothing more than the transformation of the movements which correspond to the existence of this element and which communicate to it its properties, into the movements corresponding to the existence of another element."[48]

Occasional internal strains are evident in this monograph.[49] At some points Berthelot suggested the "arrangement" explanation for isomerism, without really exemplifying it, much less exploring its heuristic power; in other places he seemed to prefer a reductionist-physicalist hypothesis, but this remained completely undeveloped; and in still other places he produced neologisms and taxonomies whose bases were not always clear. About the time this monograph was published, Berthelot turned from organic chemistry to chemical dynamics and thermodynamics. This important change in

47. Ibid., p. 126. A nearly identical statement appears on p. 17.

48. Ibid., pp. 165–166.

49. Ciancia colorfully (and, I think, accurately) depicts this monograph as a snapshot of "une attitude épistémologique prête à basculer sous le poids de ses propres contradictions" ("Isomérie," p. 54; for some examples of these contradictions, see also pp. 69, 77–78).

his career path suggests that he may no longer have felt comfortable, and perhaps he no longer felt powerful, in the field of organic chemistry. Curiously, Deville had made the same turn about 10 years before Berthelot. Jacques put the point clearly, with his customary sardonic wit: Thermochemistry required no detailed understanding of the interior of molecules. "Thus, by a very cunning maneuver, Berthelot chose a new research domain where his false ideas were without consequence. As was his wont, he would attempt to impose his imprint and his philosophy on this chapter of chemistry where there were few competitors and few precursors."[50]

Even while the Académie debate was going on, Wurtz wrote and published a substantial treatise simply entitled *La théorie atomique*.[51] It is a reasonable conjecture that he was induced to write this book by his initial confrontations with Deville and Berthelot in the spring of 1877. The book, published in 1879, was a systematic treatment of the entire subject—historical, philosophical, chemical, and physical—tracing the doctrine from Dalton to Wurtz's time. There was much blunt criticism of his opponents in the Académie debate: for example, Wurtz laid out his position against Troost's work,[52] attacked Berthelot's argument for halving certain atomic weights,[53] and contested Berthelot's assertions regarding heats of reaction.[54]

Most important, Wurtz—using isomerism as his domain—argued that Berthelot's organic chemical theories of the 1860s were scientifically sterile, sometimes even incoherent, in comparison with structure-theoretical ideas. First, Wurtz challenged Berthelot's position that isomerism could helpfully be based on a generating taxonomy. If a compound with a certain formula is produced by three distinct chemical reactions, he wrote, one of two circumstances obtains: either the three reactions produce distinct isomers of the substance, or the reactions produce the identical substance. If

50. Jacques, *Berthelot*, pp. 134–135.

51. Wurtz, *La théorie atomique* (Baillière, 1879). In 1878, Wurtz effectively retired from teaching and administration—with the exception of his lectures in the Sorbonne—leaving him sufficient leisure to write this landmark work.

52. "The experiment was inexact and the conclusion inadmissible." (ibid., pp. 83–85)

53. "M. Berthelot himself has invoked this feeble argument. . . ." (ibid., p. 132)

54. "Here the matter is irrelevant and the argument no longer has any point." (ibid., p. 153)

the former occurs, then Berthelot's scheme does not help us to understand the chemical differences between the isomers. If the latter occurs, then this is not even an instance of isomerism at all, so it cannot be used as the basis of an explanation of that phenomenon.[55]

As we have seen, Berthelot had also occasionally suggested the structuralist explanation that isomers differ because of differing arrangements of their subcomponents. In his hands this second approach to an explanation was no more satisfactory, wrote Wurtz, because he failed to proceed past this bland programmatic pronouncement. It remained pure conjecture, abstract and unapplied to particular cases. Finally, Wurtz criticized a third gambit also found in Berthelot's writings on isomerism: that changing the order of reaction sequences necessarily changed the resulting product—for example, in multiple chlorinations of a hydrocarbon. On the basis of these extraordinarily ad hoc ideas on isomerism, Berthelot had asserted that there should be (for example) hundreds of isomers of trichloropropane. Wurtz pointed out that structure theory predicted five and only five trichloro-propanes, all of which had been prepared—some only with the assistance of the theory itself. He challenged his opponent: "Where is the sixth?"[56]

Another example Wurtz used was close to his heart, namely the isomers of amyl alcohol. In the early 1860s Wurtz himself had done some of the earliest research on this subject, discovering two new isomers of the long-known "fermentation" (iso)amyl alcohol, namely secondary amyl alcohol and amylene hydrate. Soon thereafter, Wurtz had shown that the latter vaguely named compound, structurally speaking, was tertiary amyl alcohol. A student of his named Aleksandr Zaitsev found a novel secondary amyl alcohol, and another former student named Adolf Lieben prepared the normal primary amyl alcohol. Then, in 1877, A. N. Vishnegradsky prepared a sixth new isomer, a secondary alcohol with a branched carbon chain. There were still two additional isomers predicted by structure theory that had not yet been prepared. These unknown substances could even be assigned names (secondary butyl carbinol and neopentyl alcohol) on the basis of structure theory, and the theory served as a guide to their even-

55. Ibid., p. 210.

56. "Qu'on nous montre la sixième." (ibid., pp. 211–212) I have used E. Cleminshaw's translation of this sentence (Wurtz, *The Atomic Theory* (Appleton, 1881), p. 290).

tual preparation within a few years of publication of Wurtz's *La théorie atomique*. Regarding the six then-known isomers, Wurtz wrote: "The theory predicted them; experiment has brought them to light. And this fidelity of theory, this happy coincidence between predicted and observed facts, has been tested in hundreds of cases."[57]

Let us compare Berthelot's treatment of the amyl alcohols to Wurtz's. In 1872 Berthelot had published a *Traité élémentaire de chimie organique*, which appears to be a treatment in book form of the lecture material from his course at the École de Pharmacie; factual and almost completely non-theoretical, it is virtually free of polemics.[58] In this work, Berthelot nearly ignored the amyl alcohols, barely mentioning the fact of multiple isomers. His explanation for that isomerism was only implicit. He stated that there are always as many isomers among the alcohols as there are among the hydrocarbons that generate them.[59] One may presume that he meant that hydration of the five known amylenes gives rise, schematically speaking, to the five then-known amyl alcohols. Berthelot's isomer numbers worked as of 1872, but not much longer, as the sixth, seventh, and eighth amyl alcohols were successively discovered. Structure theory had always predicted precisely eight possible isomers, and indeed no ninth one has ever appeared.

Faced with the explicit challenges in Wurtz's *La théorie atomique* and with the empirical record that appeared to be working against him, Berthelot prepared a second edition of his textbook in collaboration with a gifted former student, Émile Jungfleisch.[60] Whereas the first edition barely mentioned atomic theory, a new preface was mostly devoted to discussing it, and a new chapter was inserted to respond to the attacks. "We have not believed it necessary," Berthelot wrote, "to adopt the atomic notation, regarding it as less correct, because it pursues an impossible conciliation between weight and volume relations, by taking as a fundamental principle a third order of

57. Ibid., pp. 215–216, 244–246. A fine near-contemporaneous account of all eight amyl alcohols may be found in V. Meyer and P. Jacobson, *Lehrbuch der organischen Chemie*, second edition (Veit, 1907), vol. 1, pp. 238–243.

58. Berthelot, *Traité élémentaire de chimie organique* (Dunod, 1872).

59. Ibid., pp. 148–151.

60. Berthelot and Jungfleisch, *Traité élémentaire de chimie organique*, second edition (Dunod, 1881).

relations, drawn from the specific heats of solid elements"—an order that lacks both experimental and theoretical rigor. On the other hand, he added, both systems are exactly equivalent notationally, the "positive foundation" of the science being expressed equally in both. Consequently, suggesting a slap at Wurtz, "there would be some intolerance" to ascribe an "exclusive dogmatic value" to one or the other. Nonetheless, the system using equivalent weights was indeed to be preferred, since it distinguished between hypotheses and theories. However, the new edition also included occasional atomistic formulas, so that the book could also be used by atomists. This was the first appearance of such formulas in any of Berthelot's works.[61]

Berthelot stated in the preface that his collaborator Jungfleisch had been responsible for several new features of the present edition, including a discussion of benzene theory. There is indeed a clear description of Kekulé's theory, which by 1881 had become nearly universally accepted. One of the merits of this theory had been to explain why each di-derivative of benzene has exactly three isomers. Jungfleisch and Berthelot offered an alternative explanation, however. Since Berthelot had shown that benzene could be formed from three moles of acetylene, they assumed three acetylenic groups in the benzene molecule; they then argued that two substituting groups could enter into such a molecule in exactly three different ways, hence giving rise to three distinct isomeric series. Unfortunately for this theory, there are in fact six different ways for two groups to apportion among three, and it is not at all clear why they excluded half of the alternatives.

Kekulé's theory had also predicted only one isomer of every monoderivative and one isomer of every pentaderivative of benzene. During the late 1860s and the early 1870s these predictions had not always seemed secure. For example, since 1860 Hermann Kolbe had believed that he could identify a second isomer of benzoic acid, which he called "salylic" acid. Friedrich Beilstein demonstrated the nonexistence of this compound as early as 1864, but Kolbe refused to give up on this putative refutation of his enemy's theory. However, by 1875 even Kolbe had convinced himself that there was only one benzoic acid.[62] Jungfleisch had also apparently

---

61. Ibid., vol. 1, pp. i–vii. The preface was signed by Berthelot alone.

62. For a discussion and relevant citations, see Rocke, *Quiet Revolution*, pp. 295–304.

contradicted Kekulé's prediction in 1868 by preparing a second isomer of pentachlorobenzene, and this work had been repeated and verified by R. Otto 2 years later.[63] However, Albert Ladenburg, a student of both Wurtz and Kekulé, contested the existence of this second isomer, and by 1874 he had adequately refuted the work of Jungfleisch and Otto.[64] As for the amyl alcohols, language similar to that in the first edition still appears, even though Berthelot's and Jungfleisch's counting of isomers was no longer accurate and thus their theory no longer matched the empirical record. Such examples could be multiplied; clearly there were serious strains appearing once more.[65] Structure theory, and the benzene theory based upon it, looked ever more securely established by innumerable predictions that seemed to be nearly invariably confirmed by empirical evidence.

Three years after the second edition of the *Traité* was published, in May of 1884, Wurtz died suddenly. Berthelot composed a generous éloge, stressing the "fruitful rivalry" that had characterized their relationship over 30 years. "Sustained by a common love for a science that we cultivated in parallel," he wrote, "this competition never damaged the courtesy of our personal relationship." Wurtz's "powerful school of chemistry" at the Faculté de Médecine had been one of "that brilliant Pléiade of three schools of French chemistry" (meaning Wurtz's, Deville's at the École Normale, and Berthelot's at the Collège de France). "M. Wurtz also counted among his claims to glory," Berthelot added significantly, "the influence that he exerted on the development of the doctrines and notations of the newer atomic theory."[66]

63. E. Jungfleisch, "Sur une seconde série de dérivés chlorosubstitués de la benzine," *Bulletin de la Société Chimique* 9 (1868): 346–356; R. Otto, "Ueber zwei isomere Pentachlorbenzole und Bichlorbenzolchlorid," *Annalen* 154 (1870): 182–187.

64. A. Ladenburg, "Die Pentachlorbenzole," *Berichte der Deutschen Chemischen Gesellschaft* 5 (1872): 789–790; E. Jungfleisch, "Sur les deux benzines quintichlorées," *Bulletin de la Société Chimique* 18 (1872): 531–534; Ladenburg, "Sur les benzines pentachlorées," ibid., p. 548; Ladenburg, "Die Pentachlorbenzole," *Berichte* 6 (1873): 32–33; Ladenburg, "Zur Constitution des Benzols," *Annalen* 172 (1874): 331–356. Jungfleisch refused to concede the point in his and Berthelot's *Traité* (second edition, 1881, pp. 150–154), though he added, without argument, "nous n'insisterons pas ici sur ce point."

65. *Traité*, second edition, pp. 136–140 and 192–196.

66. Berthelot, "Adolphe Wurtz," *Le Temps*, 14 May 1884, p. 1.

In the same year, Berthelot had a conversation with a fellow senator and former chemist, Alfred Naquet. When Naquet asked Berthelot why he insisted on retaining equivalents, the latter responded: "It is because I do not want to see chemistry degenerate into a religion. I don't want anyone to believe in the real existence of atoms as Christians believe in the real presence of Jesus Christ in the consecrated host." Naquet protested that atomic theory was only a useful mental model, a fruitful theory, nothing more; no one believed in the real existence of atoms. Berthelot retorted brusquely: "Wurtz has seen them."[67]

Also in 1884, the year of Wurtz's death, curricular discussions at the École Polytechnique touched on the choice of notation. A strong case was made by two professors, G. Lemoine and A. Cornu, to require equivalent notation, at least in inorganic chemistry courses. This proposal was rejected, leaving the choice of notation entirely in the hands of the instructor of each course. Lemoine and Cornu, unhappy, argued "c'est l'avenir des élèves qui est en jeu et que l'on sacrifie!"[68] In contrast to the École Polytechnique, atomic notation was kept out of all courses in the École Normale until after 1900.[69] This period was profoundly confusing to students (who often did not know which system was in favor where, or which system should be used in a given situation). The atomistically inclined Marcel Delépine recalled a student competition in 1893 in which he "took the precaution" of using both notations, each of which he "knew perfectly," for fear of otherwise "indisposing" Professor Jungfleisch were he to use atomic weights alone.[70]

Berthelot and Jungfleisch published a third edition of their textbook in 1886.[71] Atomistic formulas are more visible in this edition: there is full

---

67. A. Naquet, *Moniteur scientifique* 14 (1900): 792; quoted at length in Jacques, *Berthelot*, pp. 203–204. Naquet dated the conversation to 1884.

68. C. Kounelis, "Heurs et malheurs de la chimie," in *La formation polytechnicienne, 1794–1994*, ed. B. Belhoste et al. (Dunod, 1994), p. 261.

69. C. Zwerling, *The Emergence of the École Normale Supérieure as a Center of Science Education in Nineteenth-Century France* (Garland, 1990), pp. 159–161.

70. Delépine, quoted in A. Metz, "La notation atomique et la théorie atomique en France à la fin du XIXe siècle," *Revue d'histoire des sciences* 16 (1963), p. 235. This article principally consists of a summary of portions of Delépine's *Auguste Béhal* (Société d'Histoire de la Pharmacie, 1960).

71. Berthelot and Jungfleisch, *Traité élémentaire de chimie organique*, third edition (Dunod, 1886).

parallel presentation of formulas in both systems. To be sure, the authors changed little of substance: anti-atomistic claims still appear in chapter 2, and the Berthelot-Jungfleisch benzene theory is unaltered from the second edition. However, careful readers could detect some subtle shifts; for instance, the structuralist explanation of the nature of secondary alcohols was provided without comment, and with apparent approval.[72] In 1898 Berthelot and Jungfleisch published a final edition of this textbook. This edition is fully atomistic and structuralist—a completely altered, completely modern work of chemistry. Berthelot had fully conceded the victory to his opponents.[73]

Berthelot's actual moment of official capitulation can be fixed rather precisely.[74] In April of 1896, Berthelot returned to his instructional duties at the Collège de France after a 5-month stint as France's foreign minister. He directed his préparateur, Delépine, to erase the equivalent weights that had framed the blackboard in all his classes to that date, and to replace them with atomic weights. "Thus it happened," wrote Delépine, "that I wrote on the board $O = 16$, $S = 32$, etc., in place of $O = 8$, $S = 16$, etc." Delépine's editor commented: "This action of the préparateur, on the instructions of the most important defender of 'equivalents,' marked the end of the discussions on atomic notation."[75]

At this time, Wurtz had been dead for 12 years. It was Adolphe Wurtz's tragedy not to have survived to see this ultimate victory for atomism and structure theory. It was Marcellin Berthelot's tragedy that he did.

72. Ibid., pp. 33–37, 155, 233. Another indication: the phrase "we have not believed it necessary to adopt the atomic notation, regarding it as less correct" no longer appears in the preface (p. vi).

73. Berthelot and Jungfleisch, *Traité élémentaire de chimie organique*, fourth edition (Dunod, 1898).

74. Ciancia ("Isomérie," p. 63n.) cites a paper by Berthelot that uses atomic weights as early as 1892. However, this paper was published in the *Bulletin de la Société Chimique*, the historic redoubt of French atomists, and it is not clear what significance should be imputed to this publication.

75. Metz, "La notation atomique," p. 237.

11

Later Years

In 1868 Wurtz began the publication of a massive encyclopedia of chemistry, following the model of the famous Liebig-Wöhler-Poggendorff *Handwörterbuch der reinen und angewandten Chemie*, first published 30 years earlier in Germany, which was about to come out in a second edition. Wurtz's *Dictionnaire de chimie pure et appliquée* became an influential treatise for all French chemists, from rank beginners to accomplished researchers. As the publisher correctly noted in the opening *avertissement*, the work was more a collection of monographic articles than a dictionary. The first volume alone was the equivalent of several thousand pages of octavo print; subsequent volumes and supplements were published until 1908. Wurtz hoped that the project would enliven the French chemical community; he used it to propagate his central concerns for atomic theory and structural organic chemistry, but he tried to include at least some defenders of equivalents among the authors of articles.

"Chemistry is a French science"

The first fascicle of the *Dictionnaire* was prefaced by a substantial history of chemical theory since Lavoisier, written by Wurtz. The opening words of this history immediately became famous (and infamous): "La chimie est une science française. Elle fut constituée par Lavoisier d'immortelle mémoire." Composed in 1867 and dated 1 May 1868, the fascicle was available in bookstores by the end of that year; in 1869 the Histoire was reprinted as a monograph.[1] The first sentence created an uproar. Wurtz's apparent attempt

1. A. Wurtz, "Discours préliminaire," in *Dictionnaire de chimie pure et appliquée*, ed. Wurtz (Hachette, 1868), vol. 1; *Histoire des doctrines chimiques depuis Lavoisier*

to appropriate the entire science for his own country was taken to be the height of arrogance and blind chauvinism, and a storm of criticism followed, especially from German chemists.[2]

However, there is an alternative interpretation that I believe better fits the circumstances. Throughout the entire *Histoire*, Wurtz was concerned to demonstrate that the conceptual seeds that (in his view) had yielded such impressive scientific fruit in contemporary German chemical laboratories had been French originally. Not only had Lavoisier started the science with his famous revolution, Wurtz wrote, but the modern atomic-structural movement—a second chemical revolution, in a sense—was an outgrowth of the work of Gerhardt and Laurent. Selling this viewpoint was, I believe, a part of his long-term strategy to improve the poor fortunes in France of these new chemical ideas. The work was directed inward, not outward—toward those French colleagues who had not yet signed on to the reforms. What better way to persuade them to do so than to appeal to their patriotism? This was—if I am right—rhetorical patriotism put to a didactic cognitive purpose, not gratuitous jingoism.

To be sure, Wurtz was an ardent patriot, urging French paternity for the science of chemistry on the basis of the contributions of Lavoisier and others. However, in a deeper sense he was perhaps the most internationalist of all French chemists, for the mission of his mature years was to infuse French chemistry with what were, in fact, German theories and German pedagogical methods. Moreover, non-French chemists figured prominently in Wurtz's *Histoire*. Davy, Dalton, and Berzelius play leading roles in his narration, and Avogadro even wins precedence over Ampère

---

jusqu'à nos jours (Hachette, 1869). Readers of *The Laboratory* for 20 July 1867 (vol. 1, p. 284) learned that Wurtz's *Histoire* was in production and that publication was predicted to begin in November 1867. Wurtz sent a newly printed copy to Williamson on 25 December 1868 (Harrris Collection, Bloomsbury Science Library, University College London, Ms. Add. 356).

2. The discussion that follows is based on my paper "History and Science, History of Science" (*Ambix* 41 (1994): 20–32) and on pp. 342–349 of my book *The Quiet Revolution*. Some of the phrases in the following two paragraphs are taken directly from these two sources. A wider context is portrayed in my paper "Pride and Prejudice in Chemistry" (*Bulletin for the History of Chemistry* 13–14 (1992–93): 29–40). Readers are directed to these sources for fuller discussions of this well-known episode. For an interpretation similar to mine of Wurtz's apparently chauvinistic dictum, see B. Bensaude-Vincent, *Lavoisier* (Flammarion, 1993), pp. 393–400.

regarding the submolecularity hypothesis. It is true that Wurtz rather slighted German chemists (including, surprisingly, Wurtz's own revered teacher Liebig); but Kekulé's name was prominently displayed in the last chapter, and some other younger Germans such as Erlenmeyer were mentioned. Furthermore, Wurtz was writing a history of chemical *theory*, and many leading older German chemists, such as Bunsen, Wöhler, Kopp, Mitscherlich, and even Liebig after 1840, were staunch empiricists. Besides, once again, he was concerned that the book be influential for his particular didactic purpose among his (somewhat chauvinistic) French colleagues, and stressing the work of foreigners was not the route to rhetorical success for that audience.

It is ironic that within a year after the publication of Wurtz's monograph history France and Germany plunged into a disastrous war. Chauvinism and the deepest bitterness were then well in evidence, on both sides of the Rhine.

The Franco-Prussian War and Its Consequences

Napoleon declared war on Prussia on 19 July 1870, and mobilization began. Nearly everyone expected the French army to give effective battle, and many well-informed neutral observers gave them a decided edge over the Prussians. Nevertheless, the Prussians immediately gained the upper hand, and France was forced to abandon Alsace little more than 2 weeks after the start of hostilities. September opened with the disaster at Sedan; all could see that the war was lost, and that Paris would soon find itself besieged. Napoleon's empire fell, and on 4 September a new "government of national defense" was organized.

Wurtz, along with innumerable other Parisians of means, took measures to protect his family. He brought them to the home of his elderly friend J. B. Caventou (one of the discoverers, a half-century earlier, of alkaloids), who lived in a Normandy village near the mouth of the Somme. Back in Paris, Wurtz stood guard in his home, where he invited a student and friend, Caventou's son Eugène, to join him.

Just before the fall of the empire, a Commission Scientifique pour la Défense de Paris was established, under the leadership of Berthelot; this commission sought to use innovative science to generate useful technologies, both offensive and defensive (such as extending the food supply of the

2 million people under siege). Wurtz was not a member of this commission, though he did do research for it[3]; his time was more taken up by the Conseil Supérieur de l'Hygiène Publique and the Association de la Croix Rouge, of which organizations he was a member; his residence was converted into a hospital, and his lecture room and laboratory stood empty of activity. After the siege was lifted and peace was declared, Wurtz took up his duties as dean once more, but this was of short duration; the revolutionary government fired him in late March of 1871, and a few days later he barely escaped arrest by the new governing Commune. He traveled to Normandy to be reunited with his family.

About this time, Louis Pasteur published an article in the *Salut Public* of Lyon (where he had fled) that had the provocative title "Why did France fail to find superior men in its moment of peril?" To set the stage to answer this rhetorical question, Pasteur pursued pathological and gendered metaphors. Germany, an "arrogant, ambitious, and deceitful nation," expanding at the expense of its neighbors "like a malignant tumor," was personified as a highwayman who had armed himself in the shadows and then had ambushed his "gentle and trusting rival," France. The latter, unexpectedly assaulted and in danger of having her throat slit, might have been able successfully to extricate herself from this deadly embrace had not the "blows of her cruel adversary" come on top of her earlier indiscretions and past faults. These faults were many, Pasteur averred, yet there was an important one that had "always obsessed" him and to which he wanted now to draw attention: France had failed to find superior men in the late conflict because "for the last half-century France has been disinterested in the great works of thought, especially in the exact sciences."[4]

Pasteur meant this quite literally and directly, and he was far from alone in this opinion. "Few understand the true origin of the wonders of industry and the wealth of nations," he wrote. He related a conversation he had had with a distinguished government official. Pasteur was complaining to him of the frequent abandonment of scientific careers by promising young Frenchmen, and the minister replied that this was not surprising, for after

3. M. Crosland, "Science and the Franco-Prussian War," *Social Studies of Science* 6 (1976), p. 200.

4. L. Pasteur, "Quelques réflexions sur la science en France," as reprinted in *Moniteur scientifique* 13 (1871): 176–182 (see p. 176); dated 16 March 1871.

all "the reign of the theoretical sciences today was yielding its position to that of the applied sciences." Pasteur thought that "nothing is more erroneous nor more dangerous than this opinion." "No, a thousand times no," he thundered; "there exists no category of science to which one can give the name of 'applied' science. *There is science, and there are applications of science*; the two are tied together like fruit on the tree that bears it." It is possible, he conceded, that random, chance experience may have led to the rise of the industrial arts in the earliest stages of human history. "However, it is certain that, in modern times, chance favors invention only for minds prepared for discoveries by patient study and persevering efforts."[5]

In short, Pasteur was convinced that applications of science come only from a basis in pure science, and that France's neglect of the latter had allowed Germany to prevail in the war. Here Pasteur was deploying a rhetorical strategy that had been used in the past, and one that continues to be used by scientists today: support my research, for it will help provide the basis for national power, wealth, and well-being. But how far can Pasteur's opinion be substantiated? What role did science and technology play in the outcome of the conflict, and to what extent was the French government culpable for its defeat because it had not supported science in the preceding half-century?[6]

The Franco-Prussian War is today a rather neglected topic for historical research, but in the generation after 1870 literally thousands of reminiscences, interpretations, and assessments appeared. The causes suggested for Germany's victory and for France's defeat have been many and varied, but most analysts agree in citing the effective German mobilization, tactics, and logistics and the general failure of French leadership. The problems of Napoleon III ran deeper, too; as Michael Howard described it in his classic monograph on the subject, the regime was the victim of a "faulty military system."[7] If one focuses on science and technology, one could indeed claim

---

5. Ibid., p. 178.

6. Crosland raises this question in "Science and the Franco-Prussian War." His discussion, however, is almost entirely devoted to efforts by scientists to lift, or at least cope with, the siege of Paris in the winter of 1870–71, rather than to whether inadequate government support for science in the Second Empire prepared the way for defeat in war. See also G. Ortenburg, *Waffe und Waffengebrauch im Zeitalter der Einigungskriege* (Bernard & Graefe, 1990).

7. M. Howard, *The Franco-Prussian War* (Macmillan, 1961), p. 1.

that the unquestionably superior Prussian artillery posed an insurmountable problem for the French army. However, that technology had been developed with little input from basic research or theoretical science, and moreover it was just as available to the French as to the Prussians. In fact, just 2 years before the war, Alfred Krupp had made every effort to sell Napoleon the same steel breech-loaders that would later so butcher his army. The sale was derailed only by Minister of War Edmond Leboeuf, who had a personal interest in the Le Creusot gunworks and was not convinced of the safety of breech-loaders.[8]

With the hindsight of historical analysis, we can say that Pasteur's analysis of the cause of French defeat in the Franco-Prussian War was clearly erroneous, perhaps even disingenuously self-serving.[9] And yet there is another way to look at the question that may provide a somewhat different perspective, and a more generous judgment of Pasteur's thesis. Superior German science did not win the war in any kind of direct or immediate sense, nor did inferior French science lose it. However, there might well be a wider argument to be made about national differences regarding indirect effects of scientific education and culture.

Pasteur asserted that there was a "law of correlation between theoretical science and the life of nations," and that France, troubled by political instability, had neglected this great truth:

Germany was busy multiplying its universities, creating the most beneficial competition between them, surrounding its teachers and doctors with honor and esteem, and creating vast laboratories with the finest instruments, while at the same time France, enervated by revolutions, constantly occupied by sterile research into the best form of government, gave but desultory attention to its institutions of

8. W. Manchester, *The Arms of Krupp* (Little, Brown, 1968), p. 101.

9. It must be stressed, however, that many contemporary scientists had exactly the same naive view of the relationship between basic science and technology. A month before Pasteur's article was written, Edward Frankland was asked by a parliamentary commission whether he thought that a country's technological well-being was directly related to the health of its pure science. He responded affirmatively, noting that a country would always have the option of buying new technology abroad but would then lock itself into a state of permanent technological inferiority. (Frankland testimony, 14 February 1871, in *First and Second Reports from the Royal [Devonshire] Commission on Scientific Instruction*, British Parliamentary Papers (London, 1872), vol. 25, p. 372) The unstated axiom underlying this argument was that the *source* of all technology is science.

higher education. At the point where we have arrived at what is known as modern civilization, the culture of science in its highest expression is perhaps even more necessary to the moral state of a nation than it is to its material prosperity. [10]

This is, of course, a subtler assertion than Pasteur's earlier argument that, put crudely, technology comes only from pure or theoretical science. Pasteur wanted to be clear about the generality of the assertion. Not only science but "disinterested works of the mind of every sort" introduce into a society the "philosophical or scientific spirit, that spirit of discernment which submits everything to rigorous reason, condemns ignorance, and dispels prejudices and errors." They "raise the intellectual level and moral sentiment; through them, the divine idea itself is diffused and exalted."[11] What France had neglected, according to Pasteur, was this "culture of science." This was the crucial difference that had spelled victory for Germany and defeat for France. The problem, thought Pasteur, was not so much that Germany had better weapons, created by better science; it was that the German culture had been breeding better minds, because Germany had gained a monopoly on modern universities.

There are two notable ironies in this situation. One is that the claim Pasteur was making here was precisely the claim made a generation earlier by Justus Liebig: that the only way properly to educate the mind—for all purposes, including technological ones—was to learn the pure sciences, and to do so in a laboratory setting. In a famous polemic that appeared in his proprietary journal in 1840, Liebig proclaimed:

In the natural sciences, Germany has again taken the place that nature has accorded her. A mass of intelligence pulses through our numerous universities, in the arteries of so many states, which secure for their scientists the most complete independence. . . . From Germany has come the impulse of progress in all the natural sciences of modern times. . . .[12]

Science, affirmed Liebig, is naturally constituted "more than any other subject, to exercise influence on the culture of the mind." What is necessary, he wrote, is that the student become thoroughly familiar with the ABCs of science, just as students must learn to read and write before tackling advanced

10. Pasteur, "Quelques réflexions sur la science en France," pp. 178–179.

11. Ibid., p. 179.

12. J. Liebig, "Der Zustand der Chemie in Preussen," *Annalen der Chemie und Pharmacie* 34 (1840), pp. 100–101.

subjects. "In our lectures we familiarize students with the alphabet [of science], in our laboratories they learn the use of these symbols, they master reading the language of phenomena. . . ."[13] A scientifically educated man thus understands phenomena and is able to think; he easily outpaces the empirical technician, even in purely practical matters.

Whether or not this analysis was truly reasonable or sustainable, many intellectuals in both countries agreed: Germany prevailed in the Franco-Prussian War because of a superior intellectual culture, bred by the German universities. Just 10 days before Pasteur wrote his article, Henri Sainte-Claire Deville declared in a public meeting of the Académie des Sciences:

> In the final analysis, the liberal organization of the German universities was placed in the service of the odious passions directed against our country. And it is said everywhere, with justice, that it is through science that we have been defeated. The cause of this is in the regime that has crushed us for the last eighty years, a regime that subordinates men of science to men of politics and administration. . . .[14]

Precisely the same argument was made 3 weeks later in a public address by none other than Justus Liebig: it was Prussian scientific culture that had provided the margin of victory. And back in Paris, Ernest Renan was moved to exclaim, at a Goncourt dinner 4 days after Napoleon's surrender: "In all things that I have studied, I have always been struck by the superiority of German intelligence and work. It is not surprising that in the art of war, which is after all an inferior but complex art, they have attained this superiority. . . . Yes, gentlemen, the Germans are a superior race!"[15]

And there is a second irony. Pasteur may have been moved to make a larger, cultural argument for the applicability of pure science to the real world because of the dearth of appropriate concrete examples. To be sure, many isolated instances of basic science's leading to usable technology were ready to Pasteur's hand—including some from his own career—but nothing of the magnitude that could provide a convincing argument regarding, say, the loss of a war, or the rise of a major new industry. The second irony,

13. Ibid., pp. 104, 114–115.

14. H. Deville, "De l'intervention de l'Académie dans les questions générales de l'organisation scientifique en France," *Comptes rendus* 72 (1871), p. 238.

15. Liebig, "Eröffnungsworte . . . nach dem Friedensschluss," in *Reden und Abhandlungen* (Winter, 1874), pp. 331–333; Renan, quoted from E. and J. de Goncourt, *Journal* (Laffont, 1989), vol. 2, p. 277 (6 September 1870). I thank Natalie Pigeard for the latter reference.

then, is that the period of the Franco-Prussian War marks the approximate transition point to the time when such instances were becoming common.

The best example of this is one that is no less true for being something of a cliché: the coal-tar dye industry. An important turning point in this history was the artificial synthesis of the natural product alizarin, the bright red coloring principle of the madder plant, which is commercially the most important traditional dye next to indigo. The synthesis was achieved by Carl Graebe and Carl Liebermann in Adolf Baeyer's laboratory at the Berlin Gewerbeakademie, in 1868–69. This event transformed the young synthetic dye industry. First, it was the occasion of a gradual shift from French and English to German leadership in the new industry; second, this was the first important natural dye to yield to the synthetic chemical arts; third, many future large chemical firms established themselves with this dye; and finally, this event marked a shift from more or less empirically driven innovation, to product development that owed much to chemical theory. It was the Germans that excelled in this important stage of development, just then beginning, and by the time of the First World War Germany had a near monopoly, not only on dyes but also on many other important products of the chemical industry. Germany's ability to prosecute what was been called "the chemists' war" was thus immeasurably enhanced. Pasteur's words had finally came true then, and with a vengeance.

The Research School of the Mature Wurtz

After the Franco-Prussian War, Wurtz continued a personal research trajectory that he had established nearly 20 years earlier. A lightning review: From Alexander Williamson's ether syntheses and water type had come Wurtz's interest in the relationships of alcohols, acids, and ethers, and especially in that most interesting naturally occurring compound, glycerin. Wurtz was also influenced by the important early work of his rival Berthelot. From here, he was inspired to seek, and then successfully to find, the new dialcohol, glycol. Glycol then opened for Wurtz a treasure trove of new derivatives and polyfunctional new compounds, as did the dehydrated glycol he called "ethylene oxide." Along with his colleagues, he was powerfully struck by the strongly alkaline character of the latter substance; he developed theoretical views that placed it as the central link between

Figure 11.1
Adolphe Wurtz. Source: E. F. Smith Collection, University of Pennsylvania Library.

organic and inorganic chemistry, demonstrating along the way the essential unity of all of chemistry.

Ethylene oxide is isomeric with aldehyde (acetaldehyde). When treated with alcohol, ethylene oxide yielded the diether of glycol; treated similarly, aldehyde was long known to yield the isomeric diether known as acetal. In both the prewar and the postwar years, Wurtz and his students prepared a variety of homologous glycols and their ethers. Oxidizing each provided a route to many alpha-hydroxy acids, which could also be derivativized. Dehydrations provided means to prepare novel olefins and olefin oxides, which themselves could form the start of new reaction chains. "In view of these events," Hofmann later opined, "no one would deny that Wurtz's glycol researches were often influential and often even decisive on the development of chemical theories."[16]

Another offshoot from this program, the "aldol reaction," was even more significant, for it led to a group of synthetic techniques of enormous consequence. "Synthesis" was and is used in many different senses in chemistry,[17] but it is perhaps most common among current organic chemists to restrict the term "synthetic reactions" to those in which the carbon chain at the core of a molecule is lengthened. This category of reaction, then, alters the heart of a molecule by increasing its fundamental size, and the armamentarium of such reactions constitutes the chief synthetic repertoire of a modern organic chemist. These synthetic reactions in the narrow sense create new carbon-carbon bonds—which explains their importance for chemists.

Most of the reactions that have appeared in this book, as important as they are, do not rise to this category. The multiple-alkylation strategy described in chapter 5, pursued by Wurtz and others after 1849, was simply a way to generate innumerable novel variations on a well-understood basic molecule; so were Berthelot's millions of theoretically synthesizable artificial triglycerides; and so was the routine derivativization of acids and alcohols (e.g., formation of methyl, ethyl, propyl, etc. esters and ethers). The reaction developed by Kolbe and Frankland in 1847 whereby a halogen

16. Hofmann, "Adolph Wurtz," in *Erinnerung an vorangegangene Freunde* (Vieweg, 1888), vol. 3, p. 358.

17. C. Russell, "The Changing Role of Synthesis in Organic Chemistry," *Ambix* 34 (1987): 169–180. See the third section of chapter 8 above.

atom was replaced by nitrile, and the nitrile then hydrolyzed to carboxyl, increased the chain length by a single carbon atom and thus technically fits our specialized definition, but this was a marginal case.

At least one "synthetic" reaction in this narrow sense had been known for more than a century, namely the formation of acetone by the destructive distillation of acetates. However, in the absence of formulas this was nothing more than an empirical datum; it was not a scientifically useful reaction. Another was discovered in the 1830s, namely the formation of mesitylene from three moles of acetone; but again before structure theory there was little immediate significance of this discovery. Kolbe's electrolysis reaction of 1849 and Wurtz's similar reaction of 1855 created new carbon-carbon bonds, but Kolbe contested that interpretation for at least 15 years, and the "Wurtz reaction" was difficult and synthetically not very useful. Berthelot's discovery that acetylene trimerizes to benzene (1866) and some of his other pyrolyses were likewise interesting but not really productive. In sum: as late as the mid 1860s, despite the growing maturity of organic chemistry, carbon-bond-forming reactions were rare and not very fruitful.

It was Edward Frankland who began to change this situation. As early as 1851 Frankland had attempted, with incomplete success, to "ascend the homologous series of organic bodies"—that is, to discover synthetic reactions in this sense. He first hit pay dirt around 1863. Using his novel zinc alkyls and sodium metal, he succeeded in effecting substitutions of alkyl groups (principally methyl and ethyl) directly into the carbon chains of certain organic compounds. In a remarkable 1866 paper Frankland and B. F. Duppa reported the preparation of a number of interesting new compounds and noted almost incidentally that ethyl acetate could be induced to dimerize with loss of water under the influence of sodium metal, forming "acetoacetic ester."[18] This was the first example of a category of organic reactions that is extraordinarily important today: synthetic condensation reactions. However, Frankland's reaction did not become generally useful until after 1881, when Ludwig Claisen, a former student of Kekulé's, demonstrated how to make it work much more conveniently by using sodium hydroxide rather than sodium metal.

18. For references and discussion, see J. Partington, *A History of Chemistry* (Macmillan, 1961–1970), vol. 4, pp. 500–532; Rocke, *Quiet Revolution*, pp. 181–190, 312–315.

Shortly after the war, Wurtz picked up the work with aldehyde that branched off from his glycol work. Aldehyde had long been known to polymerize naturally, and Kekulé had found that a dehydrated condensation dimer—crotonaldehyde—could also be formed. In 1872 Wurtz showed that under the right conditions—acid catalysis at room temperature—an undehydrated intermediate could be isolated: beta-hydroxy-butyraldehyde.[19] Since this new compound was both an aldehyde and an alcohol, Wurtz called it "aldol," and the reaction was thereafter known as the "aldol reaction." This self-condensation of aldehyde was entirely analogous to Frankland's self-condensation of acetic ester; in each case, the reaction produced a new carbon-carbon bond and a doubling of the size of the carbon skeleton.

The aldol reaction, the earliest flexible and conveniently workable synthetic condensation reaction, was adaptable to innumerable special circumstances and compounds. After Wurtz's 1872 publication, there was a proliferation of "name reactions" in organic chemistry, a large number of which were synthetic condensations: the Perkin reaction, the Claisen condensation, the Claisen-Schmidt condensation, the Knoevenagel reaction, the Tollens reaction, the Wittig reaction, the Thorpe reaction, the Darzens condensation, the Erlenmeyer-Plöchl azlactone synthesis, and others. It is ironic that the founding fathers of this reaction type, Wurtz and Frankland, failed to have any of these reactions named after themselves.

The other major sense of "synthesis" is the artificial preparation of natural products, and here too Wurtz was in the vanguard. For example, in 1867 he synthesized choline, the first naturally occurring alkaloid to be prepared in a laboratory. The substance proved to be readily formed from trimethylamine and ethylene oxide; this mode of preparation also proved to be the key to Wurtz's demonstration of its structure a year later.[20]

These are only a few highlights from Wurtz's personal research of the last two decades of his life. Over the course of his career, Wurtz produced scientific work that equaled that of the best chemists of Germany or any other country. But what marks Wurtz's professional life even more strongly is his

19. Wurtz, "Sur un aldéhyde-alcool," *Comptes rendus* 74 (1872): 1361–1367.

20. Wurtz, "Synthèse de la névrine," *Comptes rendus* 65 (1867): 1015–1018; "Sur l'identité de la névrine artificielle avec la névrine naturelle," *Comptes rendus* 66 (1868): 772–776.

activity as director of a research school that was unique in France. (Institutional aspects of this school, and its character in the 1850s and the early 1860s, were described in chapters 6–8.) Édouard Grimaux later described the scene from the student's point of view:

He worked among us, having exactly as much room at his disposal as the least of his students; merry, serene, passionately pursuing his research, despite the noise, the laughter, and often the singing of the youngest. When he was asked, "Why do you not have a private room where you will not be disturbed?" he answered, "But here I am in the midst of my students." He loved his students; he loved to advise them, direct them, discuss with them. We would ask him for his opinion, or tell him an idea that we believed to be new, or give him our latest results; he would listen with affectionate grace, modestly advancing his objections or exposing his doubts, not as a dogmatic professor or an imposing master, but unaffectedly, as a more experienced comrade advising younger colleagues, ready to abandon his opinion if the arguments of his contradictors appeared to him more solid, and not afraid to learn something from those whom he had been charged to teach.[21]

Armand Gautier, one of Wurtz's best students and most loyal friends—and eventually his successor at the Faculté de Médecine—provided a telling anecdote along the same lines. In Gautier's first months in the lab, Wurtz counseled him to cease working on cyanogen derivatives, which topic he thought had been fully exhausted. Gautier declined to follow this well-intended advice, and soon thereafter made a number of significant discoveries. Wurtz then gathered his students together, explained what had happened, and said amiably: "He was right, and I was wrong. Do as he has done."[22]

Carneiro and Pigeard have recently examined Wurtz's and other French research schools in great analytical detail, and I will summarize some of their findings here.[23] Among rivals leaders of research schools, Berthelot

21. Grimaux, "Adolphe Wurtz," *République française*, 20 May 1884.

22. Gautier, "Ch.-Adolphe Wurtz, sa vie, son oeuvre, sa personnalité," *Revue scientifique* 55 (1917), p. 775.

23. A. Carneiro, The Research School of Chemistry of Adolphe Wurtz (Ph.D. dissertation, University of Kent/Canterbury, 1992), pp. 114–115 and appendix I; Carneiro, "Adolphe Wurtz and the Atomism Controversy," *Ambix* 40 (1993): 75–95; N. Pigeard, L'Oeuvre du chimiste Charles Adolphe Wurtz (1817–1884) (thèse de maîtrise, Université Paris X Nanterre, 1993), passim; Pigeard, "Un alsacien à Paris," *Bulletin de la Société Industrielle de Mulhouse* 833 (1994): 39–43; Carneiro and Pigeard, "Chimistes alsaciens à Paris au 19ème siècle," *Annals of Science* 54 (1997): 533–546.

and Deville were the only competitors who could really claim to have created and led such entities; Berthelot was therefore right in speaking of this "Pléiade" of three chemists of the mid to latter nineteenth century.[24] In 1865, when Berthelot gained his permanent appointment at the Collège de France and was given a laboratory there, he could appoint préparateurs who were interested in performing research under his direction. Three years later, the new infusion of funding and positions after the establishment of the École Pratique des Hautes Études provided a means to expand such activity. Carneiro has identified about 40 young chemists, including a few foreigners, who worked with Berthelot at the Collège in the last third of the nineteenth century. However, at any given time Berthelot's research group was always quite small, a handful of students at most. Moreover, Berthelot's research was quite disparate and was sufficiently anti-theoretical in outlook that he never developed a coherent theme or focus for the group. Finally, much evidence attests to the authoritarian character of Berthelot's direction of his students. Consequently, Berthelot's group does not compare well to Wurtz's either in total size or in reference to traditional positive models for research schools.[25]

Deville also taught at the École Normale, where he was a professor from 1851 on. (He also taught at the Sorbonne as suppléant and professor for nearly 30 years, but there were no proper labs there.) As has already been discussed, during the Second Empire Deville managed to finance his laboratory munificently by persuading Napoleon III to provide him annual allocations, in hopes of industrial applications. However, the total size of Deville's school was also small—Carneiro was able to find only about 30 students in all—and Deville's anti-theoretical attitude worked in a similar

24. Carneiro (The Research School, pp. 25–32, 38–46) also discusses the "schools" of Auguste Cahours and Louis Pasteur. However, she identifies only six students of Cahours, which does not rise to the level of a research school; furthermore, Pasteur left chemistry so early in his career that it does not mean much to speak of his school in the context of competition with Wurtz, Deville, and Berthelot.

25. Ibid., pp. 46–61. The classic literature on research schools includes the following: M. Crosland, *The Society of Arcueil* (Harvard University Press, 1967); J. Morrell, "The Chemist Breeders," *Ambix* 19 (1972): 1–46; G. Geison, "Scientific Change, Emerging Specialties, and Research Schools," *History of Science* 19 (1981): 20–40; J. Fruton, *Contrasts in Scientific Style* (American Philosophical Society, 1990); G. Geison and F. Holmes, eds., "Research Schools," *Osiris* [n.s.] 8 (1993).

fashion to Berthelot's to limit the impact of his influence. Where Deville did have significant success was in educating a new generation of chemistry teachers. His students filled the majority of the teaching posts in general and inorganic chemistry at the École Normale and the Sorbonne, and a large number in the provincial faculties of sciences, as well as in many lycées. Carneiro has argued persuasively that it was in this way that Deville—and not Berthelot—effectively prevented atomistic ideas from spreading in France in the second half of the century.[26]

In this regard something also must be said about Pelouze's laboratory school, which operated in two successive locations for a total of 19 years. One reliable estimate is that over 250 chemists received training in these two labs (Rue Guénégaud, 1838–1845; Rue Dauphine, 1845–1857), and of these, around 40 published research performed there.[27] However, Pelouze's institution was of a different type than Wurtz's. Pelouze ran a commercial school, designed especially to prepare young men for industrial-chemical careers, and those who pursued research there did so in a way quite unconnected with the bulk of Pelouze's "customers." In contrast, Wurtz's was an academic research school in the classic sense: an enterprise connected to a university and devoted to basic science, rather than technical training. One measure of this difference is the percentage of those who published research relative to the total student population. Pelouze's and Wurtz's total student numbers over their careers were probably similar, and yet the number that published from Wurtz's school was many times that of Pelouze's comparable group. It is certainly true that Pelouze was personally devoted to pure science, but he was even more averse to theory than Deville or Berthelot, if that was possible. He never developed a clear research pathway or agenda, so he had none to communicate to students.

26. Carneiro, The Research School, pp. 32–38, 94.

27. This is according to the undated, untitled, anonymous, 3300-word manuscript on Pelouze and his laboratories that is preserved in the Dossier Pelouze of the Académie des Sciences Archives in Paris. Written by a student of Pelouze (perhaps Aimé Girard) shortly after Pelouze's death in 1867, it is highly specific and well researched, and it seems authoritative. Some of Pelouze's better-known students (in a semi-chronological list) were Malaguti, Frémy, Gelis, Barreswil, F. L. Knapp, Millon, Zinin, Cloëz, Chancel, Gerhardt, Reiset, Plessy, Sobrero, Claude Bernard, F. Moldenhauer, Berthelot, Alvaro Reynoso, Aimé and Charles Girard, Dusart, de Luca, Vée, Davanne, Péan de St. Gilles, and Delaire.

Figure 11.2
Rue Dauphine 24, near the Mint. This was the site of Pelouze's laboratory school
in 1845–1857. Photograph taken by the author, March 1999.

Wurtz certainly did. Wurtz ardently believed that theory was the "soul of science," and the theories that won his full loyalty included chemical atomism, type theory, and (after 1858) structure theory. These ideas provided the conceptual vitality and heuristic guidance that informed not only his own personal research but also that of his entire research school. By both quantitative and qualitative measures, Wurtz's school dominated French chemistry during the last three decades of his life; his competitors really did not come close to his achievement.[28] The work of his students, who were often essentially independent but strongly influenced by the maître's direction and advice, ranged over all of chemistry, including inorganic, biological, analytical, physical, and industrial chemistry.

However, at the center of gravity of the laboratory always stood organic chemistry, which entered a phase of explosive growth in Germany after 1860. In this field Wurtz ran essentially the only show in town, or in the country—at least until his own students began to establish themselves as independent researchers. Wurtz's elders and contemporaries in the field avoided (or had abandoned) organic chemistry to an extraordinarily complete degree. Who were the other French organiciens? Carneiro argues that Auguste Cahours's research (some of which was significant in the organic field), and that of his few students, can almost be considered an offshoot of Wurtz's school; I think she is correct in this.[29] Of those chemists who were about Wurtz's age, Regnault, Deville, Berthelot, Frémy, and Pasteur had all begun their careers in organic chemistry; however, Regnault worked exclusively in physics after 1840,[30] Deville was not active in organic chemistry

28. In 1877 Wurtz asserted that more than 400 research publications had emanated from his laboratory (*Rapport sur l'École Pratique des Hautes Études* (Delalain, 1877), p. 40). A conservative estimate of the number of students that passed through the lab is 300.

29. Carneiro, The Research School, pp. 28–32, 131–132. Cahours operated almost as a "closet" advocate of the reformed chemistry. His excessive modesty and his instinctive conservatism and political caution led him to retain conventional equivalents until the end of his life; however, he urged his students to use atomic weights. Most of Cahours's students also considered Wurtz their mentor. See also E. Grimaux, "L'oeuvre scientifique d'Auguste Cahours," *Revue scientifique* 49 (1892): 97–101; A. Étard, "Notice sur la vie et les travaux de A. Cahours," *Bulletin de la Société Chimique* [3] 7 (1892), i–xii.

30. On Regnault's predilection for precise experimental work in physics after 1840, and its drawbacks, see M. Dörries, "Vicious Circles, or, The Pitfalls of Experimental

after about 1850, Berthelot published only occasionally in the field after about 1864, Frémy became more an editor and textbook writer than research chemist; and Pasteur shifted to biology around 1857. Of the older chemists, Dumas had ceased original scientific activity in 1848, Chevreul had migrated to other areas of work, Balard became inactive in research in the 1840s, and Pelouze's work after 1845 was of only occasional interest for organic chemistry.

In addition to the emphasis on the guiding role of theory and on organic chemistry (unusual in France at the time), Wurtz's school had many other distinctive characteristics. Although it was located physically and institutionally in the Faculté de Médecine, the Faculté offered no official sponsorship or curricular connection, nor was the teaching or research carried out there relevant to medicine except in the most distant sense. The presumed academic degree that could have been expected after a stint in Wurtz's laboratory, the medical doctorate, was rarely sought by his students. Many of his students were foreigners, for whom a French doctorat would have been of no use at home; in any case, they and their French comrades were there to learn cutting-edge chemistry, not medicine. The fact that Wurtz charged his students a monthly fee was also unique in French academic establishments. Structurally, the arrangement was that of purchasing a service[31]; one concomitant was that students could claim (and were willingly accorded) more independence than could be had in Berthelot's or Deville's lab. In those labs, students were officially sanctioned by the government, and they paid nothing for the privilege of working there (or were even given salaries as préparateurs or répétiteurs); by the same token, they did what they were required to do by the maître to gain their degree and prepare for a career in France.[32]

In this manner an extraordinary situation was established in France. In the 1860s and the 1870s, as organic chemistry mushroomed in Germany

---

Virtuosity," in *Experimental Essays—Versuche zum Experiment*, ed. M. Heidelberger and F. Steinle (Nomos, 1998).

31. Wurtz followed the practice of remitting all fees for those who were unable to pay; there were always at least a few such students present in the lab. Pelouze, in contrast, made no pretense of running other than a commercial operation. Whereas Dumas could afford to operate an academic-style lab out of his own pocket, Wurtz's financial arrangements were required for the very existence of his lab.

32. Carneiro, The Research School, pp. 127, 144.

and fueled much of the growth in academic science there that so astonished the world, there was only one place where the field flourished across the Rhine: in a makeshift, unofficial laboratory attached to the Paris medical school. That one modest academic laboratory (and its later "branches" led by former students of Wurtz) accounted for nearly all French research in classical organic chemistry that was published during the critical period 1864–1884.

In these years Wurtz's laboratory was in fact internationally recognized as one of the world's premiere "finishing schools" for prospective elite chemists, and gifted young men came from all over—Germany, Russia, Great Britain, the United States, Switzerland, and many other countries— to spend a few months at the Faculté de Médecine. Among the famous non-French chemists who claimed Wurtz as a mentor were A. M. Butlerov, A. S. Couper, Adolf Lieben, George Carey Foster, Friedrich Beilstein, James Crafts, N. A. Menshutkin, A. M. Zaitsev, Albert Ladenburg, and J. H. van't Hoff. We have also seen that Wurtz's influence was crucial for August Kekulé's intellectual growth, although Kekulé's Paris sojourn took place 2 years too early for him to have been able to benefit from actually working in the Wurtz lab. As Crafts put it, Wurtz's research school "succeeded in some measure to that of Liebig, and was visited by chemists of all nations; for European science still held to the traditions of Humboldt, and a sojourn, however short, at Paris was considered a desirable part of a scientific education."[33]

We have already seen (in chapter 6) that the percentage of foreigners among Wurtz's laboratory workers was particularly high, and that Alsatians constituted much of the French contingent. The laboratory community was, consequently, not typically French, but rather an international group. Chemistry was spoken in many different languages—including a great deal of the curious alemannic Alsatian language (Wurtz himself spoke fluent Alsatian, French, German, English, and Italian). An Austrian student during 1859–60 recalled the experience:

We formed a society of people who had come from various countries: Austria, Russia, Spain, Cuba, even India, and of course France was represented; but we always consorted in a friendly way as a "people of brothers," devoted with equal enthusiasm to the study of science, under the direction of a superb researcher who

33. Quoted in ibid., p. 129; see also p. 127.

as a teacher was accessible equally to everyone, who was always cheerful and benevolent, and who knew how to keep us earnestly at our work.[34]

Carneiro and Pigeard have shown that the "hard nucleus" of Wurtz's school was formed nearly entirely of Alsatian students, such as Charles Friedel, Auguste Scheurer-Kestner, Charles Lauth, Edmond Willm, Joseph Achille LeBel, and Willliam Oechsner de Coninck. The Alsatian community had many strongly marked characteristics: an orientation toward German ideas and culture, facility with foreign languages, internationalism, Protestant-ism, a sensitivity toward industrial connections, political liberalism, and a respect for the role of imagination, intuition, and theory in science. All these qualities characterized Wurtz, and also characterized the Alsatians at the core of Wurtz's group. After the Franco-Prussian War, Wurtz played an important role in establishing the École Alsacienne de Paris and other insti-tutions to aid his displaced countrymen; throughout his life Wurtz was both proud and self-aware of his Alsatian roots.[35]

These brief remarks on the content and character of Wurtz's research school during its mature years provide only a sketch of the essentials. For fur-ther details, the reader should refer to the outstanding studies of this subject carried out during the last few years by Ana Carneiro and Natalie Pigeard.

Administering the Faculté de Médecine

In chapter 9 we saw how Wurtz collaborated with Minister of Public Instruction Victor Duruy after 1863 on a campaign to convince the gov-ernment that the French science establishment had fallen far behind that of Germany and thus required a massive infusion of funds. On 18 January 1866, Duruy elevated his co-conspirator to the deanship of the Faculté de Médecine.[36] During the troubled deanship of Ambroise Tardieu

34. A. Bauer, "Erinnerungen," *Oesterreichische Chemiker-Zeitung* 22 (1919), p. 117. Bauer provides an extraordinary depiction of daily life among the German chemical colony in Paris at the time.

35. See the sources cited in n. 23 above, especially "Atomism Controversy" and "Chimistes alsaciens à Paris."

36. MIP to Vice-Recteur, 18 January 1866, AJ16/295. On the context of Wurtz's activities in this period, see G. Weisz, "Reform and Conflict in French Medical Education, 1870–1914," in *The Organization of Science and Technology in France 1808–1914*, ed. Fox and Weisz (Cambridge University Press, 1980).

(1864–1866), the Faculté had suffered various political and social problems layered on top of fiscal distress; in giving Wurtz the nod, Duruy remarked: "M. Wurtz, I am going to age you ten years."[37] Wurtz—being more of a chemist than a medical scientist—was not Duruy's first choice for the job, but others had refused to take on the difficulties. In the end, Wurtz proved to be the perfect man for the position. In the 9 years he held this position (including a brief suspension from duties during the Commune), he brought to the school political stability, modest expansion and renovation, and eventually even a measure of tranquillity.[38]

Not that it was easy. Under Doyen Wurtz, the Faculté de Médecine continued to be the site of republican student protests against the illiberal policies of the empire, as well as the focus of charges from conservatives that irreligious doctrines were being taught there.[39] Students disrupted the opening convocation in November of 1867 and the subsequent classes of Professors Vulpian, Baillon, and Sée; meanwhile, denunciations of the alleged materialism of Professors Axenfeld, Vulpian, and Robin were published.[40] Wurtz responded with a powerful defense of freedom of academic opinion and of the principles of disinterested scientific investigation; in the process he provided insight into his own philosophy of science. The study of medicine, he wrote to Duruy, had entered a new phase in which observation and experiment, elements of the scientific method, were paramount:

Like physics and chemistry, the science of organization [i.e., physiology] today begins by establishing facts, and, after having drawn from these facts the immedi-

37. Hofmann, "Wurtz," p. 258. Hofmann quoted the sentence in French; it is likely he had the story from his friend Wurtz.

38. N. Pigeard, "Wurtz, doyen de la Faculté de médecine de Paris, 1866–1875," *Club d'histoire de la chimie, Bulletin de liaison* 3 (1994): 24–29.

39. Carneiro, The Research School, pp. 91–93; A. Corlieu, *Centenaire de la Faculté de Médecine de Paris (1794–1894)* (Imprimerie Nationale, 1896), p. 245; Hofmann, "Wurtz," pp. 256–261; A. Prévost, *La Faculté de Médecine de Paris, ses chaires ses annexes et son personnel enseignant de 1794 à 1900* (Mailoine, 1900); R. Fox, "Positivists, Free Thinkers, and the Reform of French Science in the Second Empire," in *Science, Industry, and the Social Order* (Variorum, 1995).

40. Wurtz, annual report to Conseil Académique, 26 November 1867, AJ16/6566; Wurtz to MIP, 6 April 1868, AJ16/6494; Albert Duruy to Wurtz, 25 April 1868, AJ16/6494. See also R. Fox, "Science, the University, and the State in Nineteenth-Century France," in *Professions and the French State, 1700–1900*, ed. G. Geison (University of Pennsylvania Press, 1984), pp. 97–99.

ate and proximate consequences, it arrives at more general and elevated inductions, if the solidity of the base allows access to the heights. This is the positive experimental method, sometimes confounded, wrongly and intentionally, with positivism: these two have nothing in common. Science is free to choose the method that suits it, and of holding to its proper sphere, which is that of pure reason. Absolute independence is necessary for it to do so. The Faculté de Médecine has introduced this exact method of modern science in its instruction. This is the propensity that is being unjustly incriminated.[41]

As a direct result of Wurtz's passionate representations, the principles of academic freedom were upheld, and the protests quieted.

A second nexus of controversy at this time was the effort of women to gain entrance to courses in the Faculté. Against the will of most or all of his colleagues, Wurtz successfully appealed to Duruy to allow Mlle. Mary Putnam, "pharmacien et docteur de l'Université de Philadelphie," to register for the course of study leading to the doctorate in medicine.[42] Putnam was apparently preceded briefly at the Faculté by a French woman, Madeleine Brès, who was allowed to register. Once these exceptions had been made, a precedent was established, and gradually women were admitted to the various faculties of the Université de France. Pigeard has identified six women who studied in Wurtz's laboratory. When Wurtz died, organized groups of women students from the Sorbonne and Faculté de Médecine attended his funeral, out of respect for what he had done for them.[43]

41. Wurtz to MIP, "Sur l'Enseignement à la Faculté de Médecine," n.d. (ca. May 1868), AJ16/6494. "Ainsi que la Physique et la Chimie, la science de l'organisation commence aujourd'hui par établir les faits, et, après avoir tiré de ces faits les conséquences immédiates et prochaines, elle arrive à des inductions plus générales et plus élevées à la condition que la base affermie permette l'accès des hauteurs. Telle est la méthode expérimentale positive, quelquefois confondue, à tort et à dessein avec le positivisme: ces choses-là n'ont rien de commun. La science est libre de choisir la méthode qui lui convient et de se maintenir sur son domaine qui est celui de la raison pure. Il faut qu'elle y conserve une indépendance absolue. La Faculté de médecine a introduit dans son enseignement cette méthode exacte de la science moderne. C'est là la tendance qui est injustement incriminée."

42. Perhaps this was Mary Putnam-Jacobi, mentioned by Margaret Rossiter as "a prominent New York City pediatrician and neurologist" in 1893: *Women Scientists in America* (Johns Hopkins University Press, 1982), pp. 97–98.

43. Pigeard, "Un alsacien à Paris," p. 42; Pigeard, "Chemistry for Women in Nineteenth-Century France," in *Communicating Chemistry*, ed. B. Bensaude-Vincent and A. Lundgren (Science History Publications, 2000), pp. 403–422; Gautier, "Ch.-Adolphe Wurtz," pp. 777–778; AJ16/6255; AJ16/269.

According to one report, Wurtz waited a long time before accepting the deanship.[44] He may have laid down conditions for his acceptance. Whether or not this was the case, in March of 1866 he gained approval from his faculty and from Duruy to hire a special lecturer to supplement his own chemistry teaching. He appointed the agrégé Alfred Naquet for this task. A former student of Wurtz, Naquet was an outspoken republican, and a year earlier had published the first textbook in France that used atomic weights.[45] (When Duruy was removed from his post in 1869, so was Naquet.) In later years, under Wurtz's deanship, additional special lecturers were added in other subjects, significantly supplementing instruction throughout the Faculté. A second measure taken within months of Wurtz's accession may also have been a condition of his acceptance. In June of 1866 Wurtz finally won approval for a step he had been advocating for years: that the Faculté assume the financial details and bookkeeping for his research laboratory. The lab still had no official existence, much less a state-conferred budget, but at least Wurtz no longer had to keep the books himself.[46]

There was a third new initiative that first year, which was certainly the most significant of all. In the fall of 1866, Duruy obtained and granted sufficient funds to allow the Faculté to begin designing student practica for physiology, chemistry, and clinical medicine. Space was found by converting the dissection pavilions into summer teaching laboratories. (They were usable for dissections only in winter, anyway, when corpses had longer shelf life.) Wurtz's annual report submitted ca. November 1866 combines expressions of gratitude for this grant with his usual warnings about the utter insufficiency of the physical structures. The pavilions had not changed in 30 years; they were "poorly constructed, poorly ventilated,

44. Hofmann, "Wurtz," pp. 257–258.

45. Wurtz to MIP, 13 March 1866 and Duruy to Vice-Recteur, 26 March 1866, AJ16/295. Naquet's textbook was *Principes de chimie fondée sur les théories modernes* (Savy, 1865). Wurtz had published the first volume of his *Traité élémentaire de chimie médicale*, which used equivalents, in 1864; the first fascicles of his *Leçons élémentaires de chimie moderne*, which used atomic weights, did not begin appearing until the end of 1866. Of course, Wurtz had been using atomic weights in his research papers, monographs, and invited lectures since late 1858.

46. MIP minutes and redaction de M. Dumas, Comité de l'Inspection Générale, 6 June 1866, F17/4020.

poorly heated, crowded with tables, offering our throngs of students but a miserable facility." The standing of the Faculté was being "seriously menaced by foreign competition," and these matters still desperately needed to be addressed.[47] Wurtz's reports for 1867 and 1868 were even more pessimistic. In response, Duruy provided the Faculté a regular budget line of 15,000 francs for these medical school practica. For Wurtz, it was still far too little. In the summer of 1868, under these "deplorable" conditions, 80 students completed a chemical practicum. "And what a depressing contrast between these bargain-basement laboratories and the monumental edifices" of German universities.[48]

After the École Pratique des Hautes Études was established, Wurtz wrote to Duruy once more, asking to be officially appointed director of the research laboratory that he had run for the previous 15 years with no financial or material resources from the state. His case was all the more compelling, he thought, in view of the fact that he had found that two of the rooms of the dean's official residence were too filthy to house his young family, and instead had assigned them to be used for examinations. When the only response to this request was the grant of a dean's salary of 3000 francs, Wurtz was incredulous, for this was exactly what every dean had been paid since 1813, and what he was already getting. In April of 1869, Wurtz was at last granted a sum of 2000 francs per year as director of his laboratory; he used the money to provide the salary for his préparateur.

47. Wurtz to Rector and Conseil Académique, annual report of the Faculté de Médecine for 1865–66, AJ16/6566. "Nos pavillons de dissection mal contruits, mal ventilés, mal chauffés encombrés de tables n'offrent à nos élèves qui s'y pressent qu'une installation misérable. . . . [N]otre Faculté est sérieusement menacée par la concurrence étrangère." On this small renovation see also Wurtz to M. le Recteur, 21 December 1866, AJ16/6348. A few months later Wurtz widened the scope of his requests, but without success (Wurtz to MIP, April 1867, AJ16/6360).

48. Ibid., annual reports for 1866–67 and 1867–68 (latter quoted: "Et quel contraste douloureux entre ces laboratoires d'occasion et les constructions monumentales qui s'élèvent de toutes parts dans les Universités allemandes, et que j'ai visitées récemment"); Procès-Verbaux de l'Assemblée des Professeurs de la Faculté de Médecine, 31 October 1867, AJ16/6255. The instructional laboratory that began at the Faculté in 1867 was separate and distinct from Wurtz's research laboratory, which had been operating since 1853.

The new director's salary and the dean's stipend were in addition to his regular professorial salary of 10,000 francs.[49]

But as doyen, Wurtz consistently put the Medical School's needs above his own. Just before the war, Wurtz characterized the situation as perilous and requested permission to implement the program for reconstruction that had already been approved. Asked to prioritize his requests in the stringent environment after the war, Wurtz urged first the full reconstruction of the École Pratique de Médecine (the clinical branch of the school), estimated to cost 9.4 million francs, and next the renovation of facilities for the study of anatomy and physiology, at 4.1 million. Architectural plans were drawn up and approved, but implementation was delayed. A few months later, in an annual report, Wurtz returned to the same theme, decrying the sad physical state of his school, which he had been lamenting in reports for 20 years. "This scientific poverty in which [the Faculté] is struggling forms a striking contrast with the state of prosperity and of progress which has been noted in this regard in the faculties across the Rhine." The same lament came again in 1870: "You see, Monsieur, it is always the same note which returns and the monotony of my complaints is their best justification. Every year, I repeat [that] the status quo cannot continue, and yet it continues; it continues with an ever increasing number of students." In the winter of 1873–74, a fire in the Hôtel de Ville destroyed the architectural plans, and they had to be redrawn.[50]

49. Wurtz to MIP, 22 November 1868 and 19 March 1869; Vice-Recteur to MIP, 25 March 1869; note to MIP, 17 April 1869; unpaginated notebook labeled Wurtz, EPHE, Lab. de chimie, FM Paris, 1869–1884; all in F17/4020. Wurtz's professorial salary, like that of each of his colleagues, consisted of a base amount of 7000 francs, plus a capitation salary set at 3000; these numbers never changed throughout the Second Empire. In the 1870s the base salary was raised to 13,000 francs, then to 15,000, with no capitation (MIP, Notice Individuelle, Dossier Wurtz, AJ16/6565; Fiche d'état de service for Wurtz, F17/21890; Corlieu, *Centenaire*, p. 70).

50. Wurtz to MIP, 12 March 1870, AJ16/6348; Wurtz, "Rapport à M. le Ministre de l'Instruction publique sur l'état des bâtiments et des services matériels de la Faculté de Médecine," 1 February 1872, AJ16/6357, printed in revised form in *Revue des cours scientifiques* [2] 1 (1872): 852–854; Wurtz to Recteur, annual report, 29 November 1872, AJ16/6566 ("Cette misère scientifique où elle se débat fait un contraste saisissant avec l'état de prospérité et de progrès que l'on constate à cet égard dans les Facultés d'outre Rhin"); ibid., 4 November 1873 ("Vous le voyez, Monsieur, c'est toujours la même note qui revient et la monotonie de mes

Nothing happened; and then nothing happened some more. Proper laboratories had been promised upon the foundation of the Faculté de Médecine in 1796, but never constructed. Dumas's schemes for reconstruction of various Parisian institutions had been on the brink of success at the beginning of 1848, but then lightning had struck with the onset of revolution. After the failure of the revolution and the establishment of the empire, a full renovation had been planned and promised in 1855, but again no movement occurred. In the late 1860s success had once more seemed within reach; however, disaster had come again, this time in the form of a national military catastrophe. Wurtz had had more than enough. The renovation was, he told a correspondent, "the principal issue of my deanship," and he wanted to see it to conclusion.[51] To his old friend, former student, and now senator Scheurer-Kestner, Wurtz wrote: "The insufficiency of our laboratories and the poverty of our installation does not allow us to have many students. Tell that, dear friend, to your co-sovereigns who hold the purse-strings and the destiny of France."[52]

A crucial problem here was, of course, the chaotic political situation in the 4 years after the Commune.[53] Only with the successful "amendement Wallon" of February 1875—passed with a single-vote majority—was the form of government for France clearly indicated as a republic. In the legislative elections of early 1876, republicans took the Chamber of Deputies, while the Senate remained divided. Late that year, no longer under Wurtz's deanship, the money was finally released for full reconstruction of the

---

doléances est leur meilleure justification. Tous les ans je répète le statu quo ne peut pas durer et pourtant il dure: il dure avec un nombre d'élèves croissant sans cesse"); Wurtz to Recteur, 31 March 1874, AJ16/6360.

51. Wurtz to L. Micé, 22 February 1875, Wellcome Institute, Western Manuscripts Collection, Wurtz file. "C'était là la principale affaire de mon décanat; je voudrais la faire aboutir."

52. Wurtz to Scheurer-Kestner, 6 July 1875, Ms. 5983, Bibliothèque Nationale et Universitaire de Strasbourg: "L'exigüité de nos laboratoires et la misère de notre installation ne nous permettent pas de faire beaucoup d'élèves. Dites-cela, cher ami, à vos co-souverains qui tiennent les cordons de la bourse et les destinées de la France."

53. For an excellent analysis of the growth of the academic reform movement coincident with the rise of political institutions of the incipient Third Republic, see G. Weisz, *The Emergence of Modern Universities in France, 1863–1914* (Princeton University Press, 1983), pp. 90–133.

Faculté, and work began early the next year. Among the functions that had to be moved to allow for the new construction was Wurtz's research laboratory. At a cost of 13,000 francs, a neighboring building was purchased for this purpose, on the Rue Hautefeuille at the corner of the Rue de l'École de Médecine; the cost of conversion and outfitting over the summer of 1877 was another 50,000 francs. In December, Wurtz reported that the new lab was operating smoothly, with 21 French and 7 foreign chemists working there, but that new appropriations were necessary to keep the operation going properly. The new larger facility drew an expanded clientele: by 1881 there were 42 working there.[54]

Wurtz's new research lab was substantially more commodious and appropriate than the makeshift lab he had used for the last quarter-century. There were ten rooms, including for the first time a private lab for the director and one for his old teacher Amédée Cailliot, now a refugee from Alsace. And at this point, finally, the leaders of the new Third Republic supplied permanent budget lines to support the laboratory. From money provided from the budgets of the Faculté and the École Pratique des Hautes Études combined, Wurtz could hire two préparateurs and a préparateur adjoint, plus two orderlies; in addition, he had a materials budget of 5400 francs. (The separate chemistry teaching lab had a staff of seven and a budget of about 20,000, and served more than a hundred students per semester.) Wurtz had always hoped eventually to move his professional activity to the Sorbonne, but it was in the new lab of the Rue Hautefeuille that he performed his last scientific work.[55]

The Final Battle of the Campaign: The Sorbonne

The Sorbonne had long been the scene of fine science oratory, but no research. After the retirements of Gay-Lussac and Thenard, it was Dumas and Balard who held the chairs of chemistry, but both men were highly

54. Architect to Doyen, 24 January 1877, Vice-Recteur to Doyen, 31 July 1877, accounting sheet, ca. 1878, all in AJ16/6661; Wurtz to MIP, 10 December 1877, F17/4020; Wurtz to MIP, annual report, 11 April 1881, AJ16/6555.

55. Wurtz to MIP, 29 April 1881, F17/4020; C. Friedel, "Notice sur la vie et les travaux de Charles-Adolphe Wurtz," *Bulletin de la Société Chimique* [2] 43 (1885), pp. xvi–xvii.

cumulated and used substitute lecturers for many years.[56] In a curious maneuver in 1867, both Dumas and Balard retired from the Sorbonne, and these two chairs were given to Deville and Pasteur respectively, the latter now intended to be a chair for organic chemistry (though not named such).[57] However, Pasteur rarely gave his own lectures and lasted only 7 years in the position; furthermore, Deville's lectures were given by his protégé Louis Troost. (Deville was, of course, simultaneously professor at the École Normale, where Pasteur also retained a research connection.) In 1874, Pasteur, who had long been devoting his career to biological rather than organic-chemical research, retired from the Sorbonne, and Troost succeeded him. Neither Troost nor Deville had much interest in organic chemistry, and neither accepted the reformed chemistry based on atomic weights. The situation was now particularly bleak for the modernists; it was simply staggering that the famous Sorbonne offered nothing in the way of organic chemistry, or much of modern chemistry however defined. (Even the professor of organic chemistry at the Collège de France, Berthelot, had left that field for physical chemistry virtually the same year he had been appointed.)

Wurtz took this opportunity to approach Pasteur, seeking his support for the proposal that he might offer a "cours annexe" in modern organic chemistry at the Sorbonne. Pasteur, Deville, and Dean Milne-Edwards all supported the idea, and it was approved unanimously by the faculty; Wurtz

56. Dumas was elevated from adjoint to titulaire at the Sorbonne upon Thenard's retirement in 1841; Balard, initially appointed adjoint to replace Dumas in that role, was apparently made a second titular professor of chemistry around 1846. (Gay-Lussac's chair had been in physics, not chemistry.) These details are sketchy; a history of the science chairs at the Sorbonne is yet to be written. See P. Julien and L. Marquet, "Balard," *Revue d'histoire de la pharmacie* 24 (1977): 65–73; F. Gallais, "Balard à Paris," *Revue d'histoire de la pharmacie* 24 (1977): 28–33.

57. These changes were indirectly precipitated by the death of Pelouze. In Pelouze's place, Dumas became president of the Commission des Monnaies, a lucrative but time-consuming position; Balard took Dumas's position as Inspecteur Général de l'Enseignement Supérieur. Duruy had wanted Pasteur for the latter post, but Pasteur refused on principle to compete with his former patron. (Berthelot had no such scruples, and became a candidate, but was not successful.) See Pasteur to Duruy, 31 July and 5 September 1867, *Correspondance de Pasteur*, vol. 2, pp. 340, 347–352. The direct intercession of the emperor can be assumed for the award of the Sorbonne chairs, for both Deville and Pasteur had close ties not only to Duruy, but also directly to Napoleon.

Figure 11.3
A portrait of Deville, Balard, and Wurtz taken on the occasion of the Exposition
Universelle of 1867. Source: Autrand and Oechsner de Coninck families.

was to be paid 5000 francs.[58] A poster preserved in the Archives Nationales advertised the course to students:

Special Course on Organic Chemistry, WF 1:30. M. Wurtz, Member of the Institute, will set forth the new theories in organic chemistry. He will then treat alcohols and compounds related to or derived from them. He will conclude with the study of aromatic compounds, by indicating the applications of which they have become the object. Vice-Recteur Mourier; Dean of the Faculté des Sciences Milne-Edwards.[59]

On 22 April 1874, Mourier reported to the Ministry of Public Instruction on Wurtz's first lecture:

M. Wurtz gave today his first lecture on organic chemistry at the Faculté des Sciences; the hall was filled as it used to be in the great days when MM. Thenard and Dumas attracted to the Sorbonne a studious elite. M. Wurtz was a very great success; his lecture, presented with talent and flair and a great animation of speech, was heard with lively interest and hailed with unanimous applause. The lecture began at 1:30 and did not end until after 3:00.[60]

Dumas wrote to Wurtz on that day, apologizing for not being able to attend. Wurtz replied, "very touched," he said, by his mentor's kindness. He had had, that day, the opportunity, "before a large and sympathetic audience," to dwell upon "the guiding influence of your work." If Dumas would pay him the honor of coming to a future lecture, Wurtz continued, he would let him know in advance which would be a good one to attend; and "that day will remind both of us of an already distant past, when the roles were fortunately reversed."[61] Two weeks later Wurtz wrote to his

58. Wurtz to Dean Milne-Edwards, reproduced in transcripts of the procès-verbal of the Faculté meeting of 24 February 1874, and Milne-Ewards to MIP, 27 February 1874, both in F17/21890; original procès-verbaux of the Faculté des Sciences in AJ16/5121; also miscellaneous correspondence regarding this course in AJ16/331.

59. F17/21890.

60. Mourier to MIP, 22 April 1874, F17/21890. "M. Wurtz a fait aujourd'hui la première leçon de chimie organique à la faculté des sciences; l'amphithéâtre était plein comme aux grands jours où MM. Thénard et Dumas appelaient à la Sorbonne une élite studieuse. M. Wurtz a eu un très grand succès; sa leçon, faite avec talent et éclat et une grande animation de parole, a été écoutée avec un vif intérêt et saluée par d'unanimes applaudissements. La leçon, commencée à 1 heure 1/2, n'a fini qu'après 3 heures."

61. Wurtz to Dumas, 22 April 1874, Dossier Wurtz, Archives de l'Académie des Sciences. "J'ai été très touché de la lettre que vous m'avez adressée aujourd'hui. . . . [V]ous auriez été à la fois juge et partie, car j'ai fait ressortir, devant un auditoire

friend and former student, the structuralist organic chemist A. M. Butlerov:

You will perhaps be interested to learn that I am now giving a course in organic chemistry at the Sorbonne, where the new ideas have had such difficulty in penetrating. I hope that my efforts will help to bring about this desirable result, and to shake off the spirit of routine and torpor of some of our French professors.[62]

Wurtz did indeed have a larger goal in mind: the establishment of a third permanent chair of chemistry at the Sorbonne—a professorship that, unlike Pasteur's, could truly be dedicated to organic chemistry. It was at the Sorbonne, the Science Faculty of the University of Paris, that the teaching of modern advanced chemistry naturally belonged. That was where one could expect to find well-prepared and well-motivated students who wished to learn at the cutting edge—not at a medical school where pure chemistry could never be anything more than an ancillary study. Dean Milne-Edwards presented the proposal to add this new chair at a faculty meeting in November of 1874, and it was sent with their approval to the Conseil Académique in December. Vice-Recteur Mourier reminded the Conseil of "the brilliant success achieved by M. Wurtz, Dean of the Faculté de Médecine, in the special course which he was authorized to present" earlier that year, and suggested that he be nominated to fill the chair. Both the Conseil and the Minister of Public Instruction, the Vicomte de Cumont, then gave their endorsements.[63]

This action resulted in a bill presented to the National Assembly in February of 1875. "No one is unaware," the bill states,

that since instruction in chemistry at the Faculté des Sciences was organized a half-century ago, new theories have arisen, methods have been transformed, and discoveries that appear to offer a purely scientific interest have provided the most fertile and unexpected applications in the field of commerce and industry. To cite

---

nombreux et sympathique, l'influence dirigeante de vos travaux. . . . Ce jour nous rappellera à tous deux un passé déjà lointain, où les rôles étaient heureusement intervertis."

62. Wurtz to Butlerov, 4 May 1874, in G. Bykov and J. Jacques, "Deux pionniers de la chimie moderne, Adolphe Wurtz et Alexandre M. Boutlerov, d'après une correspondance inédite," *Revue d'histoire des science* 13 (1960), p. 128.

63. Procés-verbal of Faculté des Sciences, 21 November 1874; Procés-verbal of Conseil Académique, 8 December 1874; Vice-Recteur to MIP, 15 and 18 December 1874; all in F17/21890.

but one example, the manufacture of coal tar dyes and the artificial production of the coloring principle of madder have resulted from the most scholarly and abstract research on the substances concerned.

No matter how deep their knowledge, the text goes on, the eminent professors who currently occupy the only two chairs of chemistry at the Sorbonne (Deville and Troost) do not have the opportunity to give more than a sketch of these exciting new developments.

The interest that has been shown in the attempts already made in special lectures by a dedicated professor, demonstrates the necessity of an instruction that would initiate students into the development of the industrial arts and the most diverse scientific applications, illuminate the origin and causes of discoveries, and, finally, show by what route the high significance of science can extend our national wealth.

This is effective writing for an audience of politicians, and the bill was approved in March. The "dedicated professor" to whom it referred—Wurtz—was hired at the usual Sorbonne salary (13000 francs), and given a budget of 3000 and an orderly. He began his Sorbonne career that fall, retiring from his deanship in order to manage the load (though he did not relinquish his medical professorship). In gratitude for his extraordinary service, the Faculté de Médecine took the unusual step of appointing him doyen honoraire for life.[64]

I have already noted how tiny and miserable the Sorbonne laboratories were. Since Wurtz was the newest professor, he had no laboratory whatever, not even an empty room in which to arrange demonstrations for his lectures. (Even Deville and Troost only had a small laboratory for lecture experiments, and none for teaching or research.) Wurtz and his préparateur, Georges Salet, were compelled to transport the apparatus and materials from the Faculté de Médecine to the Sorbonne before each lecture, then cart them back after it was over. Wurtz had hoped for better; indeed, he had hoped that a proper modern research laboratory would be built for him within the precincts of the Sorbonne, so that he could transfer his research activity there. Wurtz wrote to Milne-Edwards in June of 1876, expressing this urgent request.[65]

64. Assemblée Nationale, Projet de Loi No. 2901, 23 February 1875; MIP Décret, 1 August 1875; both in F17/21890.

65. Friedel, "Wurtz," pp. xix–xxi; Hofmann, "Wurtz," pp. 306–308; Wurtz to Dean Milne-Edwards, 2 June 1876, summarized and excerpted in auction catalogue, Dossier Wurtz, Académie des Sciences. (The present location of this memo is not known.)

But even if Milne-Edwards had the will, he did not have the resources to grant Wurtz's wish. Wurtz would have to be satisfied with the new lab in the Rue Hautefeuille, built in 1877.

However, reform and reconstructions were in the air. In July of 1878 Wurtz was sent once more by the Ministry of Public Instruction on a fact-finding mission to Germany, Switzerland, and Austria-Hungary, in order to produce a second edition of his 1870 report on practical higher education in the physical and bio-medical sciences in the Germanic countries. "I fear," he wrote to Scheurer-Kestner while traveling in Switzerland, "that we are embarking on these constructions a little too lightly, and that we remain behind the Germans."[66] Wurtz's report, published in 1882, not only underlined the extent to which France was still far behind the Germans in facilities for academic science but also gave Wurtz an opportunity to express his opinions on the best ways to ameliorate the situation.

Wurtz not only advocated the German model of purpose-built scientific laboratories with all the most modern facilities and appliances; he also defended the German practice of providing dedicated academic "institutes" for each specialty field—organic chemistry, physical chemistry, pathological anatomy, experimental physiology, and so on. Wurtz thought that the advantages of this system outweighed the usual criticism of French observers that it would lead to isolation and prevent cross-disciplinary collaboration. Wurtz also strongly urged the integration, in the German (Liebigian) mold, of teaching and research laboratories; the École Pratique des Hautes Études, for all its merits, had maintained a strict separation of the two categories. Wurtz's only criticism of the laboratories he had seen was that many of them were too lavish; science facilities should be utilitarian and practical, not pointlessly luxurious.

When Jules Ferry was appointed Minister of Public Instruction in January of 1879, Wurtz and his friends knew that they had finally won the day. A republican, an advocate of secular control of education, and fully convinced of the inferiority of French higher education to German, Ferry backed total

66. Wurtz to Hofmann, n.d. [ca. 22 June 1878] and 5 July 1878, Chemiker-Briefe 118, Berlin-Brandenburgische Akademie der Wissenschaften; Wurtz to Scheurer-Kestner, 7 August 1878, Ms. 5983, Bibliothèque Nationale et Universitaire de Strasbourg ("je crains qu'on ne se lance un peu légèrement dans ces constructions et qu'on ne reste au-dessous des Allemands"); Wurtz, *Les hautes études pratiques dans les universités d'Allemagne et d'Autriche-Hongrie* (Masson, 1882).

renovation of the system, physically and organizationally. Moreover, the republicans, natural allies of educational reformers, gained control of both branches of the legislature in that year. The economy was thriving, and the budget was balanced. Accordingly, it is not surprising that 2 years later a complete reconstruction of the Sorbonne was approved by the French legislature, the Université, and the City of Paris, and an open architectural competition was held.

The architect selected, Henri-Paul Nénot, was only 29 years old, but the choice proved wise. Nénot worked closely with Wurtz and others in determining the scientific requirements for the new laboratories. Following Wurtz's lead, he even toured several (mostly Germanic) European countries to gather ideas. Hofmann was then rector of the University of Berlin; Wurtz provided the introduction for Nénot, and Hofmann spent days with the young architect going over even the smallest details. The total cost of the project was no less than 22 million francs. Ground was broken in November of 1884, and almost immediately the infamous and long-lost cornerstone of the abortive 1855 reconstruction was unearthed. The new cornerstone was laid in August of 1885, but this time there was no miscarriage. The new Sorbonne, three times the size of the old and finally properly equipped with scientific laboratories, was completed by 1894.[67] Unfortunately, Wurtz did not live to see even the groundbreaking.

Home Life and Politics

During the last 14 years of his life, Wurtz enjoyed the esteem of his colleagues both in France and abroad, a politically powerful position in the structure of French science, and an emotionally satisfying and financially stable home life. The Wurtzes had four children, Marie, Lucie, Robert, and Henri[68]; in addition, they adopted the four orphaned daughters of Mme. Wurtz's sister. In the last few years before her death in 1878, Adolphe's

67. H. Nénot, *La nouvelle Sorbonne* (Colin, 1895), pp. 1–5, 20, 37–56; O. Gréard, *Éducation et instruction*, second edition (Hachette, 1889); Hofmann, "Wurtz," p. 308.

68. Marie (1854–1930) married one of her father's students, William Oechsner de Coninck, who became professor at Montpellier. Lucie (1856–1922) married Denis de Rougemont, an artillery captain. Robert (1858–1919) became a physician. Henri (1862–1944) pursued a military career.

mother also joined the household. By all reports it was an extraordinarily happy home, filled with sociability, music, and conversation. The parents of Constance Wurtz née Oppermann were well to do, and there is evidence to suggest that the Wurtzes enjoyed at least moderate wealth from the time of their marriage in 1852.[69]

In the 1860s the family spent summers in rented villas near Paris or in the Loire Valley. When Wurtz's salary doubled upon his hire at the Sorbonne in 1875, he purchased a stately and historic mansion in the village of Juvisy, just south of what is now Orly Airport. There Wurtz expertly grew rare varieties of fruit—mostly as art and for gifts, for he did not eat much himself—and his wife grew flowers. Other activities included rowing, swimming, hunting, fishing, and hiking. From this second home Wurtz was able to commute conveniently by rail to the Latin Quarter, so that he was able to spend not just a few summer weeks but months there each year.[70] In 1882 the Wurtzes moved their city home from Rue Saint-Guillaume 27, where they had lived for 30 years, to a new residence a short distance away, at Boulevard Saint-Germain 176.

Meanwhile, honors accumulated for Wurtz in the postwar years, including many of the highest international honorary memberships, medals, and awards. In 1874 and again in 1878, he served as president of the Société Chimique, as he had also done in 1864. In 1875 he became mayor of the Seventh Arrondissement. Chevalier of the Légion d'Honneur since 1850, he became successively Officier, Commandeur, Grand Officier, and member of the Council after the war. In 1879 he added the presidency of the Comité Consultatif d'Hygiène Publique to his other duties. In 1880 he was vice-president of the Académie des Sciences, and he served as president 2 years later. Upon Pelouze's death, in 1867, Wurtz was brought into the senior editorial collective that ran the important journal *Annales de chimie*.

---

69. Shortly before the marriage, Hofmann visited Wurtz's newly rented residence in the Rue St. Guillaume. The "magnificent" apartment, Hofmann commented, looked nothing like a bachelor's digs, and Hofmann guessed the reason why ("Wurtz," p. 245). Other indications of wealth: Wurtz to Charles Laileumy [?] in Bordeaux, 12 June 1866, regarding a chalet Wurtz wished to build there (AJ16/6565); Wurtz to L. Micé, 31 July 1871: "J'ai éprouvé de grandes pertes par l'incendie de deux de mes forêts . . ." in Gascony, south of Bordeaux (Wellcome Institute, Western Manuscripts Collection, Wurtz file).

70. Hofmann, "Wurtz," pp. 311–321; Bauer, "Erinnerungen," p. 118.

Within about 4 years he became, de facto, the chief editor. In that capacity he was able to highlight the contributions of structural organic chemists by continuing his practice of publishing translated extracts of foreign (mostly German) papers, as he had been doing since 1852.[71]

Shocked by the war and still concerned about France's inferiority to Germany, Wurtz played a leading role in founding a French scientific society that was meant to be analogous to the Gesellschaft Deutscher Natur-forscher und Aerzte and to the British Association for the Advancement of Science: the Association Française pour l'Avancement des Sciences. The goals were to promote "the progress and diffusion of science," to increase interactions between the pure and applied sciences, to lobby the government for additional support, and to invigorate science in the French provinces. Meetings would be held every year in a different French city. The society was officially inaugurated in Paris in April of 1872, and the first annual meeting took place that September in Bordeaux.[72]

By the end of the 1870s, Wurtz had effectively retired from teaching at the Faculté de Médecine, though not at the Sorbonne. He was anxious for wider influence, and as the republican tide waxed he decided to seek office. His former student Scheurer-Kestner, already a permanent senator, was helpful in this effort.

Seventeen letters written by Wurtz to Scheurer-Kestner between 1877 and 1881 document Wurtz's route to political success.[73] Wurtz explained his motivation as follows: ". . . at 62, I aspire to a situation that is a little less militant than that which I have occupied in science and in education for so many years."[74] In 1881 Wurtz was named permanent senator—in the same

71. See M. Letté, "Les Annales de chimie et de physique, la quatrième série (1864–1873)," *Sciences et techniques en perspective* 28 (1994): 218–286, esp. pp. 249–255.

72. On this society, see especially Carneiro, The Research School, pp. 239–245; Carneiro and Pigeard, "Chimistes Alsaciens," pp. 544–545; R. Fox, "The Savant Confronts His Peers," in *The Organization of Science*, pp. 272–276; Fox, "Science, the University, and the State," pp. 107–111.

73. Wurtz to Scheurer-Kestner, letters 34–50 (ff. 473–493), Bibliothèque Nationale et Universitaire de Strasbourg.

74. Wurtz to Scheurer-Kestner, 20 January 1880: ". . . à 62 ans j'aspire à une situation un peu moins militante que celle que j'occupe depuis tant d'années dans la science et dans l'enseignement."

year that Berthelot received the same honor—and joined the center-left faction of the Senate. In his letter thanking Scheurer-Kestner for help and defending himself against an anonymous charge of being politically illiberal, Wurtz wrote: "I am a sincere republican, and, as a Strasbourgeois, I would be ashamed to be anything else."[75]

Wurtz was feeling his age. Reluctantly declining an invitation from Alexander Williamson, he wrote: "As for me, your old friend, I am no longer young and I feel tired. At the moment I am overwhelmed with work: senator, mayor of the 7th Arrondissement, professor, examiner, president of the Hygiene Committee. I must take care of myself."[76] Wurtz had a robust constitution, but he was troubled by symptoms of diabetes, which increasingly sapped his strength. After an exhausting winter semester of 1883–84, Wurtz was clearly in physical distress. He spent March of 1884 in Cannes, where his daughter and son-in-law (and, coincidentally, Dumas) were wintering.[77] Having returned to Paris only partially recuperated, Wurtz was shocked to learn of Dumas's sudden death. He gave the funeral oration, in the name of the Académie des Sciences, on 14 April.[78]

For many years Dumas had held the position of Permanent Secretary of the Académie, the most prized honor in French science. Upon Dumas's death, this post was offered to Wurtz, who after some hesitation decided to accept it. However, by this time he was seriously ill. He gave his first lecture of the summer semester on 27 April; it is said that he had his usual fire and eloquence, but by the end of the class he was almost in a faint. For

75. Ibid., 26 June 1881: "Je suis un républicain sincère, et, en ma qualité de Strasbourgeois, je serais honteux d'être autre chose." Wurtz's political silence during the Second Empire was apparently held against him by those on the left.

76. Wurtz to Williamson, 5 November 1881, Harris Collection, Bloomsbury Science Library, Ms. Add. 356: "Et moi, votre vieil ami, je ne suis plus jeune et je me sens fatigué. Dans ce moment je suis accablé de besogne, sénateur, Maire du VIIième Arrondissement, professeur, examinateur, président du Comité d'Hygiène. J'ai le devoir de me ménager."

77. Hofmann, "Wurtz," pp. 321–326; H. Gregor, "Académie des Sciences, Séance du 12 mai, Présidence de M. Rolland: M. Adolphe Wurtz," unidentified newspaper tearsheet, 15 May 1884, Académie des Sciences, Archives, Dossier Wurtz. Friedel stated that Wurtz suffered (also?) from bladder and prostate conditions ("Wurtz," p. xxxii).

78. "Discours de M. Wurtz," *Comptes rendus*, 98 (1884): 940–944.

the first time in his life, he had to cancel classes owing to illness. On the morning of 12 May, 4 weeks to the day after he had given the funeral oration at Dumas's grave, he succumbed. Word was brought instantly to the Académie, which was in session; members were "stupefied" at the news, and the meeting was immediately adjourned. The funeral took place 3 days later at the Temple de la Redemption, and Wurtz was laid to rest in Père Lachaise Cemetery.

12

A Summing Up

From the time of his full conversion to the new chemistry of Laurent, Gerhardt, and Williamson (ca. 1854), Wurtz strove to gain acceptance in France for atomic weights, atomic theory, and, slightly later, structure theory. Thirty years of this campaign did not suffice to gain victory, and it is reasonable to inquire why. This inquiry will lead to a concluding discussion of broader questions about the course of nineteenth-century French science, and to comparisons with German science.

Celebrity Culture

One factor had to do with the exercise of influence and power by individual actors, and how personalities and styles interacted with contexts to create differentials in that power. To a degree that was arguably more pronounced than in other countries, there was a celebrity culture in French academic institutions, and Wurtz played the game less skillfully, or perhaps simply less enthusiastically, than others. The system must also be understood in the context of the ability of the celebrities to cumulate professional positions in Parisian institutions. All these statements require careful definition and qualification.

The first stipulation is that Wurtz's failure cannot be imputed to any lack of personal success or eminence on the scientific stage. By any measure, Wurtz was one of the greatest French scientists of the nineteenth century. His research spanned the entire science of chemistry and was notable for its volume, significance, and influence. He led France's finest laboratory-based teaching-and-research school in any field of science during these 30 years.[1] He was virtually the only renowned specialist in organic chemistry in France, and this at a time when organic chemistry had attained high importance both scientifically and technologically. His laboratory was internationally recognized as one of the world's premiere "finishing schools" for prospective elite research chemists. James Crafts cannot be faulted for regarding Wurtz's lab as, in a sense, the direct successor of Liebig's; the timing was just about right, too, for Wurtz started up his laboratory in the Faculté de Médecine just one year after Liebig transferred to Munich and abandoned laboratory teaching.[2]

Wurtz's eminence and unique position in French science were recognized by many of his contemporaries. In his detailed 1868 report to Napoleon urging the establishment of the École Pratique des Hautes Études, Minister of Public Instruction Victor Duruy asserted the importance of laboratory-based teaching and research schools in the German pattern and pointed out that such entities were very nearly absent in France. The raw material (fine scholars and capable students) was not lacking, only the institutional resources:

Gifted professors who are devoted to science sometimes discover these tenacious inclinations and encourage them. Thus we have had for the last thirty years a school of chemistry which has given French chemistry such a high reputation in the scholarly world. If we were to have such schools in other sciences, we would obtain the same results.[3]

In view of Duruy's precise wording ("thirty years," the implied continuation to the present, and the use of the singular in the phrase "a school of chemistry"), it can be assumed that he was referring to the laboratory

1. One might reasonably argue that Pasteur's research group was equal or superior to Wurtz's in eminence. However, Pasteur's laboratory in the École Normale was devoted exclusively to research. From 1862 on it was adequately funded, and from 1874 on it was richly endowed by special government grants. After the Franco-Prussian War, Pasteur did no teaching anywhere.

2. See first section of chapter 6; see also chapter 11, passim.

3. Duruy, "Rapport à l'Empereur à l'appui de deux projets de décret relatifs aux laboratoires d'enseignement et de recherches et à la création d'une école pratique des hautes études," 14 July 1868, p. 8, in F17/13614. "Des maîtres habiles et dévoués à la science découvrent parfois ces vocations opiniâtres et les encouragent. C'est ainsi que nous avons depuis trente ans une école de chimie qui a donné à la chimie française un rang si élevé dans le monde savant. Ayons des écoles semblables pour les autres sciences et nous obtiendrons les mêmes résultats."

school of Dumas (1838–1848) and its successor led by Dumas's student Wurtz (from 1853 on). As we have seen, Dumas and Wurtz both consciously modeled their labs after the Giessen pattern. This and additional evidence provided in chapter 9 suggests that when Duruy designed the École Pratique des Hautes Études after German models he was also thinking of Wurtz's laboratory school.

Wurtz's high contemporary standing is also evident from the results of elections to the Académie des Sciences. (To be sure, these elections were always influenced by politics, arm twisting, and log rolling; however, as Maurice Crosland has argued, the degree to which they reflected a genuinely fair-minded, meritocratic, and consensual assessment of scientific attainments should by no means be underestimated.[4]) The first election in the Académie's Chemistry Section after Wurtz's arrival in Paris was in 1857, after the death of Baron Thenard. Wurtz was deeply disappointed that Edmond Frémy was chosen. However, in view of the fact that Frémy was older than Wurtz, had been resident in Paris far longer, was a professor in a leading grande école and in a leading research institution, and had the proper connections (especially through Pelouze), and since Wurtz's research was only just then beginning to take off, the result should not have been a surprise.

The next vacancy in the Académie's Chemistry Section occurred when Pelouze died in 1867. Wurtz won the election over Berthelot by a vote of 46 to 3. The same year, an anonymous ministerial report addressed to Napoleon assessing candidates for Dumas's vacated office of Inspecteur Général de l'Enseignement Supérieur was frankly critical of Berthelot. The most eminent chemists in France, the writer declared, clearly were Pasteur, Deville, and Wurtz, and Balard was senior to all of them (Balard was chosen).[5] In 1868 another vacancy was created in the Académie's Chemistry Section, upon Dumas's elevation to Secrétaire Perpétuel. For a third time Berthelot was a candidate, and for a third time he lost, this time to Cahours. Berthelot finally won election to the Académie in the Physics

4. M. Crosland, "Assessment by Peers in Nineteenth-Century France," *Minerva* 24 (1986): 413–432; *Science under Control* (Cambridge University Press, 1992), pp. 203–241.

5. L. Velluz, *Vie de Berthelot* (Plon, 1964), pp. 119–120. The writer was probably Duruy.

Section, when a vacancy occurred in 1873. Even then he was third on the section's nomination list, tied with nine others; nonetheless, he prevailed in the general election.[6]

The perceived eminence of Wurtz over Berthelot and other rivals was also evident in how their respective protégés fared. A vacancy in the Chemistry Section occurred upon the death of Regnault in 1878, and the leading candidates were Wurtz's former student Charles Friedel and Berthelot's former student Louis Troost. Troost had several advantages: at age 53, he was senior to the 46-year-old Friedel. Both were professors at the Sorbonne, but Troost had seniority there, and his chair (in chemistry) was more prestigious than Friedel's (in mineralogy). And Troost had some impressive backers working for him behind the scenes. It was thus with considerable relief that Wurtz wrote to Hofmann, in a postscript scrawled in a large triumphant hand: "Victory! Friedel was chosen and the grand coalition of Dumas, Deville, Berthelot in favor of Troost produced [only] 14 votes."[7]

Adolphe Wurtz lived a modest life and died a modest death. His mausoleum in Père Lachaise Cemetery is small and undistinguished, crowded cheek-by-jowl with its neighbors, and marked only by the simple inscription "Famille Wurtz." A short and entirely undistinguished street in the Thirteenth Arrondissement was named for him 10 years after his death, and there is no other Wurtz memorial in Paris.[8] A statue of Wurtz stands in front of his father's church of St.-Pierre-le-jeune in Strasbourg. The carved inscriptions on the base, altered by the German authorities around 1900, have been restored, but not well.

6. J. Jacques, *Berthelot 1827–1907* (Belin, 1987), pp. 105–111; A. Carneiro, The Research School of Chemistry of Adolphe Wurtz (Ph.D. dissertation, University of Kent/Canterbury, 1992), p. 69. Deville had been elected in the mineralogy section of the Académie as early as 1861; Pasteur was elected to the same section in 1862.

7. Wurtz to Hofmann, 5 July 1878, Chemiker-Briefe, Berlin-Brandenburgische Akademie der Wissenschaften. "Victoire! Friedel est nommé et la grande coalition Dumas, Deville Berthelot en faveur de Troost a produit 14 voix."

8. Measured in square footage of Parisian asphalt named for chemists, the Rue Wurtz is beaten by eleven others, including the rather obscure "nominations" for Rue [Charles] Moureu and Rue [Charles] Lauth (Wurtz's own student); see next note and Jacques, "Test," p. 131. Jacques comments that Rue Wurtz actually exceeds Place M. Berthelot in square footage, but only by a fluke of measurement.

Figure 12.1
The statue of Adolphe Wurtz outside the church of Saint-Pierre-le-Jeune in Strasbourg, in whose parsonage he was born. Photograph taken by the author, 1992.

Figure 12.2
An inscription on the statue. Note the alteration of the letter u. During the
German period (1870–1918) an umlaut was added in order to Germanize the
name; in the interwar period, the orthography was corrected by the French
authorities. Photograph taken by the author, 1992.

Figure 12.3
An inscription on the statue, quoting Wurtz's controversial claim. Photograph
taken by the author, 1992.

Marcellin Berthelot, in contrast, lived and died larger than life. The fiftieth anniversary of his first publication (1901) was celebrated by 3000 invited guests in the Great Hall of the new Sorbonne. His state funeral, 6 years later, was marked by speeches by the president and the prime minister of the republic, and hundreds of other dignitaries were in attendance. Berthelot and his wife were interred in the Panthéon—an unprecedented honor made possible by a special legislative act. A statue was erected prominently in the Rue des Écoles, in the square opposite the Collège de France—a square that now bears Berthelot's name. The centenary of his birth, in 1927, was the occasion of elaborate celebrations, including the preparation of a sumptuous commemorative volume whose dimensions and weight can scarcely be imagined by anyone who has not hefted it personally. No fewer than 50 schools in France, including six lycées, are now named in his honor.[9]

The modesty of Wurtz's profile extends into the modern secondary literature on French science. So eminent a specialist as Harry Paul appears to have a blind spot for Wurtz, despite having written early in his career an outstanding treatment of "the French scientist's image of German science" in which Wurtz necessarily figures. When in a recent work Paul cites Duruy's 1868 statement concerning the vigor of "the French school of chemistry," he concludes that Duruy must have been referring to either Deville or Berthelot, rather than (as I assumed) to Wurtz. In a pathbreaking article published in the same year as his monograph on the French-German scientific relationship, Paul omits Wurtz from a list of directors of leading academic chemical laboratories (he names Berthelot, Deville, Frémy, and Le Châtelier), and later he cites him as a prominent "medical scientist" in the French community.[10] I do not mean to single out Paul's fine work for criticism; rather, I use him representatively, for most other specialists seem

9. Jacques, *Autopsie*; Jacques, "Un test de la popularité de Berthelot" and B. Javault, "Berthelot, héros de la IIIème République," both in *Marcelin Berthelot*, ed. J. Dhombres and B. Javault (SFHST, 1992). According to Jacques, the only scientists who have had more French city streets named for them than Berthelot are Pasteur and the Curies.

10. H. Paul, *The Sorcerer's Apprentice* (University of Florida Press, 1972), pp. 7–11; *From Knowledge to Power* (Cambridge University Press, 1985), p. 47; "The Issue of Decline in Nineteenth-Century French Science," *French Historical Studies* 7 (1972), p. 418n. and p. 430.

to have the same problem. Fortunately, this situation seems to be changing; one contemporary senior scholar who has written with sophistication on Wurtz is Robert Fox, and, as I have already mentioned, two younger scholars, Ana Carneiro and Natalie Pigeard, have begun to produce fine detailed studies.

Wurtz was not a "star" in the French scientific firmament. Berthelot was. So was Pasteur, for whom a new research institute was named long before his death. So was Claude Bernard, in whose honor the French government conducted the first state funeral for any scientist (1878). In an earlier generation, so were Thenard, Gay-Lussac, Dumas, Arago, Biot, and Cuvier. What made a scientist a "star"? Not just contemporary research renown, for Wurtz had that. Conversely, Thenard fell far short of the research productivity of Gay-Lussac, Dumas, or Wurtz. What one needed in addition was a collection of certain other human qualities: enterprise, networking skills, political savvy (not necessarily in the wider political culture, but in the politics of science), self-confidence, showmanship, and ambition. And good fortune never hurt.

Cumul, Textbooks, and Control of Pedagogy

Celebrity status in France during the July Monarchy and Second Empire was indicated less by peer-judged elections such as those at the Académie des Sciences, and more by success in securing professional positions, which were largely controlled by the central political authorities—and indirectly by powerful collegial patrons. I will begin my analysis of this situation by presenting a diachronic overview of the structure of the community, by positions held, from about 1820 until 1870.

In the 1820s and the 1830s, French academic chemistry was controlled by Gay-Lussac and Thenard. Gay-Lussac held professorships at the Sorbonne and the École Polytechnique, and also was assayer at the Mint; in 1832 he traded the École Polytechnique for the Muséum d'Histoire Naturelle. Thenard likewise held professorships at the Sorbonne and the École Polytechnique, while simultaneously occupying the chemical chair at the Collège de France.

Slowly the Gay-Lussac–Thenard monopoly was transferred to Dumas, Pelouze, and Balard. From footholds in the École Polytechnique and the

Sorbonne, Dumas won professorships in both locations, although he traded the École Polytechnique for the Faculté de Médecine in 1838. Pelouze gained footholds in the Mint and the École Polytechnique, then succeeded Dumas in the chemical chair of the latter institution in 1838. He then traded the Polytechnique chair for the professorship at the Collège de France, when Thenard left that institution in 1846 to become Chancelier of the Université de France. Meanwhile, Balard was appointed to chairs at the Sorbonne and the École Normale. In 1851 Balard inherited Pelouze's chair at the Collège when Pelouze simultaneously gained the top job at the Mint and Gay-Lussac's lucrative consulting position at Saint-Gobain.

Enter the next generation. Wurtz succeeded Dumas at the Faculté de Médecine in 1853, while Deville succeeded both Balard at the École Normale and Dumas at the Sorbonne (only unofficially at first, as suppléant). Berthelot benefited from Balard's patronage, by being named for new professorships, created expressly for him, at the École de Pharmacie and then at the Collège de France. When Balard and Dumas simultaneously retired from the Faculté des Sciences in 1867, the two chemistry professors at the École Normale, Deville and Pasteur, added the Sorbonne chairs to their responsibilities. Lesser cumulards also appeared. Victor Regnault succeeded Gay-Lussac at the École Polytechnique in 1840, and added a chair in physics at the Collège de France in 1841. Edmond Frémy succeeded his maître and patron Pelouze at the École Polytechnique in 1846, then garnered Gay-Lussac's old position at the Muséum d'Histoire Naturelle 3 years later. Auguste Cahours won positions at the École Polytechnique and at the Mint.

As complicated as all this may sound, it is much simplified from the actual situation. For instance, I have not mentioned in this account the École Centrale des Arts et Manufactures, where Dumas, Pelouze, Wurtz, Frémy, Cahours and others also worked; nor have I fully indicated the complex of political, administrative, and consulting positions that were also traded among leading cumulards. The academic "standard load" appears to have been two or three (and sometimes even four) full professorships, not counting consulting and administrative positions. Wurtz and Pasteur were the only significant exceptions to this rule (although each of them added a Sorbonne chair to that of their home institution for a few years). Indeed, as I argued in chapter 5, Wurtz's position at the Faculté de Médecine was in some measure the opposite of cumul, for the position he accepted in February of 1853 was altered 10 months later to include responsibilities from two different professorships, organic chemistry and medical chemistry. A pattern can also be seen whereby senior cumulards eventually divested themselves of their professorships in favor of lucrative consulting and administrative posts. This was clearly the case for Thenard, Gay-Lussac, Dumas, Balard, and Pelouze, though it was less true for the next generation.

My diachronic review can be complemented by a synchronic summary. During the heart of Wurtz's career (from the early 1850s to the late 1870s), the cast of characters at the principal Parisian institutions was surprisingly stable. At the École Polytechnique were Frémy, Cahours, and Regnault, and at the École Normale were Pasteur and Deville. Pasteur and Deville also held the chemistry chairs at the Sorbonne after Dumas and Balard left these chairs in 1867 (Deville, suppléant for Dumas, had been the de facto chemistry professor there since 1853). Deville's protégé Troost succeeded to Pasteur's Sorbonne chair in 1874. Balard and Regnault were professors at the Collège de France, and a chair for Berthelot was added in 1863. Frémy and Chevreul held the two chemical chairs at the Muséum d'Histoire Naturelle throughout this period. Finally, Wurtz was professor at the Faculté de Médecine and, in the last few years of this period, also at the Sorbonne.

These were the leading personnel in the Parisian chemical world in the third quarter of the century. This "staffing" context related directly to the vicissitudes of the reformed chemistry. Recall that Berthelot, Deville, and Troost were distinctly inimical to the reforms; that Pasteur, Regnault, and Chevreul, although potentially favorable to the new ideas, were unengaged because they were active in unrelated fields, and perhaps also constitutionally too cautious; that Cahours was an atomist at heart but not in public; and that Dumas, Balard, and Frémy were, each for their own reasons, dogmatically uninterested in taking a position one way or the other. Consequently, throughout the heart of Wurtz's career, personalities opposed or indifferent to the reformed chemistry controlled the Sorbonne, the principal grandes écoles, and the leading research institutions of Paris. Moreover, the four most adamantly anti-atomistic chemists of all—Berthelot, Deville, Troost, and Frémy—covered, between them, every institution named. The only exception was the Faculté de Médecine, which as a medical school was structurally peripheral to pure research in a physical science such as chemistry. Such was the power of centralization of the academic community, and the complementary power of cumul: from the middle of the century until the late 1890s, a small handful of like-minded scholars held a virtual lock on chemical doctrines in Parisian institutions of higher education and research.

But what exactly was the means by which this control was exercised? Reading the literature of the day, one occasionally encounters the complaint by atomists that the teaching of atomistic chemistry was banned in official French science education. This appears to be overstated. However, de facto it was true that almost no atomic theory was taught in France between about 1850 and 1890; this was a direct or indirect consequence of certain structural characteristics. Degrees in higher education were all conducted by examination, and examinations were based largely on the study of a standardized syllabus of printed works; this was so because French university study before 1868 had no practica, nor even an expectation that students needed faithfully to attend lecture courses. Neither hands nor ears were the sanctioned route to knowledge, but eyes on the printed page. This extended from the licence and agrégation all the way down through baccalauréat to the lycées and collèges of France (the doctorat, in contrast, was a research degree, requiring a dissertation).

The history of nineteenth-century French science textbooks has only recently begun to be studied.[11] As far as chemistry instruction at the highest level is concerned, from the 1810s until the 1840s, three Francophone writers dominated the literature: Thenard, Dumas, and Berzelius.[12] Although initially atomistically inclined, the first two of these authors always wrote

11. A. Lundgren and B. Bensaude-Vincent, eds., *Communicating Chemistry* (Science History Publications, 2000); B. Belhoste, *Les sciences dans l'enseignement secondaire en France* (INRP, 1995); A. Choppin, *Les manuels scolaires* (Hachette, 1992); J. Dhombres, "French Textbooks in the Sciences, 1750–1850," *History of Education* 12 (1984): 153–161.

12. L. Thenard, *Traité de chimie élémentaire, théorique et pratique* (Crochard, 1813–1816); J. Dumas, *Traité de chimie appliquée aux arts*, 8 vols. (Béchet, 1828–1846); J. Berzelius, *Traité de chimie*, 8 vols. (Didot, 1829–1833), second edition, 1845–1850. A. Garcia Belmar and J. Sanchez ("French Chemistry Textbooks, 1802–1852," in *Communicating Chemistry*) have shown what a profusion of books there was at the secondary level.

cautiously, and from their increasingly marked empiricism as well as from increasing disinterest in scientific theory, neither man could be considered an atomist after 1840. As for Berzelius, after 1840 his work was increasingly (and accurately) viewed as old-fashioned. In any case, there was no uniformity of atomic weight systems in this early period, so that no pedagogical uniformity in chemistry existed. A sense of disunity was only increased by the arrival on the scene of aggressive atomists such as Laurent, Gerhardt, Gaudin, and Baudrimont. All these men wrote textbooks, but none was significantly accepted in lycées, faculties, or grandes écoles—with the possible exception of Gerhardt's *Traité de chimie organique* (1853–56), which ironically was written using conventional equivalents. None of these men had much success in entering the academic establishment of their day.

In 1837, the official government syllabus for the curriculum leading to the baccalauréat in physical science was altered to eliminate any mention of the atomic theory; the word "atom" did not reappear in these syllabi until 1902. Catherine Kounelis argues that Dumas and Thenard were probably both complicit in this change.[13] (In principle, between 1793 and 1865 every French textbook required a "prior authorization" from the government to ensure conformity with the official syllabus before it could be adopted in the classroom, but this system was not uniformly implemented.[14]) Coincidentally across the Rhine, several leading German chemists agreed privately among themselves in 1838 to campaign for the general acceptance of chemical equivalents; Liebig wrote to Berzelius and to Pelouze in order to internationalize the movement (see the fourth section of chapter 3 above for a discussion of these circumstances). Berzelius was unsympathetic to Liebig's entreaty, but Pelouze's reaction must have been enthusiastic.

The next year, an unnamed savant—probably Pelouze—argued in the Conseil de Perfectionnement of the École Polytechnique that the school must systematically adopt equivalents and exclude atoms, a proposition that "has received general consent among scientists of the north."[15] It is

13. C. Kounelis, "Atomism in France," in *Communicating Chemistry*.

14. Garcia Belmar and Sanchez, "French Chemistry Textbooks, 1802–1852," p. 25.

15. École Polytechnique, Conseil de Perfectionnement, Procès-verbaux, vol. 6, p. 144, quoted in M. Crosland, *Gay-Lussac* (Cambridge University Press, 1978), p. 155. Crosland notes that, in addition to Pelouze and Gay-Lussac, Thenard and

uncertain what reception this proposition received in 1839, but by 1846 the Polytechnique faculty included two others with preferences similar to Pelouze's: his former student Edmond Frémy, and Victor Regnault.

Whatever may have occurred behind the scenes, uniformity in the French textbook literature was finally achieved with the publication of works by Pelouze, Frémy, and Regnault that attained great popularity for a generation after 1847.[16] These authors, though not positivists in the strict sense, were always extremely cautious in tone. Their textbooks were dogmatically non-theoretical, and used conventional equivalents throughout. The first textbooks accepted in French educational circles that used atomic weights arrived only in the mid-1860s, from the hands of Naquet and Wurtz; although Wurtz's textbook appears to have been successful, it was very much a rear-guard action.[17] The net result of all of this is that for a period of 40 years beginning about 1837 French lycée students learned only one kind of chemistry: that based on equivalents and suffused with theoretical caution, if not outright positivism. This created a cycle that was hard to break, especially in the centralized system of the Université de France, for advanced textbooks that used atomic weights had little market appeal to customers who had been raised on equivalents.[18]

There were additional factors involved in the equivalentists' control of chemical curricula. Deville was quite fortunate to be positioned at both the Sorbonne and the École Normale. The former institution was dedicated to

---

Chevreul were on the Conseil, and he speculates that it was Chevreul who made this proposition. Crosland describes the proposal as implying "penetrating criticism" of Gay-Lussac, but it need not be interpreted that way. The two professors of chemistry at that time were Gay-Lussac and Pelouze; Regnault was hired in 1841 and Frémy in 1846.

16. V. Regnault, *Cours élémentaire de chimie à l'usage des facultés, des établissements d'enseignement secondaire, des écoles normales et des écoles industrielles* (Masson, 1847–1849); Regnault, *Premiers éléments de chimie à l'usage ...* (Masson, 1850); J. Pelouze and E. Frémy, *Traité de chimie générale* (Masson, 1847–1850); Pelouze and Frémy, *Notions générales de chimie* (Masson, 1853), second edition, 1855; Pelouze and Frémy, *Abrégé de chimie, conforme au programme officiel de l'enseignement dans les lycées (section des sciences)* (Masson, 1848–1850).

17. For a discussion, see the last section of chapter 8 above.

18. B. Bensaude-Vincent ("Atomism and Positivism," *Annals of Science* 56 (1999): 81–94) defines a movement, which she calls "textbook positivism," that accurately characterizes the point of view of these books.

pure science, while the latter had begun life with the mission to train future teachers at the lycées throughout France. Although the mission of the École Normale broadened considerably toward scientific research in the middle years of the century, due largely to Pasteur's remarkable energy and commitment, still Deville was able to exercise considerable influence on French chemical pedagogy from this foothold.

In particular, many of Deville's students went on to teaching careers in exactly the kind and level of chemistry (lower-level general, analytical, and inorganic chemistry) that was most valuable in maintaining, throughout the entire French scientific establishment, Deville's own particular orientation—empirical, non-theoretical, and experimentalist.[19] By contrast, organic chemistry was an intrinsically "advanced" field, and as such it was by its very nature strategically ill positioned to exert influence at the crucial lower levels of the pedagogical pyramid, especially in the lycées. Paradoxically, it was through organic chemistry that the decisive evidence for the reformed chemistry had come; but once in possession of this new gospel, the modernist organic chemists found themselves with less power to evangelize than they would have wanted. The best example of this is Wurtz, well boxed off at the Faculté de Médecine. Wurtz wrote Kekulé in 1864: "The old notation has been imposed on me by the nature of my teaching [at the Faculté], which is addressed to students whose chemical education has already taken place in the secondary schools. You cannot believe how much this notation has hampered me."[20]

I have not yet mentioned a final factor in the dominance of equivalentist chemistry in France, and the one that is most often cited in secondary

19. Carneiro, The Research School, pp. 34–38, 94. John Servos makes a parallel argument on pp. 87–99 of *Physical Chemistry from Ostwald to Pauling* (Princeton University Press, 1990). Early-twentieth-century American physical chemists, Servos contends, were successful in expanding their numbers dramatically not because of the scientific worth of their research but because of the demand for chemistry teachers at the booming universities in the elementary areas of general, inorganic, and analytical chemistry. Organic chemists could not, in general, fit this bill.

20. Wurtz to Kekulé, 25 August 1864, August-Kekulé-Sammlung, Institut für Organische Chemie, Technische Hochschule, Darmstadt. "L'ancienne notation m'était imposée par la nature de mon enseignement, qui s'adresse à des élèves dont l'éducation chimique est déjà faite dans les collèges. Vous ne sauriez croire combien cette notation m'a gêné."

accounts: the putatively dominant influence of Berthelot. Berthelot was president of the section of physical sciences of the EPHE, Inspecteur Général de l'Enseignement Supérieur (1876–1888), member and later vice-president of the Conseil Supérieur de l'Enseignement Publique (from 1880), Sénateur Inamovible (from 1881), Ministre de l'Enseignement Publique (1886–87), Secrétaire Perpétuel de l'Académie des Sciences (from 1889), and Foreign Minister of France (1895–96). However, I believe that his influence against atomism and modernist chemistry, though real enough, has been consistently overestimated. As I have argued, during the 1850s and the 1860s Berthelot's influence was not as strong as it may have appeared. Many competent observers had doubts about the worth of parts of his scientific corpus, and in his personal politics he was out of step. It was only after the Third Republic was well established that his power began to increase substantially. During the earlier period, it was perhaps the support and reinforcement that he gave Deville—as well as the increasing influence of his students—that constituted his most effective resistance to modernist chemistry.

The younger generation of atomists who looked back on the old days of the 1880s and the 1890s tended to remember Berthelot with little fondness, which probably accounts for his exaggerated reputation as the principal reason for the suppression of atomism. (Deville had died in 1881, even before Wurtz.) The Nobel Prize in chemistry for 1912 was given jointly to Victor Grignard and Paul Sabatier, both of whom had been youthful provincial atomists (at Lyon and Toulouse, respectively) during the "Babylonian Captivity" of atomic theory. Grignard recollected his initial dislike of chemistry as a student of mathematics:

It was in 1894 [at the Faculté des Sciences in Lyon] and we had not yet emerged from the period where the influence of Berthelot was being exercised despotically on secondary education. It hindered the atomic theory from replacing that of equivalents. At [the École Normale Spéciale in] Cluny [ca. 1890], where even then teaching was on a high level, they had treated mineral chemistry in equivalents and organic chemistry in atoms. I had an impression of it as an incoherence and mnemonism that frightened me.[21]

Similarly, a former student of Sabatier's wrote that his study of equivalentist chemistry in the lycée had been "grim" and "unprofitable," but that

21. Quoted in M. Nye, *Science in the Provinces* (University of California Press, 1986), p. 166.

Sabatier's lectures in the 1890s made the science "logical" and "deductive" for him.[22]

In summary: The patronage exerted by Gay-Lussac, Thenard, Balard, and Dumas led to the accession to academic power of Pelouze, Deville, and Berthelot, whose students in turn—Frémy and Troost, in particular—continued a kind of dynasty until very late in the century. The one commonality in the first four chemists named—at least in the period after 1840—was a profoundly positivistic[23] alignment, an orientation strongly in evidence among the last five. These five chemists were able to dominate French chemical pedagogy in the manner described above, until the turn of the century.

French and German Laboratories

It is not quite true, as some have asserted, that there were no chemical laboratories in institutions supported by the French state during the first two-thirds of the nineteenth century. Certain of the grandes écoles always had some sort of lab facilities. For instance, Dumas, Boullay, Pelouze, Frémy, and others performed chemical research in the École Polytechnique in the 1820s and the 1830s,[24] as did Regnault and certain répétiteurs such as Berthelot in the Collège de France, in the new building erected in the 1830s. There was also a chemical lab at the Muséum d'Histoire Naturelle under the control of Chevreul and later Frémy. At the Sorbonne, a small disused room was converted into a meager chemical laboratory during the 1830s. There was always an assay lab at the Mint, and there were other mission-oriented applied chemistry labs at such institutions as the Arsenal and the Conservatoire des Arts et Métiers.

22. Paul Dop (1930), cited in ibid., p. 288, n. 125. Nye's book is indispensable for understanding the relationship between Paris and the provinces in the two generations after 1860.

23. None was a positivist in any proper sense. On the influence of "positivism" in French chemistry (or the lack thereof), see Bensaude-Vincent, "Atomism and Positivism."

24. "M. Frémy, jeune homme qui travaille à mon laboratoire . . ." (Pelouze to Liebig, 12 February 1834, Liebigiana IIB, Bayerische Staatsbibliothek). This could only have been at the École Polytechnique, for the letter was written nearly a year before Pelouze's appointment as assayer at the Mint. Dumas, too, had occasional collaborators in his research at the Polytechnique.

But this enumeration must be placed in perspective. Gay-Lussac had pro-fessorships at the Sorbonne and the École Polytechnique, and was assayer at the Mint. Where did he choose to set up his research lab? At the Arsenal, where he consulted. In the next generation, Pelouze was professor at the Polytechnique, the École Centrale, and the Collège de France. Where did he perform his research? At the Mint, where he bypassed the in-house assay lab to create an altogether new laboratory with his own resources. Regnault was professor at the Polytechnique and the Collège de France, but set up his research at the state porcelain factory in Sèvres, where he consulted; the government provided apparatus by special grants. Dumas was Dean of the Sorbonne and Professor at the Faculté de Médecine, but chose to work in his private lab in the Rue Cuvier. Clearly, leading scientists systematically shunned the facilities in state institutions.

Detailed information on these academic laboratories is difficult to find.[25] What the qualitative (uniformly derogatory) descriptions that are available tell us is that they were utterly insufficient for the purpose of research; indeed, at least some of them were fugitive jury-rigged affairs. It seems that they were generally used only as a pis-aller, when a sufficiently inventive and energetic scholar managed to cobble together enough appropriate equip-ment in constricted quarters to lead a precarious existence as a researcher. Not one had a state-derived operating budget. Pasteur's assistant Émile Duclaux summarized: "When [French scientists] wanted to work, they exer-cised their wits; they skimped on the essentials in order to have something left over; they set themselves up as best they could in the quiet corner of a lecture room, always ready to clear away their apparatus at lecture times."[26]

We saw in chapter 4 how two leading chemists (and eventually bitter rivals), Dumas and Pelouze, took matters into their own hands by simulta-neously creating private labs of their own in 1838. Both labs were designed

25. I have never been able to find detailed descriptions, plans, dimensions, equip-ment, etc., for the early-nineteenth-century chemical labs at the Collège de France or the Sorbonne. A small amount of information on the early-nineteenth-century chemical laboratory of the École Polytechnique can be found in the sources cited in note 3 to chapter 4.

26. Quoted in R. Fox, "Scientific Enterprise and the Patronage of Research in France, 1800–70," *Minerva* 11 (1973), p. 459. This article, which has much on lab-oratories (esp. on pp. 458–469), is one of the most valuable sources in the secondary literature for this topic. See also chapters 2, 4, and 9 above.

to support the scientific research of the director and of a small coterie of advanced independent researchers; both were modeled after Liebig's lab in Giessen; and both had the same capacity of about ten students; but there were also differences between them. Pelouze's Rue Guénégaud lab was designed to be financially self-supporting, or even to make a profit, by taking in pupils who wanted to learn enough chemistry to operate comfortably in private industry. The center of gravity was technical training, and pure science was done on the side. In contrast, Dumas's Rue Cuvier lab was the reverse: there were some industrial chemists in training, but the raison d'être was research, and Dumas bore all the costs himself. However, neither facility was an "academic" laboratory in any proper sense, for both were free-floating private concerns.

In 1845 Pelouze expanded his business by building the Rue Dauphine facility, which could accommodate up to 30 students at a time. Just when it appeared that there might be a change of government policy to provide for better support for state academic labs, the Revolution of 1848 occurred. Dumas was forced to close his laboratory, and the plans for renovation of the state facilities for scientific research collapsed. Absent the competition from Dumas, two more entrepreneurs then entered the market, Wurtz (Rue Garancière, end of 1850) and Gerhardt (Rue Monsieur-le-Prince, May 1851); these concerns, like Pelouze's, were oriented to industrial chemical training, and were intended to make money, or at least to break even. Both were short-lived affairs, probably because Pelouze managed to soak up too much of the limited customer demand with his larger and more famous establishment.

Meanwhile, ambitious would-be scientists, faced with nearly non-existent state laboratories and too few private facilities oriented toward basic research, did what they could. Anyone who hoped for a scholarly career had to produce a doctoral thesis based on original research, which meant access to a laboratory. Those who had private means often provided their own facilities, usually in their lodgings. For example, Deville's small lab in the garret of his rented house in the Rue de la Harpe (1839–1845) was still large enough to accommodate the occasional guest worker, such as Cahours. Boussingault, Fizeau, and Foucault all did productive scientific work in their homes. Even foreign visitors to Paris sometimes set up modest facilities in their lodgings; for example, Alexander Williamson worked at home (8, Rue des Francs Bourgeois) during his stay in Paris from 1846 to 1849.

318   L'ILLUSTRATION, JOURNAL UNIVERSEL

Figure 12.4
The president of the republic, Adolphe Thiers, visiting the laboratory of Deville and Debray at the École Normale on 6 May 1873. From *L'Illustration, Journal Universel*, 61 (1873), 318. Source: Photograph Collection, Deutsches Museum, Munich.

The situation changed both for better and for worse in the 1850s. Wurtz created the first teaching-and-research lab actually located on the premises of a Parisian academic institution, even though officially it was unsanctioned by the Faculté de Médecine and most of his students were not studying for a medical degree at all. In the new premises of the École Normale in the Rue d'Ulm, Deville and Pasteur gradually found funding for adequate laboratories by appealing directly to the emperor—Deville from 1854 and Pasteur from 1858. These gains were counterbalanced by the permanent closure of Pelouze's facility in 1857; the private labs begun by Gerhardt and Wurtz likewise no longer existed by this time. Moreover, the major renovations of the Facultés promised by Napoleon's regime in 1854–55 were never funded.

This was the state of play at the time of the creation of the École Pratique des Hautes Études, which brought a modest increase of state financial support and provided for the foundation of a few additional academic laboratories (chapter 9). At the end of the 1860s it looked once more as if the road might be clear for major improvements in French establishments for academic science. However, the advent of war created another hiatus. It was only in the new environment of the Third Republic, beginning in the late 1870s and reaching fruition in the 1880s and the 1890s, that a definitive change of direction can be discerned.

Many historians have noted the apparently sudden upwelling, after about 1860, of public complaints regarding the miserable state support for laboratory research; the liberalization of Napoleon's policies at this time and the appointment of the sympathetic minister Victor Duruy soon thereafter have been cited as reasons for this sudden spate of protests. There were other reasons, as well. The intensity of dissatisfaction must have been increased by the collapse of the system of private labs, as inadequate as they had been even when healthy. And even more to the point, German academic laboratories were just then experiencing explosive growth.

A quick review: When Liebig came to Paris to finish his scientific education (1822–1824), he found a new world of experimentalist precision that he had not seen in Germany, and fell in love with it. However, German universities, which had been hidebound, poor, and narrow in the eighteenth century, were already in flux at this time. New universities were founded early in the century in Bonn and Berlin; these served as models for German states outside of Prussia. The Prussian university reforms were associated with the rise of a new research mandate, associated with late-Romantic German Neohumanism. The newly invigorated orientation toward research gradually took off in the decentralized German states. Since every state had its own university and its own ambitions for national and international reputation, the resulting competition drove a self-reinforcing cycle.[27]

27. There is a good literature on these questions. See especially R. Turner, "The Growth of Professorial Research in Prussia, 1818–1848—Causes and Contexts," *Historical Studies in the Physical Sciences* 3 (1971): 137–182; Turner, "Justus Liebig versus Prussian Chemistry," *Historical Studies in the Physical Sciences* 13 (1982): 129–162; C. McClelland, *State, Society, and University in Germany, 1700–1914* (Cambridge University Press, 1980).

I recounted Liebig's rise to fame in the first four chapters. The apparatus for organic analysis that he invented in 1830 proved an excellent fit with his evolving strategy to create an institution that combined high-level science pedagogy with a novel group-oriented approach to research.[28] He argued passionately that knowledge of the pure sciences was fully as essential to the cultured intellect as were humanist subjects such as classical philology, and that such scientific knowledge could only properly be obtained in a laboratory setting. This was how Liebig harmonized his views with neohumanist doctrines, but he also was careful to emphasize that those who had learned the science properly would be more successful in applied as well as theoretical contexts.

By such rhetoric Liebig was soon granted state sponsorship for his efforts, and a rising customer demand. Dumas's anxious rivalry with his German colleague was described in chapter 4. In 1831 and 1832 both he and Pelouze wrote dismissively to Liebig of their countrymen's efforts in chemistry, and lamented their lack of facilities for research. Meanwhile, like Dumas and Pelouze, other German chemists were following Liebig's lead, but unlike the French they increasingly had state support. At the University of Göttingen, where he was appointed professor in 1836, Liebig's close friend Friedrich Wöhler developed a chemical practicum that was similar to what Liebig was doing, and he also began attracting sizable numbers of students after about 1840. At the University of Marburg, Robert Bunsen did the same from the time of his appointment there in 1839. Professors in other disciplines also followed the same pattern, somewhat later than for chemistry: the physicists Wilhelm Weber (Göttingen), Franz Neumann (Königsberg), and Gustav Magnus (Berlin) and the physiologists Johannes Müller (Berlin) and Jakob Henle (Heidelberg).

Liebig was therefore not alone in mounting intensive laboratory-based education for university science students at this time. Nor in fact was he the first to do so: for instance, it is well known that Wöhler's predecessor at Göttingen, Friedrich Stromeyer, conducted university-sponsored practica for many years. Indeed, Ernst Homburg has recently demonstrated how greatly German chemistry was influenced by Stromeyer, and by an associated revolution in inorganic and analytical chemistry that has hitherto

---

28. F. Holmes, "The Complementarity of Teaching and Research in Liebig's Laboratory," *Osiris* 5 (1989): 121–164.

remained nearly invisible.[29] Nonetheless, there were distinct and novel elements to the Liebig organic-chemical phenomenon. Liebig insisted on the need for laboratory experience for his entire classes, not only for selected professionalizing chemistry students, and he did so on the basis of a fundamental pedagogical conviction. This was the beginning of a system of mass state-supported laboratory education replacing a system of elite private patronage. Furthermore, he regularly used his advanced students in his research program. He was phenomenally successful in attracting practicum students, far more than any other science professor in Germany or anywhere else. He was a committed—and gifted—proselytizer for his methods and for the intellectual and practical value of chemical study. And finally, in contrast to Stromeyer, his work was focused on organic chemistry, which was entering a period of explosive growth. For all these reasons, the traditional emphasis on Liebig's pedagogical innovation is not misplaced.

By 1842, Liebig's French students had included Jules Gay-Lussac, Pelouze, Regnault, Oppermann, Gerhardt, and Wurtz; together with his major influence on Dumas, Liebig was a far more important figure in France than has hitherto been appreciated—even if these men had great difficulty in reproducing the Giessen model in Paris.[30] Liebig also made his influence felt in Britain. Liebigians such as William Gregory, Lyon Playfair, A. W. Hofmann, Edward Frankland, and Alexander Williamson succeeded in reforming English science education in the 1840s and the 1850s, with the warm advocacy of well-placed individuals in government, including the (German-born) Prince Consort.

In Germany, a kind of takeoff in chemistry took place after the revolution of 1848. In a landmark study, Peter Borscheid concluded that the burst of state support for German academic chemistry after 1850 was directly related to the crop failures, famines, and social unrest of the 1840s—and to

29. E. Homburg, "The Rise of Analytical Chemistry and Its Consequences for the Development of the German Chemical Profession (1780–1860)," *Ambix* 46 (1999): 1–32.

30. Two scholars who have perceptively explored Liebig's influence in France in the late 1830s and the early 1840s are F. Holmes and Mi Gyung Kim. See Holmes's biography of Liebig in the *Dictionary of Scientific Biography* and Kim, "Constructing Symbolic Spaces," *Ambix* 43 (1996): 1–31.

the influence of Justus Liebig.[31] Borscheid argued that the rulers of the various German kingdoms and principalities became converts to Liebig's prescriptions for academic and agricultural chemistry in order to raise agricultural productivity and thus prevent future famines and revolutions. Later historical work has cast doubt on the simple version of this scenario, without destroying its general thrust. In any case, the fact is that a number of German states moved aggressively after 1850 to strengthen the study of chemistry and the other laboratory sciences.

The process began when the University of Heidelberg sought to attract either Liebig, Hofmann, or Bunsen to its vacant chair of chemistry. Liebig played Heidelberg against Munich, and ended up by accepting a munificent offer from the Bavarian government, which included no obligation to teach laboratory classes. Hofmann was still firmly ensconced in London, so Bunsen was the ultimate choice for Heidelberg (1852). Just a year earlier, Prussia had enticed Bunsen to transfer from Marburg to Breslau, by offering him a new chemical institute. In addition to the Breslau building, the Prussians built new chemical institutes at the Universities of Greifswald, Königsberg, and Halle during the 1850s. Heidelberg had a new building ready for Bunsen by 1854, and the Bavarian government built one for Liebig.

Beginning in 1861, the Prussian government engaged in a long series of conversations and negotiations with Hofmann, hoping to attract him back to his homeland. In 1863 Hofmann accepted a call to the University of Bonn, replete with promises of princely emoluments and a large new chemical institute. Before he could transfer there, however, another position came open at Berlin, and the Prussians also offered him this chair, and once again a large laboratory to be built to his preferences. Hofmann's leverage both in Berlin and in London was so great that he was allowed to hold an exclusive option on all three positions (including the Royal College of Chemistry) until 1867; but in 1865 he did indeed begin his new career at the University of Berlin. Bonn then went after Kolbe, who had recently transferred from Marburg to Leipzig with the promise of a large new institute in his pocket. When Kolbe turned this offer down, Bonn called Kekulé,

31. P. Borscheid, *Naturwissenschaft, Staat und Industrie in Baden (1848–1914)* (Klett, 1976).

Figure 12.5
An elevation of the Munich chemical laboratory, built shortly after Liebig's
transfer there in 1852. From August von Voit, *Das chemische Laboratorium der
Königlichen Akademie der Wissenschaften in München* (Braunschweig: Vieweg,
1859). Source: Photograph Collection, Deutsches Museum, Munich.

Figure 12.6
A photograph of an etching showing Liebig in his private study in Munich.
Source: E. F. Smith Collection, University of Pennsylvania Library.

another student of Liebig; Kekulé took possession of the palatial chemistry building originally intended for Hofmann. Frankland, who had strong associations with Bunsen, Liebig, and Kolbe, then succeeded Hofmann at the Royal College of Chemistry. In addition to the ten German universities named in this and the previous paragraph, Giessen, Göttingen, Erlangen, Würzburg, and Tübingen also built new chemical institutes before the Franco-Prussian war. Similar building programs began in Austria-Hungary and Switzerland.[32]

The new German chemical institutes of the 1850s and especially the 1860s were built on a scale and with amenities and instrumentation that the world had never seen. All of a sudden, it seemed, these palaces of science—for physics, physiology, and the other disciplines shared in the wealth—were mushrooming across the German landscape. The boom in German academic science was a complicated affair, involving pressures of German industrialization, economic prosperity, enlightened modernization policies, a substantial rise in student enrollments, and competition engendered by the decentralized German polities. Once they were made aware of what had transpired, French scientists were fairly terrorized by these developments, and with good reason. Consequences of this were discussed in chapter 9.

The Issue of Decline in French Science

Did the Germans truly surpass the French in science during the middle decades of the nineteenth century? Opinion in both countries during the 1860s and the 1870s was uniform: almost without exception, informed observers believed that they assuredly had. Granted, French scientists could have had a political agenda to deliberately underestimate the quality of their science (in order to appeal for enhanced support from their government), just as there may have been external reasons for Germans to overestimate their own (prestige, patriotism, chauvinism). On the other hand, the French were by no means immune to chauvinism, nor did the Germans ever hesitate to poor-mouth their own assets in order to beg for more. So, to the extent that the historian should consider the views of the

32. J. Johnson, "Academic Chemistry in Imperial Germany," *Isis* 76 (1985): 500–524.

principals themselves in assessing the situation, the uniformity of opinion must at least be taken seriously.

And so it was, until a generation ago. In a landmark essay on the patronage of French science in the nineteenth century (1973), Robert Fox posited, in common with previous scholars, that France had suffered "loss of the supremacy in scientific research" to Germany. He placed the beginning of the process of decline around 1830, and its culmination by the time of the Franco-Prussian war.[33] However, just a year earlier Harry Paul published a stimulating article that powerfully challenged the declinist thesis.[34] Paul's view was that previous scholars had too often exhibited an awestruck "fetishism" toward German science and toward the "icon" before which historians of science had always had to "genuflect," Liebig; moreover, "the year 1840 [had] exercise[d] a neo-Pythagorean fascination for the devotees of decline." Despite a few minor concessions, Paul rejected all the arguments hitherto deployed to show decline, and concluded that "it is by no means clear that any system differing significantly from the system that actually developed in France would have been any better for the 'advance of science.'"[35]

Paul's paper represented a significant contribution that cast a revealing new light on the complicated question of decline. But Paul's paper is subject to some of the same criticisms that he used against the declinists: it is often anecdotal, qualitative, and appears to argue to a foregone conclusion. Intent on demonstrating the high quality of French science, Paul suppressed chronology by using examples from throughout the nineteenth century, and even a few that date well into the twentieth. (Not even the most convinced declinist has denied that there was a powerful and efficacious movement of support for French academic science from about 1880 on.) He was also content to make broad claims about the comparison to Germany, without

33. Fox, "Scientific Enterprise," p. 445.

34. Paul, "The Issue of Decline." Paul's essay provides a convenient review of earlier historiography on this issue, as does the introduction to *The Organization of Science and Technology in France 1808–1914*, ed. Fox and Weisz (Cambridge University Press, 1980).

35. Paul, "The Issue of Decline," pp. 418, 450. Paul's later work *From Knowledge to Power*, esp. pp. 15–59, is more sympathetic to the claim of decline. "It is difficult to contest," he says on pp. 19–20, "although necessary to modify in some respects, [Antoine] Prost's designation of the period 1800–80 as 'the long stagnation' of the University. . . ."

actually citing any comparative sources or data from across the Rhine; but such a comparison is necessary, for the essential declinist question is a relative rather than an absolute one. Putting the best possible light on French deficiencies, Paul argued that "the [French] system for producing [only] a small number of scientists . . . was a perfectly reasonable policy for a society that required only a small number of scientists"; at another point he argued that because the EPHE began in 1868 to provide budget lines for certain laboratories, these laboratories "were not restricted by a lack of funds."[36]

Since the publication of this essay, historians of France have tended to adopt aspects of Paulian revisionism, or to take on a multivalent eclecticism, or to deny the very possibility of determining the existence or nonexistence of relative decline. "A perverse positivistic reader," wrote Paul in a 1991 essay, "may peevishly ask, well, was Germany superior or not? A generation ago the answer was clearly, yes; now it is obscurely no."[37] In an important 1980 essay, Robert Fox and George Weisz accepted significant elements of the anti-declinist viewpoint, suggesting finally that the entire question needed to be phrased not in terms of the fate of science as a whole, but rather of different fields of science that experienced different trajectories.[38]

About the same time, Terry Shinn studied the research productivity of the entire French science faculty system in the "long" nineteenth century.[39] Shinn's study provided evidence that the quantity as well as the quality of

36. Paul, "The Issue of Decline," pp. 429, 430. Paul's first point was later echoed by Craig Zwerling ("The Emergence of the École Normale Supérieur as a Centre of Scientific Education in the Nineteenth Century," in *The Organization of Science and Technology in France 1808–1914*, ed. Fox and Weisz, p. 60): "In short, the French economy got the level of science it needed and could support." For a sharply dissenting analysis, see F. Leprieur and P. Papon, "Synthetic Dyestuffs," *Minerva* 17 (1979), p. 218; Leprieur, "La formation des chimistes français au XIXe siècle," *La recherche* 10 (1979), p. 737.

37. H. Paul, "The Role of German Idols in the Rise of the French Science Empire," in *Einsamkeit und Freiheit' neu besichtigt*, ed. G. Schubring (Steiner, 1991), pp. 195–196.

38. Fox and Weisz, "Introduction." However, Fox's position is clearly declinist in his later essay "Science, the University, and the State in Nineteenth-Century France," in *Professions and the French State, 1700–1900*, ed. G. Geison (University of Pennsylvania Press, 1984).

39. T. Shinn, "The French Science Faculty System, 1808–1914," *Historical Studies in the Physical Sciences* 10 (1979): 271–332.

French scientific publications plunged during the generation from the mid 1840s to the mid 1870s, but he also argued that such productivity trends could not be correlated to variations in such factors as centralization or decentralization of the nation's science. However, as has been the case for most of the quantifiers,[40] Shinn's evidentiary basis was to some degree problematic.[41]

About the same time as Shinn's study, François Leprieur urged the need for a "nuanced" view of the declinist thesis. Leprieur argued for the continued vitality of French organic chemistry until about 1860, citing as evidence the parade of foreign chemists to Wurtz's laboratory, and he cogently disputed the claim that French weakness was a heritage of the "Napoleonic university system" established at the beginning of the century. However, he did agree that there was a significant decline vis-à-vis Germany after 1870, especially in the science-based industries; this was due to many factors, he argued, including the absence of a close relationship between academic and industrial chemists in France.[42]

One of the lessons that can be drawn from previous attempts at quantitative studies of trends in national research productivity is how important it is to specify the parameters of study with precision, and to keep the

40. For an excellent critique of quantitative studies of decline, see M. Nye, "Scientific Decline," *Isis* 75 (1984): 697–708.

41. First of all, it would have been better to have used the *Royal Society Catalogue of Scientific Papers* rather than Poggendorff as a database for articles, for the former is considerably more complete. (Poggendorff also includes book publications, which is presumably why Shinn used it.) Second, Shinn's finding that research success does not correlate to centralization or decentralization loses force because of intrinsic uncertainties of historical analysis; it could well be, e.g., that centralization did stifle productivity, but that this factor was counterbalanced by other concurrent trends not being measured; nor is centralization/decentralization a binary choice. Third, Shinn did not attempt to compare French productivity to concurrent changes in Germany—which would be necessary if one wished to inquire regarding *relative* decline, the crucial point at issue. Finally, Shinn's data concerned only facultés des sciences, not grandes écoles or research institutions; for our story, that is only the Sorbonne. It is not clear what one can truly say about the fate of French chemistry if one excludes, as Shinn necessarily does here, such men as Pelouze, Regnault, Chevreul, Berthelot, Cahours, and Frémy.

42. F. Leprieur, Les conditions de la constitution d'une discipline scientifique (thèse de IIIème cycle, Université de Paris I); Leprieur, "Formation"; Leprieur and Papon, "Synthetic Dyestuffs."

analysis confined within that restricted domain. As Nye has commented, many if not most of these studies stretch across the entire century, and some even incorporate mid-twentieth-century French declinist angst into the mix.[43] Along these lines, it is obviously misleading to cite, as evidence that nineteenth-century French science did not decline, the fact that a great infusion of funds took place after 1878; and yet this implication can be found in more than one analysis. Professor Paul justified his judgment expressed in the quotation cited above (p. 399) by referring to a comparative study of the health of physics in the year 1900; but this has little bearing on the questions raised in this book (whose chronology relates to chemistry in the period from circa 1820 to circa 1880).[44]

In addition to scrupulous attention to the fine structure of chronology, one must also be careful to distinguish between different scientific fields, as Fox and Weisz have rightly argued. The quality of scientific contributions and questions of national style also must be carefully considered; evaluation of such matters is highly subjective, and is usually not even attempted.[45] In addition, the issues relating to pure science on the one hand, and technology and industrial development on the other, are by no means the same; these issues are too often implicitly or even explicitly commingled. Even purely quantitative measures, such as numbers of article published, are subject to problematic judgments: What qualifies as an article? How are they to be located and enumerated? How is the database of authors determined? What fields of science are included? How should one choose the time period? And so on.

43. Nye, "Scientific Decline," p. 697. Among the studies that draw upon recent generations of pessimists are the following: R. Gilpin, *France in the Age of the Scientific State* (Princeton University Press, 1968); Paul, "The Issue of Decline."

44. For a demonstration that quantitative data do not suggest any significant inequalities in the state of health of the science of physics in Germany, France, Britain, and the United States around 1900, see P. Forman, J. Heilbron, and S. Weart, "Physics circa 1900," *Historical Studies in the Physical Sciences* 5 (1975): 1–185.

45. But see Shinn for such an attempt, and Nye for a methodological discussion. Along these lines, Fox comments that "French ways of doing things were certainly different from those of their apparently more successful rivals, but the mere fact that they were so cannot be used, at least in isolation, as a basis for assessment" (*Science, Industry, and the Social Order in Post-Revolutionary France* (Variorum, 1995), p. xii). Within limits this is eminently reasonable, but one should guard against an essentialism that would prevent any comparative analysis whatever.

Let me then pursue the following delimited question: Between about 1820 and 1880, how did the robustness of French chemical science fare in comparison with German chemistry? The aggregate details in the present volume have much bearing on the answer to this question. The evidence is not, in the main, quantitative or statistical, but I think that it is no less revealing for that. The qualitative judgment to which I have come, considering this aggregate evidence, is the following: in chemistry, France exhibited a dramatic decline, vis-à-vis Germany, in both quantity and quality of published research in this time period. The decline probably began in the 1820s (but by 1830 at the latest), and continued for at least the next half century. It accelerated after the 1860s, but was partially redressed at the end of the period in question.

For those readers who prefer quantitative measures, here are a couple of samples. A few years ago I carried out a survey of numbers of chemical papers published during the year 1841 in the principal scientific journals of Britain, France, and Germany. This survey suggested that the British community was lagging seriously behind the French, but that the French were already far behind the Germans.[46] A quarter-century later the discrepancy was much greater. Edward Frankland reported to a British government commission that during the year 1866, 445 German chemical authors published 777 "original investigations," whereas 170 French chemists published 245 papers, and only 97 British chemists published only 127 papers. Furthermore, many of the British authors were actually expatriate Germans![47] Focusing on the French-German comparison, Frankland's figures suggest that the number of German chemists actively publishing in that year was between two and three times the number of French chemists, and the number of articles was more than three times greater. The populations of the two countries were similar in that year.

Another approach with a much larger database has been provided by Christoph Meinel, who cataloged for study all papers (determined from the *Royal Society Catalogue of Scientific Papers*) published by each of the 200

46. A. Rocke, "Chemical Knowledge in 1841," unpublished conference paper, Royal Society of Chemistry Sesquicentennial Congress, 9 April 1991.

47. Testimony of Frankland, 14 February 1871, in *First and Second Reports from the Royal [Devonshire] Commission on Scientific Instruction*, British Parliamentary Papers (London, 1872), vol. 25, p. 372.

nineteenth-century chemists whose biographies appear in the *Dictionary of Scientific Biography*. The relative accounting of original chemical papers published annually in France and Germany is revealing. At the beginning of the century until the early 1820s, the French DSB chemists were publishing papers at close to twice the rate of the German DSB chemists, but by the late 1820s the latter had already surpassed the French in rate of production. From the mid 1820s until the early 1860s the ratio was roughly constant, the French usually at about three-fourths of the German rate. Then suddenly the German publication rate mushroomed, so that in the late 1860s the German rate was more than twice the French; in the last interval studied, the early 1870s, it was more than triple.[48] Since Meinel's data set includes only DSB chemists, that is, those of significant reputation, these ratios probably understate the German advantage, since more scientists of second and third rank worked in Germany than in France. This study is entirely consistent with Frankland's one data point, but what it seems to show is that in Frankland's year of 1866 the edge which the Germans had long enjoyed over the French had suddenly become a chasm.[49]

These samples can be supplemented by aggregate comparative institutional statistics for all higher education—with the important qualification that it is difficult to find true comparability in entities measured across national boundaries. In 1865, there were 823 teaching positions in French higher education, and about 1500 in Germany.[50] In 1876, this

48. C. Meinel, "Structural Changes in International Scientific Communication," in *Atti del V convegno di storia e fondamenti della chimica* (Accademia Nazionale delle Scienze, 1993), p. 53. Meinel does not provide his actual numbers, but rather summarizes them in a bar graph. I thank Professor Meinel for further clarification of these numbers, communicated privately in May 1999.

49. For yet another interesting study, see B. Willink, "On the Structure of a Scientific Golden Age," *Berichte zur Wissenschaftsgeschichte* 19 (1996): 35–49. Willink likewise documents the great burst in German science in the mid to late 1860s; interestingly, Willink also notes a temporary relative dip in German scientific power during the 1850s. This result is consistent with Meinel's figures.

50. G. Weisz, *The Emergence of Modern Universities in France, 1863–1914* (Princeton University Press, 1983), pp. 28–29. This includes, for France, all professeurs, agrégés, maîtres de conférence, chargés de cours, and répétiteurs, in facultés, grandes écoles, and research institutions, but not "laboratory personnel" (presumably janitors and aides); and for Germany, all Ordinarien, Extraordinarien, and Privatdozenten.

time comparing France with Prussia alone, considering only higher education in science, and calculated on a per capita basis, Prussia had 28 percent more instructional science positions than France, and two and a half times the number of science students.[51] If these numbers of French science professors were to be adjusted for cumul, the comparison would be even more disadvantageous for France, for the number of science teaching personnel there was smaller than the number of teaching positions. Also in 1876, there were only 293 students registered in the facultés des sciences throughout France; this was about the same number of students studying for doctorates in the various natural sciences at a single German institution, the University of Leipzig.[52]

Let us now move from aggregate science statistics to the particular specialty field with which we have been most concerned here. Here is another sample. Of 66 original papers published in the Annales de chimie in 1864, 14 were on organic chemistry—a respectable fraction. But of these 14 papers, three were written by Wurtz, three by Berthelot, and five by foreign authors. Consequently, noting that Berthelot was soon to leave the field, the organic-chemical output of France that year (other than Wurtz) consisted of one paper by Victor de Luynes and two by Cahours.[53] The contrast to other countries, and particularly to Germany, is provided by examining the foreign chemical abstracts in the Annales, for which Wurtz was responsible until his death in 1884. A breakdown of these abstracts has been done for the decennial years 1870, 1880, 1890, and 1900. In aggregate terms, about 70 percent of these abstracts came from German

51. P. Lundgreen, "The Organization of Science and Technology in France," in *The Organization of Science and Technology in France 1808–1914*, ed. Fox and Weisz, pp. 327–330. Counted are French professeurs (in facultés and grandes écoles, including the Muséum) versus Prussian Ordinarien and Extraordinarien only.

52. Weisz, *The Emergence of Modern Universities in France*, p. 22; Rocke, *Quiet Revolution*, p. 267. Weisz's figures are partly derived from F. Ringer's *Education and Society in Modern Europe* (Indiana University Press, 1979).

53. A. Rocke, "Adolphe Wurtz and the Development of Organic Chemistry in France," *Bulletin de la Société Industrielle de Mulhouse* 833 (1994), p. 31. This sample is limited by taking only papers published in the *Annales*, and I have not attempted a full survey of French scientific journals. It is, however, true that *Annales* was by far the most important journal for the physical sciences in France, and that most significant contributions in this field found their way into its pages.

journals, and within organic chemistry the proportion was 80 or even 90 percent.[54]

Yet another concordant result is shown by Leprieur's and Papon's accounting of the nationalities of authors of papers published in the Bulletin de la Société Chimique, and focusing on the category of aromatic compounds. (Aromatic chemistry was central both to the rising coal-tar dye industry, as well as to the purely theoretical development of structural ideas.) Between 1864 and 1867, the percentage of organic-chemical Bulletin papers devoted to aromatics increased from 14 to 38 percent. In the same 3-year period, the percentage of German authorship of these papers rose from 35 to 85 percent. By 1870 Germans were writing no less than 96 percent of all aromatic-chemical papers published in the Bulletin.[55]

Such examples are striking. In view of all the evidence, it appears that French chemistry, so powerful internationally at the beginning of the century, had badly deteriorated in comparison to German rivals by the time of the Franco-Prussian war. France's weakness relative to Germany in the years ca. 1830–1860 seems indicated by most of the data; virtually all of it points to a dramatic worsening of the gap in the 1860s and the 1870s.[56] Indisputably for organic chemistry, most probably for chemistry as a whole, and likely for science more broadly defined, French science had deteriorated by comparison to Germany.

The Causes of Decline

If one accepts the claim that there was a comparative decline, the next step is to inquire into the reasons for this trend. Many causes have been proposed and discussed in the literature. For convenience, these can be separated into the overlapping categories of socio-cultural, institutional, and field-dependent internal factors. Since I am concerned with analyzing French science in relative and not absolute terms, the following discussion,

54. U. Fell, "The Chemistry Profession in France," in *The Making of the Chemist*, ed. D. Knight and H. Kragh (Cambridge University Press, 1998), p. 33.

55. Leprieur and Papon, "Synthetic Dyestuffs," p. 210.

56. I suspect that Kekulé's theory of aromatic compounds, so aggressive pursued in German academic labs, played a role in this sudden trend.

like the preceding one, will incorporate continuous references to Germany. Technology will not be included in this analysis.

One constellation of trends relates to an apparent difference in the prevailing varieties of late Romanticism in the two countries. After about 1820, a kind of Romantic oratorical culture established itself in France, extending throughout the highest educational and even research institutions, including in the sciences. This "declamatory science" was particularly visible at the Sorbonne and the Collège de France, where huge audiences consisting mostly of interested laypeople were attracted to the lectures. This situation developed because there was no attendance requirement for registered students, and contrariwise there was much interest in elevated subject matter among the educated public. Despite the absence of financial incentives, prestige and self-esteem provided effective motivations for professors to try to attract large audiences. Most professors therefore paid great attention to thespian skills and sought elaborate rhetorical effects to bring in these non-academic elites. Education was theater, applause was both sought and expected, and such thespian skills sometimes verged on superficial panache. Even the best researchers (Wurtz is a good example) used these techniques enthusiastically. There seems to have been astonishingly little contact between professors and students, except in the degree examinations themselves.[57] In contrast, German university classes consisted almost entirely of serious students studying for doctoral degrees, and consequently a sober and systematic lecture style was more common. Professors supervised the practica, conducting the labs with the help of Assistenten, and usually ensuring personal contact of the entire teaching staff with the students. The most technically advanced lecture courses tended to be taught by the youthful Privatdozenten themselves, for the more basic courses were larger and therefore more lucrative, thus tended to be monopolized by the professors.

Already noticeable and commented upon as novel in the 1820s, by the time of Duruy's ministry these cultural practices were regarded by reformers as deleterious to the future of French science; scholars, it was thought, should be sober rationalists (in the model of Gay-Lussac) rather than per-

57. See Arago's description of almost complete isolation between professor and students in the École Polytechnique, cited in Crosland, *Gay-Lussac*, pp. 152–153.

Figure 12.7
A lecture at the Collège de France (date uncertain). Note the size of the audience and the presence of women. Source: Collection Iconographique Maciet, Bibliothèque des Arts Décoratifs, Paris.

formers (like Thenard). Meetings of the Académie des Sciences had similarly gone in a theatrical direction, which critics such as Biot and Libri regarded as damaging to the institution's raison d'être—pure science and serious research.[58]

Another aspect of late Romantic French intellectual culture is the rise of varieties of positivist philosophies.[59] There is a well-developed literature on this subject, and exemplification is not lacking in the pages of this book.

58. The best discussions of this French Romantic oratorical culture are found on pp. 452–458 of Fox, "Scientific Enterprise," and on pp. 81–84 of Fox, "Science, the University, and the State." See also A. Rocke, "History and Science, History of Science," *Ambix* 41 (1994): 20–32. For an eyewitness comparison of Gay-Lussac's and Thenard's lectures in 1820, see the first section of chapter 1 above.

59. Bensaude-Vincent, in "Atomism and Positivism," rightly insists on the multiplicity of "positivisms" in France. She distinguishes Comte's own ideas, which were surprisingly receptive to the heuristic use of hypotheses, from Baconian inductivism. She also distinguishes Littréan positivism, "textbook positivism," and the "positivist vulgate."

Although the inductivist-conventionalist-positivist legacy could be salutary for the development of scientific thought—for example, in the careful distinction between empirical and non-empirical information—most analysts point to developments related to these ideas that can only seen as injurious. In particular, the narrow scientism that was characteristic of many positivists (and, ironically, not very noticeable in the writings of Auguste Comte himself) placed far too little value on the heuristic utility of hypotheses, theories, and abstract ideas.[60] God is an abstract idea, too, and the association of positivism with materialism and irreligion was strong in the generation after Comte. It is significant that perhaps the three greatest French chemists of the nineteenth century, Dumas, Wurtz, and Pasteur, were all deeply religious men—as markedly different as their convictions in religion, politics, and science were. None of the three was attracted to Comtean or Littréan positivism.

After its initial early phase, Romanticism took a different turn in Germany. Speculative idealist philosophy, symbolized best by Schelling's *Naturphilosophie*, was yielding already around 1810 to a more empirical Kantian sense-intuitionism, especially at the University of Berlin, a model reformist institution. Varieties of positivism spread to Germany, too, but they did not gain there the kind of ideological and even pseudo-religious force that they did in France. In addition, the thoroughly idealist features of German Neohumanism were leavened, ironically, by a perceptible undercurrent of ideas derived ultimately from the French Enlightenment: empirical, practical, and utilitarian commitments. The apparent contradictions between Romantic Neohumanism and Enlightenment practicality in the German intelligentsia could be bridged by skillful rhetoric, and no one was more skilled at this art than Liebig. The result helped to condition the rise of the German university *Praktikum* in the sciences, and especially in chemistry, from the mid 1830s on. At the same time, Romantic qualities of iconoclasm and individualism (as opposed to Enlightenment materialistic universalism) helped to condition the formation of a powerful research ethic in the reformed universities of the Neohumanist era. Liebig, Wöhler, Bunsen, and others then grafted the pedagogy of systematic practical labo-

60. For valuable discussions, see Paul, *From Knowledge to Power*, pp. 60–92; Fox, "Positivists, Free Thinkers, and the Reform of French Science in the Second Empire," in *Science, Industry, and the Social Order*.

ratory education to the ideology of the research ethic, and a potent combination was born.[61]

As anti-declinists have often and correctly insisted, the research ethic was alive and well also in France, all during the period in which others have seen decline in French science.[62] A career in pure science began with a research dissertation for the doctorat, and further advancement was predicated upon the publication of important research. Appointments both to academic positions as also to the Académie des Sciences were judged largely by research renown. However, the system of examinations for the baccalauréat, licence, and agrégation, and emphasis on memorization and oral performance in the competitions, may well have tended to underscore superficial glibness at the expense of originality and depth. The German system, in contrast, required from a would-be academic scientist not one but two dissertations, the second necessary for "habilitation" as a Privatdozent, and both contributions needed to push the research frontier. Teaching was important, too, but not in the performative sense that was prevalent in France.

Moreover, as Fox has well noted, the acknowledged respect for research renown in France was not so strong as to keep some of the most prominent scientists in the field; too often, as with Gay-Lussac, Thenard, Dumas, Balard, and Pelouze, they were attracted into administration and politics, and ended up abandoning their research.[63] Part of the attraction was purely financial: when Dumas became senator in 1852, he began earning 30,000 francs per year,[64] six times the salary of a Sorbonne professor.[65] In the middle

---

61. For a more detailed summary with full references to the secondary literature (for most of these ideas come from other scholars), see Rocke, *Quiet Revolution*, pp. 9–34.

62. E.g., M. Crosland, "Scientific Credentials," *Minerva* 19 (1981): 605–631.

63. Fox, "Scientific Enterprise," pp. 469–470; Fox, "Science, the University, and the State," p. 83. In the former, Fox argues compellingly that the absence of individual initiative was a critical lacuna leading to the decline of French science. Such a process was possible only in a centralized system where a few individuals could make a crucial difference; this factor can also provide a partial understanding why certain fields fared so much better than others.

64. Pay voucher, Dumas file, Sammlung Darmstaedter, Staatsbibliothek zu Berlin Preussischer Kulturbesitz.

65. For salaries at the Paris Faculté des Sciences, see AJ16/189 and AJ16/323.

of a *defense* of the French system, Berthelot in 1858 mentioned in a letter to Renan the familiar "ambitious preoccupations that are the undoing of all of our scholars once they attain maturity."[66] This happened with German scientists, too, but less frequently. In various ways, then, the research imperative was more thoroughly constitutive in German than in French institutions.

In an important essay, Fox has demonstrated how much French scientists were drawn, especially during the July Monarchy and Second Empire, into the role of obedient state functionaries, a process which was ultimately deleterious to their role as independent researchers and which only began to be corrected toward the end of the century.[67] French scientists, in short, were more thoroughly assimilated to the state than was healthy for independent basic research. This was most notoriously the case during what I regard as the most crucial decade in the development of nineteenth-century science, the 1850s. German academic scientists were also subject to state controls; the infamous example of the "Göttingen seven," fired from their posts in 1837 for political reasons, comes readily to mind. However, German rulers did not normally interfere with academic autonomy; Lehrfreiheit, if not an absolute principle, was more than a simple slogan. Even when governments did exert pressure or even countermand faculty preferences on hires, it was often to ensure selection of the highest quality candidate for excellence in research, against the narrow thinking of the faculty.[68]

66. Berthelot to Renan, August 1858, quoted in L. Velluz, *Vie de Berthelot* (Plon, 1964), pp. 45–46.

67. Fox, "Science, the University, and the State."

68. In 1865, Saxon Kultusminister Paul von Falkenstein insisted on hiring the superb scientist Hermann Kolbe against the initial desires of the Leipzig University faculty, who preferred the undistinguished local candidates Heinrich Hirzel and Wilhelm Knop. Also revealing in this instance is the fact that Hirzel and Knop were oriented toward technological applications (pharmacy and agriculture, respectively); the interest of the government was operative here in support of high-quality basic science rather than mediocre applied research. For details, see Rocke, *Quiet Revolution*, pp. 265–278. A similar case a generation earlier was the first appointment of Justus Liebig at Giessen in 1824, which was pushed by Humboldt and Schleiermacher against faculty preferences. Turner (see n. 27 above) has emphasized the positive state role in ensuring excellence in the German professoriate of the Vormärz, especially regarding research renown.

To these elements must also be added an increasing sense of nationalist isolation of the French scientific community, an inclination whose inception is datable to the 1820s, approximately the same time as other trends associated with late Romanticism. Once again, examples are well represented in the present volume, and again, Robert Fox has a good discussion of this issue.[69] To a greater degree than their colleagues in Germany, French scientists neglected the study of foreign languages and, often, the study of the works of foreign colleagues; they withdrew from cooperative international projects; and they taught few foreigners in their classes and rarely went for postdoctoral or study trips abroad.[70]

In the first third of the nineteenth century, at least 20 German chemists spent study tours in Paris, including Stromeyer, F. G. Gmelin, C. G. Gmelin, L. Gmelin, G. Rose, H. Rose, Liebig, Mitscherlich, Runge, Schönbein, and Bunsen. With the exception of Liebig's earliest French students, there were no French chemists who spent study time in Germany during these years. Throughout much of the century, few French scholars could even read German, much less speak it. It is true that German chemists eventually stopped coming to Paris in such numbers; by the 1830s and the 1840s they no longer considered it necessary. Alexander von Humboldt's permanent return from Paris to Berlin in 1827 can be considered a milepost in this shift. However, it continued to be the case, at least until late in the century, that Germans profited far more from their knowledge of what their neighbors were doing than the French did.[71]

Of course, this was not true for the entire French professoriate; Gay-Lussac, Pelouze, Dumas, and especially Wurtz are representative cosmopolitan figures from chemistry (all four of whom complained about the parochialism of their colleagues in the early 1830s). Other French scholars sensitive to foreign (especially German) academic work included the historian Jules Michelet, the philologist Ernest Renan, and the philosopher

---

69. Fox, "Science, the University, and the State," pp. 93–96. See also the last section of chapter 2 above.

70. According to Meinel ("International Scientific Communication, pp. 52–55), during the nineteenth century German chemical papers were translated into French less than half as frequently as French papers were translated into German.

71. K. Kanz, *Nationalismus und internationale Zusammenarbeit in den Naturwissenschaften* (Steiner, 1997), pp. 38, 54–58, 107–120, 141–146, 228–232.

Victor Cousin. But the fact remains that, on the whole, the French research community was disturbingly insular, almost xenophobic, until the reform movements of the 1860s. Naturally this worked to inhibit adoption of foreign ideas and foreign models.

Disparate French institutional structures also often worked against a vigorous research establishment. As Robert Gilpin has rightly argued, the "functional fragmentation of French scientific and intellectual life into three separate sets of institutions"—lecture-based higher education (facultés), technical training (grandes écoles), and scientific research (Collège de France, Muséum, etc.)—"had a detrimental effect on French science and technology."[72] Some specialists have argued that, put simply, the French had failed to develop a "true university system."[73] Crosland has pointed out that the better science students who (in general) attended the grandes écoles had no access to the careers in teaching that the faculties trained for.[74] A tripartite structure similar to that of France, inherited from eighteenth-century institutions, can also be seen in early nineteenth-century Germany. For instance, Eilhard Mitscherlich, professor of chemistry at the University of Berlin in the 1820s, had his laboratory not at the university but at the Akademie der Wissenschaften, where he was a member. However, Liebig and allies such as Wöhler, Bunsen, Kolbe, and Hofmann worked for an effective fusion of these functions as early as the 1830s. In France there was also a degree of functional fusion, but only through the artificial means of the practice of cumul, a practice that surely did more harm than good.

In France, university lectures were generally free, whereas in Germany they charged by the head. Critics of the French system argued that this allowed little opportunity for younger scholars to enter careers at the bottom. This deficiency was partially addressed by the government's practice of hiring low-paid agrégés who did occasional substitute teaching, and by the system of suppléance, whereby a senior professor could remit half of his salary to a semi-permanent replacement lecturer. In Germany, Privat-

---

72. Gilpin, *France in the Age of the Scientific State*, pp. 86–87.

73. See, e.g., R. Anderson, *Education in France, 1848–1870* (Oxford University Press, 1975), p. 227. For a contrary viewpoint, see Leprieur, Les conditions de la constitution d'une discipline scientifique.

74. M. Crosland, "The Development of a Professional Career in Science in France," *Minerva* 13 (1975), p. 56.

dozenten were certified by the university (after proving their worth as researchers and teachers), and paid directly by students electing their classes. Whether or not the system provided more career opportunity—and I suspect that it did—it was certainly a more fluid and market-sensitive institution than that using agrégés.

Perhaps the most common reproach made against French academic institutions was that of excessive centralization, exacerbated by the monopolistic practice of cumul. During the period with which I have been dealing, there is no question but that the intellectual life of France centered largely on the capital.[75] To a great extent, the scientific life of France was controlled by (and nearly identified with) a few dozen important men living in Paris. This circumstance was made even more monopolistic by the power structure of the Académie des Sciences. Academicians, allied (and often identical) with the highest administrators of the Université, had effective control over the entire system; and residency in Paris was required for admission as a regular member to the Académie. We saw in the first two sections of this chapter how this system allowed a small group of similarly minded individuals (Dumas, Balard, Pelouze, Deville, Berthelot, Frémy, and Troost) to call the shots in their field for two generations. The system damaged diversity, inhibited healthy competition, and hurt the career prospects of younger scientists.

It has been argued, in defense of cumul, that only by accepting multiple positions could a senior scientist attain something approaching a bourgeois existence. Exhaustion from overwork was avoided by farming out at least one and sometimes two of the professorships to youthful suppléants. This enabled the chairholder to retain control of the position (and half the salary) for no expenditure of effort, and also provided employment for young scholars. Furthermore, only in the system of cumul could a science professor at the Sorbonne (with no laboratories) gain access to the means to perform his research. But this defense does not significantly weaken the arguments criticizing monopolistic practice, and fails in the effort to make a virtue of necessity. The rebuttal only draws further attention to the scandalously poor

75. This focus weakened in the next generation; see Nye, *Science in the Provinces.* On the tendency toward centralization of science in Paris in the early nineteenth century, see N. and J. Dhombres, *Naissance d'un pouvoir* (Payot, 1989), pp. 217–218.

financial support for academic science by successive political regimes during the 60 years in question. The solution that would have solved many of these problems definitively was as simple as it was out of reach until ca. 1880: much more money for higher salaries, and much more money for laboratories and other physical facilities.

In Germany, too, academic scientists fought tenaciously with their governments for funding. During the Vormärz (1815–1848), the battles often did not come out well for the scientists. However, the German states represented a diverse and competitive pseudo-international system, and in academia the research mandate became continuously stronger. Over a dozen German universities began to compete for the finest researchers and the biggest names. Salaries rose, new positions were created, and new or renovated facilities were built to attract the most famous professors. After the Revolution of 1848, science began to be viewed by ruling elites as an integral part of economic modernization and social stability. Career opportunities increased for graduates, and enrollments rose. As we have seen, a competitive entrepreneurial fever caught hold in most of the German states, bidding wars began, and governments began to spend money on academic science in a truly lavish fashion.

Furthermore, the principle of Lernfreiheit enabled students to choose from the best institutions and the best scholars at will, for there were never any barriers against transferring from one university to another in a different German state. It was not uncommon for a student to take all his courses at one university and then go to another to pass the doctoral examinations. The same principle of Lernfreiheit was advantageous for the professors, as well, for they had as a result a very large pool of potential student candidates, derived from all the German states, from which to select advanced Praktikanten and to recruit Assistenten—both of which groups were important for the director's research program. These factors reinforced one another. The virtues of a dynamic decentralized system suddenly became impressively apparent, and the French government finally began to awake from its slumbers in the 1860s. But reveille was played to a slow tempo.

As Fox and Weisz have noted, there was a large difference in the comparative standing of the various fields of French science. Mathematics and experimental physics, for example, fared reasonably well in international comparisons, while theoretical physics and chemistry, particularly organic chemistry, did not. This divergence in the paths of various scientific disciplines in nineteenth-century France should not be surprising. Superadded to the inevitable natural variation in the national standing of different disciplines were the structural circumstances discussed in this chapter, which allowed particularly energetic individuals, or the actions of a small group of similarly minded scientists, to do either major good or major ill to a discipline. But the present volume suggests that few fields of French science suffered more, in comparison to Germany, than organic chemistry did, and I want to introduce an additional field-specific factor that pertains to both institutional history and to the internal developmental characteristics of this particular science.

Chemistry and physics are both laboratory-based physical sciences, but there are differences between them that I believe are relevant for our analysis here. I want to suggest that chemistry, and particularly organic chemistry, is intrinsically laboratory-intensive in a way and to a degree that experimental physics is not. If this thesis is granted, then the shocking dearth of laboratory facilities in nineteenth-century France was necessarily far more damaging to the health of (organic) chemistry than it was to the health of physics.

The history of physics is replete with well-known instances of single[76] experiments that are intended to provide definite answers to theoretical issues. The Michelson-Morley ether drift experiment is one well-known example, but there are innumerable others. Just taking a few classic nineteenth-century French examples, one could cite the experimental test by Poisson and Arago of Fresnel's mathematical theory of diffraction, Le Verrier's discovery of Neptune, the Foucault pendulum experiment, Foucault's comparison of the speed of light in water and air, his measurement of the absolute speed of light, and Fizeau's measurement of the difference of the speed of light in water moving in opposite directions. These were all highly difficult and sophisticated experiments bearing crucial relevance to debated theories, but they were also singleton tests, and as such were typical for the discipline of physics.

Chemistry had singleton tests, too. The discovery of a new element is one such generic example, and I have argued that Alexander Williamson's

---

76. I hesitate to say "crucial," for that loaded word suggests a philosophical position with which I do not necessarily agree.

asymmetric ether synthesis was a good model for what could be called a crucial experiment. However, organic chemistry increasingly relied on tests that could only be made by examination of a large number of (often novel artificially produced) substances. Perhaps the best, though by no means the only, example of such work is the massive testing of Kekulé's benzene theory after 1865. The chemical equivalence of the six carbon atoms (likewise, the equivalence of the six hydrogen atoms) of benzene, a cardinal postulate of the theory, could hardly be demonstrated by a single experiment, nor even by any small group of them. However, this thesis had empirical consequences that could indeed be examined. According to the theory, every monoderivative of benzene should have a single isomeric form, while diderivatives should have three isomers each. There are many possible monoderivatives, and innumerable possible diderivatives. Evidence in favor of these predictions is often negative in character, namely the non-existence of a second isomer of a monoderivative, or the non-existence of a fourth isomer of a diderivative. This negative evidence can only be inductively confirmed through extensive cumulative experience.[77]

Positive evidence in chemistry is also often examined only by massive testing. The structure theory itself generated literally infinite numbers of "small" predictions about what compounds ought to exist and what compounds might well be impossible to prepare. Organic chemists possessed of the structure theory were quite literally off and running, and even mediocre practitioners could spin off literally hundreds of derivatives and other publishable novelties. One did not even have to accept structure theory to generate these infinite numbers of compounds, as Marcellin Berthelot himself repeatedly stressed. But it was the structuralists who really soaked up the laboratory time. Within 3 months of publishing his first brief article on benzene theory, Kekulé wrote to a friend, "A great deal is in the works;

77. "In a sense, the discovery of *each* of the tens of thousands of new aromatic compounds in the first generation after 1865 constituted a potential (weakly) verifying or falsifying instance." (A. Rocke, "Kekulé's Benzene Theory and the Appraisal of Scientific Theories," in *Scrutinizing Science*, ed. A. Donovan et al. (Kluwer, 1988), pp. 53–54) For a thorough examination of the role of novel prediction in the elaboration of Kekulé's benzene theory, see S. Brush, "Dynamics of Theory Change in Chemistry," *Studies in the History and Philosophy of Science* 30 (1999), pp. 21–79 and 263–302. Curiously, Brush concludes that the "novel prediction criterion" was far more important for chemists than it has been for physicists.

the plans are unlimited, for the aromatic theory is an inexhaustible treasure trove. Now when German youths need dissertation topics, they will find plenty of them here."[78] Structural organic chemistry was inexhaustible in the context of the 1860s in a way that physics was not. Furthermore, structural theories were productive not only for massive amounts of new science, but also important new technologies, as well.

So this is my final suggestion for an important contributing cause of French backwardness in organic chemistry vis-à-vis Germany. When organic chemistry took off as a science around 1860—and it did so with astonishing suddenness and force—it was only the Germans who were possessed of the laboratories that were so needed to support the development of this field. Even if French chemists had been more enthusiastic toward the atomic and structural ideas at the heart of the new organic chemistry, they lacked the material resources, namely, the chemical laboratories, to keep anywhere close to the pace set by the Germans. The work of physicists such as Fizeau and Foucault, who managed to maintain a tradition of excellence in French experimental physics by working out of their residences, was not nearly so sensitive to laboratory capacity.[79] And what this example demonstrates once more is the need to pay close attention to details, sometimes even fine ones, and over several independent axes—social- and cultural-historical, chronological, institutional, cognitive, field-specific, and even non-systematic contingencies—when exploring the comparative development of national science communities.

Robert Kohler, an eminent senior practitioner of the social-institutional style of history of science, has recently compared the methods of the

78. "Angefangen ist noch viel, die Pläne haben kein Ende, da die aromatische Theorie eine unerschöpfliche Fundgrube ist. Wenn jetzt deutsche Jünglinge Dissertationsarbeiten nöthig haben, so finden sie hier Themate genug." (Kekulé to Baeyer, 10 April 1865, August-Kekulé-Sammlung, Darmstadt)

79. France was by no means unique in this. Frank James, citing some examples of the many physicists who set up private laboratories in their lodgings, notes that, despite the example of the Royal Institution, throughout much of the nineteenth century "very few [British] institutions had laboratories in which physics experiments could be conducted" ("Introduction," in *The Development of the Laboratory*, ed. F. James (American Institute of Physics, 1989), p. 3). See also S. Schaffer, "Physics Laboratories and the Victorian Country House," in *Making Space for Science*, ed. C. Smith and J. Agar (Macmillan, 1998).

"strong programme" in the movement known as the "social construction of knowledge" to one of their own most striking images, a "black box," that is, a closed system whose axiomatic foundation and mode of operations are not to be questioned. In a largely favorable essay review of Jan Golinski's fine summary analysis *Making Natural Knowledge: Constructivism and the History of Science* (Cambridge University Press, 1998), Kohler points out that constructivist work "is best done in microstudies of marginal episodes than in large-scale investigations of dominant traditions," and that "for this reason, constructivists' methodological black boxes may hinder more than help historians who wish to move beyond microstudies to macrohistory."[80]

And this appears to be the case in the present macrohistorical study, that is, in viewing events over the course not just of a few years, but over several decades. One example is nationalist bias, which we see present and operating in every chapter of this book; chauvinist predilections worked in vain in each country against the direction that the country's scientists eventually took: in Germany unsuccessfully to oppose the "French" reformist chemistry of Gerhardt, Laurent, and Wurtz, and in France unsuccessfully to oppose the importation of Liebigian laboratory

80. R. Kohler, "The Constructivists' Tool-Kit," *Isis* 90 (1999), p. 330. A. Pang ("Visual Representation and Post-Constructivist History of Science," *Historical Studies in the Physical and Biological Sciences* 28 (1997): 139–171) argues that a "post-constructivist history of science" is already alive and well, though until Pang coined the phrase it had remained unnamed. I believe Pang is right. Here is how he defines post-constructivist history: "First and most important, it takes as a starting-point the work of Kuhn, the Edinburgh School, and science studies, but modulates their tone and claims. None of its creators advocate a return to logical positivism. They see great virtues in viewing science as a locally situated practice. Second, post-constructivists begin where sociology of science ends. They do not seek to demonstrate that science is social, but to show how people work through epistemological uncertainties, technical difficulties, and theoretical conundrums to produce knowledge that seems trustworthy. As a consequence, they can take for granted positions still debated in the 1970s and the 1980s: that social interests help shape scientific ideas, that standards of evidence and explanation are historically contingent and local, that theories are underdetermined by fact, [and] that the relationships between theory and experiment are complex. . . . Third, the new approach reserves a large place for materials and nature. . . . In sum, this work seeks to preserve the qualities of skepticism and intellectual daring that make constructivism attractive [and to] correct its flaws and excesses without discarding its basic premises."

methods (which happened by 1868) and Kekulean structuralism (which happened by the end of the century). Wurtz's defeat was, after all, only temporary, for the force of evidence regarding these disputed issues was still as efficacious in the long term for Frenchmen as it had been for Germans. In this episode the Germans were not cleverer than the French; they were simply fortunate enough to have been the inheritors of a more dynamic, eclectic, and flexible scientific establishment.

Bibliography

Locations of Manuscripts Cited

Archives Nationales, Paris

Archives de l'Académie des Sciences, Paris

Archives du Bas-Rhin, Strasbourg

Correspondance Scheurer-Kestner, Bibliothèque Nationale et Universitaire de Strasbourg

Liebigiana IIB, Handschriftenabteilung, Bayerische Staatsbibliothek, Munich

Chemiker-Briefe, Berlin-Brandenburgische Akademie der Wissenschaften, Berlin

Sammlung Darmstaedter, Staatsbibliothek zu Berlin Preussischer Kulturbesitz, Berlin

August-Kekulé-Sammlung, Institut für Organische Chemie, Technische Hochschule, Darmstadt

Archiv, Vieweg Verlag, Wiesbaden

Archives, University College London, Bloomsbury Science Library, London

Western Manuscripts Collection, Wellcome Institute, London

Biographical Literature on Adolphe Wurtz

Anonymous. "Adolphe Wurtz and His Chemical Work." *Nature* 30 (1884): 170–172.

Bauer, Alexander. "Erinnerungen." *Oesterreichische Chemiker-Zeitung* 22 (1919): 116–118.

Berthelot, Marcellin. "Adolphe Wurtz." *Le temps* (Paris), 14 May 1884, p. 1.

Brooke, John H. "Adolphe Wurtz." In *Dictionary of Scientific Biography*, volume 14, 1976.

Bykov, G. V., and J. Jacques. "Deux pionniers de la chimie moderne, Adolphe Wurtz et Alexandre M. Boutlerov, d'après une correspondance inédite." *Revue d'histoire des sciences* 13 (1960): 115–134.

Carneiro, Ana. The Research School of Chemistry of Adolphe Wurtz. Ph.D. dissertation, University of Kent/Canterbury, 1992.

Carneiro, Ana. "Adolphe Wurtz and the Atomism Controversy." *Ambix* 40 (1993): 75–95.

Carneiro, Ana, and Natalie Pigeard. "Chimistes alsaciens à Paris au 19ème siècle: un réseau, une école?" *Annals of Science* 54 (1997): 533–546.

Duran, M. "Charles Adolphe Wurtz: Sa vie et son oeuvre, rappelées à l'occasion du 150e anniversaire de sa naissance." *Revue générale des sciences* 75 (1968): 33–45.

Figuier, L. "Nécrologie scientifique: Ad. Wurtz." *L'Annales des sciences et industrie* 28 (1884): 539–554.

Friedel, Charles. "Discours prononcés aux funerailles de M. Wurtz." *Comptes rendus* 98 (1884): 1199–1203.

Friedel, Charles. "Notice sur la vie et les travaux de Charles-Adolphe Wurtz." *Bulletin de la Société Chimique* [2] 43 (1885): i–lxxx.

Gault, Henry. "Adolphe Wurtz, sa vie, son oeuvre." *Bulletin de la Société des sciences, agriculture, et arts du Bas-Rhin* 50 (1921): 58–81.

Gautier, Armand. "Adolphe Wurtz." *Revue scientifique* 34 (1884): 641–648.

Gautier, Armand. "Ch.-Adolphe Wurtz, sa vie, son oeuvre, sa personnalité." *Revue scientifique* 55 (1917): 769–779.

Gregor, Henri. Académie des Sciences, Séance du 12 mai, Présidence de M. Rolland: M. Adolphe Wurtz. Unidentified newspaper tearsheet, 15 May 1884, Académie des Sciences, Archives, Dossier Wurtz

Grimaux, Édouard. "Adolphe Wurtz." *République française*, 20 May 1884.

Haller, Albin. *Inauguration de la statue de Adolphe Wurtz à Strasbourg, le mardi 5 juillet 1921*. Paris: Gauthier-Villars, 1921.

Hanriot, Maurice. "Le centenaire d'Adolphe Wurtz." *Revue scientifique* 55 (1917): 779–781.

Hjelm-Hansen, N. "Une lettre inédite de A. Wurtz à J. B. Dumas." *Revue d'histoire des sciences* 28 (1975): 259–265.

Hofmann, August Wilhelm. "Erinnerungen an Adolph Wurtz." In Hofmann, *Zur Erinnerung an vorangegangene Freunde*, volume 3. Braunschweig: Vieweg, 1888.

Perkin, W. H. "C. A. Wurtz." *Journal of the Chemical Society* 30 (1884): 328–329.

Pigeard, Natalie. L'Oeuvre du chimiste Charles Adolphe Wurtz (1817–1884). Thèse de maîtrise, Université de Paris X Nanterre, 1993.

Pigeard, Natalie. "Un alsacien à Paris: C. A. Wurtz, son école, ses laboratoires." *Bulletin de la Société Industrielle de Mulhouse* 833 (1994): 39–43.

Pigeard, Natalie. "Wurtz, doyen de la Faculté de médecine de Paris, 1866–1875." *Club d'histoire de la chimie, Bulletin de liaison* 3 (1994): 24–29.

Rocke, Alan J. "Adolphe Wurtz: L'historien et le chimiste." *Club d'histoire de la chimie: Bulletin de liaison* 2 (1993): 20–27.

Rocke, Alan J. "Adolphe Wurtz and the Development of Organic Chemistry in France: The Alsatian Connection." *Bulletin de la Société Industrielle de Mulhouse* 833 (1994): 29–34.

Rocke, Alan J. "History and Science, History of Science: Adolphe Wurtz and the Renovation of the Academic Professions in France." *Ambix* 41 (1994): 20–32.

Tiffeneau, Marc, et al. "Le centenaire de deux grands chimistes à Strasbourg: Les alsaciens Charles Gerhardt et Adolphe Wurtz." *Revue scientifique* 59 (1921): 573–602.

Tissandier, Gaston. "Adolphe Wurtz." *La Nature* 12 (1884): 401–403.

Tourette, G. de la. "Adolphe Wurtz." *Le progrès médical* 12 (1884): 394–395.

Urbain, Georges. "J. B. Dumas et C. A. Wurtz—Leur rôle dans l'histoire des théories atomiques et moléculaires." *Bulletin de la Société Chimique* [5] 1 (1934): 1425–1447.

Williamson, A. W. "Charles Adolphe Wurtz." *Proceedings of the Royal Society* 38 (1885): xxiii–xxxiv.

Yoshida, Akira. "C. A. Wurtz et la théorie atomique." *Japanese Studies in the History of Science* 16 (1977): 129–135.

Textbooks and Monographs by Adolphe Wurtz

Société Chimique de Paris, *Leçons de chimie professées en 1860 par MM. Pasteur, Cahours, Wurtz, Berthelot, Deville, Barral, et Dumas*. Paris: Hachette, 1861.

Éloge de Laurent et de Gerhardt. Paris: Lahure, 1862.

Société Chimique de Paris, *Leçons de chimie professées en 1863 par MM. Adolphe Wurtz, A. Lamy, Louis Grandeau*. Paris: Hachette, 1864.

Leçons de philosophie chimique. Paris: Hachette, 1864 (separate publication of Wurtz's monograph contained in preceding entry). *An Introduction to Chemical Philosophy, According to the Modern Theories* (London: Dutton, 1867). *Lektzii po nikotorim voprosam teoreticheskii khimii* (St. Petersburg, 1865).

Cours de philosophie chimique, fait au Collège de France par M. Ad. Wurtz. Paris: Typographie de Renou et Maulde, 1864.

Traité élémentaire de chimie médicale, two volumes. Paris: Masson, 1864–65. Second edition, 1873.

Leçons élémentaires de chimie moderne, two volumes. Paris: Masson, 1867–68. Second edition, 1871; third edition, 1875; fourth edition, 1879; fifth edition, 1884; sixth edition, 1892; seventh edition, 1894. *Elements of Modern Chemistry* (Philadelphia: Lippincott, 1878); second edition, 1884; third edition, 1887; fourth edition, 1892; fifth edition, 1895; sixth edition, 1900; final printing 1902. *Elementarnii uchebnik khimii* (Kiev, 1867). *Lecciones elementales de química moderna* (Barcelona: Verdaguer, 1873). *Lezioni elementari di chimica inorganica [organica] moderna* (Naples, 1882–85). *Kimiya-yi uzvî-yi tibbî* (Istanbul, 1882).

Dictionnaire de chimie pure et appliquée, three volumes in five (Paris: Hachette, 1868–1878). First supplement, one volume in two, 1880–86; second supplement, seven volumes, 1892–1908.

Histoire des doctrines chimiques depuis Lavoisier jusqu'à nos jours. Paris: Hachette, 1869. *A History of Chemical Theory From the Age of Lavoisier to the Present Time* (Macmillan, 1869); rpt. Arno, 1981. *Geschichte der chemischen Theorien seit Lavoisier bis auf unsere Zeit* (Berlin: R. Oppenheim, 1870); rpt. Wiesbaden: Sändig, 1971.

Les hautes études pratiques dans les universités Allemandes. Paris: Imprimerie Impériale, 1870.

La théorie des atomes dans la conception générale du monde. Paris: Masson, 1875. *La teoria de los atomos en la concepcion general del mundo*. Montevideo: Globo, 1875.

Progrès de l'industrie des matières colorantes artificielles. Paris: Masson, 1876.

La théorie atomique. Paris: Baillière, 1879. Second edition, 1879; third edition, 1880; fourth edition, 1886; fifth edition, 1889; eighth edition, 1898; ninth edition, 1904; tenth edition, 1911. *The Atomic Theory* (London: Kegan Paul, 1880); second edition, 1880; third edition, 1881; fourth edition, 1885; seventh edition, 1898; eighth edition, 1910; rpt. London: Routledge, 1998; several American editions also exist. *Die atomistische Theorie* (Leipzig: Brockhaus, 1879). *La teoria atomica* (Milan: Dumolard, 1879). *Gipotezi i razvitie atomov* (Kiev, 1889).

Traité de chimie biologique, two volumes. Paris: Masson, 1880–1885. *Tratado de quimica biologica* (Madrid, 1892).

Les hautes études pratiques dans les universités d'Allemagne et d'Autriche-Hongrie. Paris: Masson, 1882.

Introduction à l'étude de la chimie. Paris: Masson, 1885.

Abhandlung über die Glycole oder zweiatomige Alkohole und über das Aethylenoxyd als Bindeglied zwischen organischer und Mineralchemie. Oswald's Klassiker Nr. 170 (edited translation of major articles published in 1859 and 1862). Leipzig: Engelmann, 1909.

Secondary Sources

Alvin, Général. *L'École polytechnique et son quartier*. Paris: Gauthier-Villars, 1932.

Anderson, Robert D. *Education in France, 1848–1870*. Oxford University Press, 1975.

Anschütz, Richard. *August Kekulé*, two volumes. Berlin: Verlag Chemie, 1929.

Balpe, Claudette. "L'enseignement des sciences physiques: Naissance d'un corps professoral." *Histoire de l'éducation* 73 (1997): 49–85.

Belhoste, Bruno. "Un modèle à l'épreuve: L'École polytechnique de 1794 au Second Empire." In *La formation polytechnicienne, 1794–1994*. Paris: Dunod, 1994

Belhoste, Bruno, et al., eds. *La formation polytechnicienne, 1794–1994*. Paris: Dunod, 1994.

Belhoste, Bruno, et al., eds. *Le Paris des polytechniciens: Des ingénieurs dans la ville, 1794–1994.* Paris: Délégation à l'action artistique de la ville de Paris, 1994.

Belhoste, Bruno, et al., eds. *Le Paris des X: Deux siècles d'histoire.* Paris: Economica, 1995.

Ben-David, Joseph. "The Issue of Decline in Nineteenth-Century France as a Scientific Centre." *Minerva* 8 (1970): 161–179.

Ben-David, Joseph. *The Scientist's Role in Society: A Comparative Study.* Prentice-Hall, 1971.

Bensaude-Vincent, Bernadette. "Une mythologie révolutionnaire dans la chimie française." *Annals of Science* 40 (1983): 189–196.

Bensaude-Vincent, Bernadette. "Karlsruhe, septembre 1860: l'atome en congrès." *Relations internationales* 62 (1990): 149–169.

Bensaude-Vincent, Bernadette. "Between History and Memory: Centennial and Bicentennial Images of Lavoisier." *Isis* 87 (1996): 481–499.

Bensaude-Vincent, Bernadette. "Atomism and Positivism: A Legend About French Chemistry." *Annals of Science* 56 (1999): 81–94.

Bensaude-Vincent, Bernadette, and Anders Lundgren, eds. *Communicating Chemistry: Textbooks and their Audiences.* Canton, Mass.: Science History Publications, 2000.

Blondel, Christine, and Matthias Dörries, eds. *Restaging Coulomb: Usages, controverses et réplications autour de la balance de torsion.* Florence: Olschki, 1994.

Blondel-Mégrelis, Marika. *Dire les choses: Auguste Laurent et la Méthode chimique.* Paris: Vrin, 1996.

Bloor, David. *Knowledge and Social Imagery*, second edition. University of Chicago Press, 1991.

Borscheid, Peter. *Naturwissenschaft, Staat und Industrie in Baden (1848–1914).* Stuttgart: Klett, 1976.

Bradley, Margaret. "The Facilities for Practical Instruction in Science during the Early Years of the École Polytechnique." *Annals of Science* 33 (1976): 425–446.

Brock, William. *Justus von Liebig: The Chemical Gatekeeper.* Cambridge University Press, 1997.

Brooke, John H. "Organic Synthesis and the Unification of Chemistry—A Reappraisal." *British Journal for the History of Science* 5 (1971): 363–392.

Brooke, John H. "Chlorine Substitution and the Future of Organic Chemistry." *Studies in the History and Philosophy of Science* 4 (1973): 47–94.

Brooke, John H. "Laurent, Gerhardt, and the Philosophy of Chemistry." *Historical Studies in the Physical Sciences* 6 (1975): 405–429.

Brooke, John H "Avogadro's Hypothesis and its Fate: A Case-Study in the Failure of Case-Studies." *History of Science* 19 (1981): 235–273.

Brooke, John H. *Thinking About Matter: Studies in the History of Chemical Philosophy.* Aldershot: Variorum, 1995.

Brush, Stephen G. "Dynamics of Theory Change in Chemistry. Part 1. The Benzene Problem, 1865–1945." *Studies in the History and Philosophy of Science* 30 (1999): 21–79.

Brush, Stephen G. "Dynamics of Theory Change in Chemistry. Part 2. Benzene and Molecular Orbitals, 1945–1980." *Studies in the History and Philosophy of Science* 30 (1999): 263–302.

Camberousse, C. de. *Histoire de l'École Centrale des Arts et Manufactures*. Paris: Gauthier-Villars, 1879.

Carnot, Paul. *Hippolyte Carnot et le Ministère de l'instruction publique de la IIe République*. Paris: Presses Universitaires de France, 1948.

Chaigneau, Marcel. *J. B. Dumas, chimiste et homme politique: Sa vie, son oeuvre (1800–1884)*. Paris: Guy le Prat, 1984.

Charle, Christophe. *Les hauts fonctionnaires en France au XIXe siècle*. Paris: Galimard, 1980.

Charle, Christophe. *Dictionnaire biographique des universitaires aux XIXe et XXe siècles*, two volumes. Paris: Éditions du CNRS, 1985–86.

Charle, Christophe. *Histoire sociale de la France au XIXe siécle*. Paris: Éditions du Seuil, 1991.

Charle, Christophe. *La République des universitaires, 1870–1940*. Paris: Éditions du Seuil, 1994.

Charle, Christophe, and Régine Ferré, eds. *Le personnel de l'enseignement supérieur en France aux XIXe et XXe siècles*. Paris: Éditions du CNRS, 1985.

Ciancia, Guy. "Marcelin Berthelot et le concept d'isomérie (1860–1865)." *Archives internationales d'histoire des sciences* 36 (1986): 54–83.

Coleman, William. "The Cognitive Basis of the Discipline: Claude Bernard on Physiology." *Isis* 76 (1985): 49–70.

Coleman, William, ed. *French Views of German Science*. Arno, 1981.

Coleman, William, and Frederic L. Holmes, eds. *The Investigative Enterprise: Experimental Physiology in Nineteenth-Century Medicine*. University of California Press, 1988.

Colmant, P. "Querelle à l'Institut entre équivalentistes et atomistes." *Revue des questions scientifiques* [5] 33 (1972): 493–519.

Corlieu, A. *Centenaire de la Faculté de Médecine de Paris (1794–1894)*. Paris: Imprimerie Nationale, 1896.

Costa, Albert. *Michel Eugène Chevreul: Pioneer of Organic Chemistry*. State Historical Society of Wisconsin, 1962.

Craig, John E. *Scholarship and Nation Building: The Universities of Strasbourg and Alsatian Society, 1870–1939*. University of Chicago Press, 1984.

Crosland, Maurice. *The Society of Arcueil: A View of French Science at the Time of Napoleon I*. Harvard University Press, 1967.

Crosland, Maurice. "Science and the Franco-Prussian War." *Social Studies of Science* 6 (1976): 185–214.

Crosland, Maurice. "The Development of a Professional Career in France." In *The Emergence of Science in Western Europe*, ed. M. Crosland. Science History, 1976.

Crosland, Maurice. "Aspects of International Scientific Collaboration and Organization before 1900." In *Human Implications of Scientific Advance*, ed. E. Forbes. Edinburgh University Press, 1977.

Crosland, Maurice. "History of Science in a National Context." *British Journal for the History of Science* 10 (1977): 95–113.

Crosland, Maurice. *Gay-Lussac: Scientist and Bourgeois*. Cambridge University Press, 1978.

Crosland, Maurice. "Scientific Credentials: Record of Publications in the Assessment of Qualifications for Election to the French Académie des Sciences." *Minerva* 19 (1981): 605–631.

Crosland, Maurice. "Assessments by Peers in Nineteenth-Century France: The Manuscript Reports on Candidates for Election to the Académie des Sciences." *Minerva* 24 (1986): 413–432.

Crosland, Maurice. "The Emergence of Research Grants within the Prize System of the French Academy of Sciences." *Social Studies of Science* 19 (1989): 71–99.

Crosland, Maurice. *Science under Control: The French Academy of Sciences, 1795–1914*. Cambridge University Press, 1992.

Crosland, Maurice. *Studies in the Culture of Science in France and Britain since the Enlightenment*. Aldershot: Variorum, 1995.

Crosland, Maurice. "The Organization of Chemistry in Nineteenth-Century France." In *The Making of the Chemist*, ed. D. Knight and H. Kragh. Cambridge University Press, 1998.

Daumas, Maurice. "L'école des chimistes français vers 1840." *Chymia* 1 (1948): 55–65.

Dhombres, Jean, and Bernard Javault, eds. *Marcelin Berthelot: Une vie, une époque, un mythe*. Paris: Société Française d'Histoire des Sciences et des Techniques, 1992.

Dhombres, Nicole, and Jean Dhombres. *Naissance d'un pouvoir: sciences et savants en France (1793–1824)*. Paris: Payot, 1989.

Dictionary of Scientific Biography, 18 volumes. Scribner, 1970–1990.

Digeon, Claude. *La crise allemande de la pensée française*. Paris: Presses Universitaire de France, 1959.

Dörries, Matthias. "La standardisation de la balance de torsion dans les projets européens sur le magnétisme terrestre." In *Restaging Coulomb*, ed. C. Blondel and M. Dörries. Florence: Olschki, 1994.

Dörries, Matthias. "Easy Transit: Crossing Boundaries between Physics and Chemistry in mid-Nineteenth Century France." In *Making Space for Science*, ed. C. Smith and J. Agar. Macmillan, 1998.

Dörries, Matthias. "Vicious Circles, or, The Pitfalls of Experimental Virtuosity." In *Experimental Essays—Versuche zum Experiment*, ed. M. Heidelberger and F. Steinle. Baden-Baden: Nomos, 1998.

Duhem, Pierre. *La science allemande*. Paris: Hermann, 1915. In English: *German Science*. Open Court, 1991.

Dumas, Général J.-B. *La vie de J.-B. Dumas, 1800–1884*. Paris: privately published, 1924.

Dumas, Jean-Baptiste. *Discours et éloges académiques*, two volumes. Paris: Gauthier-Villars, 1885.

Duruy, Victor. *L'administration de l'instruction publique de 1863 à 1869*. Paris: Delalain, 1870.

Duruy, Victor. *Notes et souvenirs*, two volumes. Paris: Hachette, 1901.

Fauque, D., and G. Bram. "Le réseau alsacien." *Bulletin de la Société Industrielle de Mulhouse* 833 (1994): 17–20.

Fell, Ulrike. "The Chemistry Profession in France: The Société Chimique de Paris/de France, 1870–1914." In *The Making of the Chemist*, ed. D. Knight and H. Kragh. Cambridge University Press, 1998.

Fell, Ulrike. *Disziplin, Profession und Nation: Die Ideologie der Chemie in Frankreich vom Zweiten Kaiserreich bis in die Zwischenkriegszeit*. Leipziger Universitätsverlag, 2000.

Fox, Robert. *The Caloric Theory of Gases: From Lavoisier to Regnault*. Oxford University Press, 1971.

Fox, Robert. "Scientific Enterprise and the Patronage of Research in France, 1800–1870." *Minerva* 11 (1973): 442–473.

Fox, Robert. "The Rise and Fall of Laplacian Physics." *British Journal for the History of Science* 4 (1975): 89–136.

Fox, Robert. "The Savant Confronts his Peers: Scientific Societies in France, 1815–1914." In *The Organization of Science and Technology in France, 1808–1914*, ed. R. Fox and G. Weisz. Cambridge University Press, 1980.

Fox, Robert. "Science, the University, and the State in Nineteenth-Century France." In *Professions and the French State, 1700–1900*, ed. G. Geison. University of Pennsylvania Press, 1984.

Fox, Robert. "Science, Industry, and the Social Order in Mulhouse, 1798–1871." *British Journal for the History of Science* 17 (1984): 127–168.

Fox, Robert. "The View over the Rhine: Perceptions of German Science and Technology in France, 1860–1914." In *Frankreich und Deutschland*, ed. Y. Cohen and K. Manfrass. Munich: Beck, 1990.

Fox, Robert. *The Culture of Science in France, 1700–1900*. Aldershot: Variorum, 1992.

Fox, Robert. *Science, Industry, and the Social Order in Post-Revolutionary France*. Aldershot: Variorum, 1995.

Fox, Robert. "Positivists, Free Thinkers, and the Reform of French Science in the Second Empire." In Fox, *Science, Industry, and the Social Order in Post-Revolutionary France*. Aldershot: Variorum, 1995.

Fox, Robert, and George Weisz. "Introduction: The Institutional Basis of French Science in the Nineteenth Century." In *The Organization of Science and Technology in France, 1808–1914*, ed. R. Fox and G. Weisz. Cambridge University Press, 1980.

Fox, Robert, and George Weisz, eds. *The Organization of Science and Technology in France, 1808–1914*. Cambridge University Press, 1980.

Franklin, Allan. *The Neglect of Experiment*. Cambridge University Press, 1986.

Fruton, Joseph S. "The Liebig Research Group—A Reappraisal." *Proceedings of the American Philosophical Society* 132 (1988): 1–66.

Fruton, Joseph S. *Contrasts in Scientific Style: Research Groups in the Chemical and Biochemical Sciences*. American Philosophical Society, 1990.

Galison, Peter. *Image and Logic: A Material Culture of Microphysics*. University of Chicago Press, 1997.

Gay, Jules. *Henri Sainte-Claire Deville: Sa vie et ses travaux*. Paris: Gauthier-Villars, 1889.

Geison, Gerald R. "Scientific Change, Emerging Specialties, and Research Schools." *History of Science* 19 (1981): 20–40.

Geison, Gerald R. *The Private Science of Louis Pasteur*. Princeton University Press, 1995.

Geison, Gerald R, ed. *Professions and the French State, 1700–1900*. University of Pennsylvania Press, 1984.

Gerhardt, Charles. *Correspondance de Charles Gerhardt*, ed. M. Tiffeneau, two volumes. Paris: Masson, 1918–25.

Gernez, D. J. B. *Notice sur Henri Sainte-Claire Deville*. Paris: Gauthier-Villars [1894].

Gilpin, Robert. *France in the Age of the Scientific State*. Princeton University Press, 1968.

[Girard, Aimé?]. [Untitled history of Pelouze's laboratories, 17-page handwritten manuscript]. Dossier Pelouze, Archives de l'Académie des Sciences, Paris.

Golinski, Jan. *Making Natural Knowledge: Constructivism and the History of Science*. Cambridge University Press, 1998.

Gooday, Graeme. "The Premisses of Premises: Spatial Issues in the Historical Construction of Laboratory Credibility." In *Making Space for Science: Territorial Themes in the Shaping of Knowledge*, ed. C. Smith and J. Agar. Macmillan, 1998.

Gooding, David. "History in the Laboratory: Can We Tell What Really Went On?" In *The Development of the Laboratory*, ed. F. James. American Institute of Physics, 1989.

Gooding, David. "'In Nature's School': Faraday as an Experimentalist." In *Faraday Rediscovered*, ed. D. Gooding and F. James. American Institute of Physics, 1989.

Gooding, David. *Experiment and the Making of Meaning*. Kluwer, 1990.

Görs, Britta. "Chemie und Atomismus im deutschsprachigen Raum." *Mitteilungen, Gesellschaft Deutscher Chemiker, Fachgruppe Geschichte der Chemie* 13 (1997): 100–114.

Görs, Britta. *Chemischer Atomismus: Anwendung, Veränderung, Alternativen im deutschsprachigen Raum in der zweiten Hälfte des 19. Jahrhunderts*. Berlin: ERS, 1999.

Graebe, Carl. "Marcelin Berthelot." *Berichte der Deutschen Chemischen Gesellschaft* 41 (1908): 4805–4872.

Gréard, Octave. *Éducation et instruction: Enseignement supérieur*, second edition. Paris: Hachette, 1889.

Gréard, Octave. *Nos adieux à la vieille Sorbonne*. Paris: Hachette, 1893.

Grimaux, Édouard, and Charles Gerhardt Jr. *Charles Gerhardt: Sa vie, son oeuvre, sa correspondance, 1816–1856*. Paris: Masson, 1900.

Gustin, Bernard H. The Emergence of the German Chemical Profession, 1790–1867. Ph.D. dissertation, University of Chicago, 1975.

Heidelberger, Michael, and Friedrich Steinle, eds. *Experimental Essays—Versuche zum Experiment*. Baden-Baden: Nomos, 1998.

Hillairet, Jacques. *Dictionnaire historique des rues de Paris*, two volumes, second edition. Paris: Éditions de Minuit, 1964.

Hofmann, August Wilhelm. *Zur Erinnerung an vorangegangene Freunde*, three volumes. Braunschweig: Vieweg, 1888.

Hofmann, August Wilhelm. "Zur Erinnerung an J. B. A. Dumas." In Hofmann, *Zur Erinnerung an vorangegangene Freunde*, , volume 2. Braunschweig: Vieweg, 1888.

Holmes, Frederic L. "Justus Liebig." In *Dictionary of Scientific Biography*, volume 8. 1973.

Holmes, Frederic L. *Claude Bernard and Animal Chemistry: The Emergence of a Scientist*. Harvard University Press, 1974.

Holmes, Frederic L. "The Complementarity of Teaching and Research in Liebig's Laboratory." *Osiris* [2] 5 (1989): 121–164.

Holmes, Frederic L., and Trevor H. Levere, eds. *Instruments and Experimentation in the History of Chemistry*. MIT Press, 2000.

Homburg, Ernst. "Two factions, One Profession: The Chemical Profession in German Society, 1780–1870." In *The Making of the Chemist*, ed. D. Knight and H. Kragh. Cambridge University Press, 1998.

Homburg, Ernst. "The Rise of Analytical Chemistry and its Consequences for the Development of the German Chemical Profession (1780–1860)." *Ambix* 46 (1999): 1–32.

Horvath-Peterson, Sandra. *Victor Duruy and French Education: Liberal Reform in the Second Empire*. Louisiana State University Press, 1984.

Jacob, Andrée, and Jean-Marc Léri. *Vie et histoire du VIe arrondissement*. Paris: Hervas, 1986.

Jacques, Jean. *Berthelot 1827–1907: Autopsie d'un mythe*. Paris: Belin, 1987.

James, Frank A. J. L., ed. *The Development of the Laboratory: Essays on the Place of Experiment in Industrial Civilization*. American Institute of Physics, 1989.

Johnson, Jeffrey. "Academic Chemistry in Imperial Germany." *Isis* 76 (1985): 500–524.

Jordan, David P. *Transforming Paris: The Life and Labors of Baron Haussmann.* Free Press, 1995.

Jungfleisch, Émile. "Notice sur la vie et les travaux de Marcellin Berthelot." *Bulletin de la Société Chimique* [4] 13 (1913): i–cclx.

Kanz, Kai Torsten. *Nationalismus und internationale Zusammenarbeit in den Naturwissenschaften: Die deutsch-französischen Wissenschaftsbeziehungen zwischen Revolution und Restauration, 1789–1832.* Stuttgart: Steiner, 1997.

Kersaint, Georges. "L'école de chimie de Frémy." *Revue d'histoire de la pharmacie* 52 (1964): 165–172.

Keylor, William R. *Academy and Community: The Foundation of the French Historical Profession.* Harvard University Press, 1975.

Kim, Mi Gyung. "The Layers of Chemical Language I: Constitution of Bodies v. Structure of Matter." *History of Science* 30 (1992): 69–96.

Kim, Mi Gyung. "The Layers of Chemical Language II: Stabilizing Atoms and Molecules in the Practice of Organic Chemistry." *History of Science* 30 (1992): 397–437.

Kim, Mi Gyung. "Constructing Symbolic Spaces: Chemical Molecules in the Académie des Sciences." *Ambix* 43 (1996): 1–31.

Klein, Ursula. "Paving a Way Through the Jungle of Organic Chemistry—Experimenting within Changing Systems of Order." In *Experimental Essays—Versuche zum Experiment* , ed. M. Heidelberger and F. Steinle. Baden-Baden: Nomos, 1998.

Klein, Ursula. Experimente, Modelle, Paper-Tools: Kulturen der organischen Chemie im 19. Jahrhundert. Habilitationsschrift, Universität Konstanz, 1999.

Klein, Ursula. "Paper-Tools and Techniques of Modelling in Nineteenth-Century Chemistry." In *Models as Mediators,* ed. M. Morgan and M. Morrison. Cambridge University Press, 2000.

Klosterman, Leo. Studies in the Life and Work of J. B. Dumas (1800–1884): The Period up to 1850. Ph.D. dissertation, University of Kent/Canterbury, 1976.

Klosterman, Leo. "A Research School of Chemistry in the Nineteenth Century: Jean Baptiste Dumas and His Research Students." *Annals of Science* 42 (1985): 1–80.

Knight, David, and Helge Kragh, eds. *The Making of the Chemist: The Social History of Chemistry in Europe, 1789–1914.* Cambridge University Press, 1998.

Kohler, Robert E. "The Constructivists' Tool Kit." *Isis* 90 (1999): 329–331.

Kounelis, Catherine. "Heurs et malheurs de la chimie: La réforme des années 1880." In *La formation polytechnicienne, 1794–1994*, ed. B. Belhoste et al. Paris: Dunod, 1994.

Langins, Janis. "The Decline of Chemistry at the École Polytechnique." *Ambix* 28 (1981): 1–19.

Launay, Louis de. *Une grande famille de savants, les Brongniart.* Paris: Rapilly, 1940.

Leprieur, François. Les conditions de la constitution d'une discipline scientifique: la chimie organique en France, 1830–1880. Thèse de IIIème cycle, Université de Paris I, 1977.

Leprieur, François. "La formation des chimistes français au XIXe siècle." *La Recherche* 10 (1979): 732–740.

Leprieur, François, and Pierre Papon. "Synthetic Dyestuffs: The Relations between Academic Chemistry and the Chemical Industry in Nineteenth-Century France." *Minerva* 17 (1979): 197–224.

Lesch, John E. "Conceptual Change in an Empirical Science: The Discovery of the First Alkaloids." *Historical Studies in the Physical Sciences* 11 (1981): 307–328.

Lesch, John E. *Science and Medicine in France: The Emergence of Experimental Physiology, 1790–1855*. Harvard University Press, 1984.

Letté, Michel. "Les Annales de chimie et de physique, la quatrième série (1864–73): Un journal au service de la science officielle." *Sciences et techniques en perspective* 28 (1994): 218–286.

Leuilliot, Paul. *L'Alsace au début du XIXe siècle*, three volumes. Paris: S.E.V.P.E.N., 1959–60.

Liard, Louis. *L'enseignement supérieur en France*, two volumes. Paris: Colin, 1894.

Lundgreen, Peter. "The Organization of Science and Technology in France: A German Perspective." In *The Organization of Science and Technology in France, 1808–1914*, ed. R. Fox and G. Weisz. Cambridge University Press, 1980.

Mauskopf, Seymour. *Crystals and Compounds: Molecular Structure and Composition in Nineteenth Century French Science*. Philadelphia: American Philosophical Society, 1976.

McClelland, Charles. *State, Society, and University in Germany, 1700–1914*. Cambridge University Press, 1980.

McCosh, F. W. J. *Boussingault: Chemist and Agriculturist*. Reidel, 1984.

Meinel, Christoph. "Nationalismus und Internationalismus in der Chemie des 19. Jahrhunderts." In *Perspektiven der Pharmaziegeschichte*, ed. P. Dilg. Graz: Akademische Druck- und Verlagsanstalt, 1983.

Meinel, Christoph. "*Artibus Academicis Inserenda*: Chemistry's Place in Eighteenth and Early Nineteenth Century Universities." *History of Universities* 7 (1988): 89–115.

Meinel, Christoph. "Structural Changes in International Scientific Communication: The Case of Chemistry." In *Atti del V convegno di storia e fondamenti della chimica*. Perugia: Accademia Nazionale della Scienze, 1993.

Meinel, Christoph, ed. *Fachschrifttum, Bibliothek und Naturwissenschaft im 19. und 20. Jahrhundert*. Wiesbaden: Harrassowitz, 1997.

Metz, André. "La notation atomique et la théorie atomique en France à la fin du XIXe siècle." *Revue d'histoire des sciences* 16 (1963): 233–239.

Miles, Ashley. "Reports by Louis Pasteur and Claude Bernard on the Organization of Scientific Teaching and Research." *Notes and Records of the Royal Society* 37 (1982): 101–118.

Morrell, J. B. "The Chemist Breeders: The Research Schools of Liebig and Thomas Thomson." *Ambix* 19 (1972): 1–46.

Munday, Pat. "Social Climbing through Chemistry: Justus Liebig's Rise from the Niederer Mittelstand to the Bildungsbürgertum." *Ambix* 37 (1990): 1–19.

Nénot, H. P. *La nouvelle Sorbonne*. Paris: Colin, 1895.

Novitski, Marya. *Auguste Laurent and the Prehistory of Valence*. Chur: Harwood, 1992.

Nye, Mary Jo. "The Nineteenth-Century Atomic Debates and the Dilemma of an 'Indifferent Hypothesis.'" *Studies in the History and Philosophy of Science* 7 (1976): 245–268.

Nye, Mary Jo. "Berthelot's Anti-Atomism: A 'Matter of Taste'?" *Annals of Science* 38 (1981): 585–590.

Nye, Mary Jo. *The Question of the Atom*. Los Angeles: Tomash, 1984.

Nye, Mary Jo. "Scientific Decline: Is Quantitative Evaluation Enough?" *Isis* 75 (1984): 697–708.

Nye, Mary Jo. *Science in the Provinces: Scientific Communities and Provincial Leadership in France, 1860–1930*. University of California Press, 1986.

Nye, Mary Jo. *From Chemical Philosophy to Theoretical Chemistry: Dynamics of Matter and Dynamics of Disciplines, 1800–1950*. University of California Press, 1993.

Pang, Alex. "Visual Representation and Post-Constructivist History of Science." *Historical Studies in the Physical and Biological Sciences* 28 (1997): 139–171.

Papillon, Fernand. "Les laboratoires en France et à l'étranger." *Revue des deux mondes* [2] 94 (1871): 594–609.

Parias, Louis-Henri, ed. *Histoire générale de l'enseignement et de l'éducation en France*. Paris: Nouvelle Librairie de France, 1981.

Partington, J. R. *A History of Chemistry*, volume 4. Macmillan, 1964.

Pasteur, Louis. *Louis Pasteur: Pour l'avenir de la science française*. Paris: Éditions Raison d'Être, 1947.

Paul, Harry W. "The Issue of Decline in Nineteenth-Century French Science." *French Historical Studies* 7 (1972): 416–440.

Paul, Harry W. *The Sorcerer's Apprentice: The French Scientist's Image of German Science, 1840–1919*. University of Florida Press, 1972.

Paul, Harry W. "La science française de la seconde partie du XIXe siècle vue par les auteurs anglais et américains." *Revue d'histoire des sciences* 27 (1974): 147–163.

Paul, Harry W. *From Knowledge to Power: The Rise of the Science Empire in France, 1860–1939*. Cambridge University Press, 1985.

Paul, Harry W. "The Role of German Idols in the Rise of the French Science Empire." In *'Einsamkeit und Freiheit' neu besichtigt*, ed. G. Schubring. Stuttgart: Steiner, 1991.

Pessard, Gustave. *Nouveau dictionnaire historique de Paris*. Paris: Rey, 1904.

Pestre, Dominique. "La pratique de reconstruction des expériences historiques, une toute première réflexion." In *Restaging Coulomb*, ed. C. Blondel and M. Dörries. Florence: Olschki, 1994.

Petrel, Jacques. *La négation de l'atome dans la chimie du XIXe siècle: Cas de Jean-Baptiste Dumas*. Paris: Centre National de Recherche Scientifique, 1979.

Pigeard, Natalie. "Chemistry for Women in Nineteenth-Century France." In *Communicating Chemistry*, ed. B. Bensaude-Vincent and A. Lundgren. Canton, Mass.: Science History Publications, 2000.

Pigeard, Natalie, and Ana Carneiro. "Atomes et équivalents devant l'Académie des Sciences." *Comptes rendus de l'Académie des Sciences* 323 (1996): 421–424.

Prévost, A. *La Faculté de Médecine de Paris, ses chaires, ses annexes, et son personnel enseignant de 1794 à 1900*. Paris: Maloine, 1900.

Prost, Antoine. *Histoire de l'enseignement en France, 1800–1967*. Paris: Colin, 1968.

Rheinberger, Hans-Jörg. *Toward a History of Epistemic Things*. Stanford University Press, 1997.

Ringer, Fritz. *Education and Society in Modern Europe*. Indiana University Press, 1979.

Ringer, Fritz. *Fields of Knowledge: French Academic Culture in Comparative Perspective, 1890–1920*. Cambridge University Press, 1992.

Rivé, Philippe, ed. *La Sorbonne et sa reconstruction*. Lyon: La Manufacture, 1987.

Rocke, Alan J. "Subatomic Speculations and the Origin of Structure Theory." *Ambix* 30 (1983): 1–18.

Rocke, Alan J. *Chemical Atomism in the Nineteenth Century: From Dalton to Cannizzaro*. Ohio State University Press, 1984.

Rocke, Alan J. "Hypothesis and Experiment in the Early Development of Kekulé's Benzene Theory." *Annals of Science* 42 (1985): 355–381.

Rocke, Alan J. "Kekulé's Benzene Theory and the Appraisal of Scientific Theories." In *Scrutinizing Science*, ed. A. Donovan et al. Kluwer, 1988.

Rocke, Alan J. "Pride and Prejudice in Chemistry: Chauvinism and the Pursuit of Science." *Bulletin for the History of Chemistry* 13–14 (1992–93): 29–40.

Rocke, Alan J. *The Quiet Revolution: Hermann Kolbe and the Science of Organic Chemistry*. University of California Press, 1993.

Rocke, Alan J. "Organic Analysis in Comparative Perspective: Liebig, Dumas, and Berzelius, 1811–1840." In *Instruments and Experimentation in the History of Chemistry*, ed. F. Holmes and T. Levere. MIT Press, 2000.

Roque, G., B. Bodo, and F. Viénot, eds. *Michel-Eugène Chevreul: Un savant, des couleurs.* Paris: Muséum National d'Histoire Naturelle, 1997.

Russell, Colin A. *The History of Valency.* Leicester University Press, 1971.

Russell, Colin A. "The Changing Role of Synthesis in Organic Chemistry." *Ambix* 34 (1987): 169–180.

Russell, Colin A. *Edward Frankland: Chemistry, Controversy and Conspiracy in Victorian England.* Cambridge University Press, 1996.

Schaffer, Simon. "Physics Laboratories and the Victorian Country House." In *Making Space for Science,* ed. C. Smith and J. Agar. Macmillan, 1998.

Scheurer-Kestner, Auguste. *Souvenirs de la jeunesse.* Paris: Charpentier, 1905.

Schnitter, Claude. "Le développement du Muséum national d'histoire naturelle de Paris au cours de la seconde moitié du XIXe siècle: 'Se transformer ou périr.'" *Revue d'histoire des sciences* 49 (1996): 53–97.

Shinn, Terry. "The French Science Faculty System, 1808–1914: Institutional Change and Research Potential in Mathematics and the Physical Sciences." *Historical Studies in the Physical Sciences* 10 (1979): 271–332.

Shinn, Terry. "Orthodoxy and Innovation in Science: The Atomist Controversy in French Chemistry." *Minerva* 18 (1980): 539–555.

Shinn, Terry. *Savoir scientifique et pouvoir social: École polytechnique et les polytechniciens, 1794–1914.* Paris: Presses de la Fondation Nationale des Sciences Politiques, 1980.

Shortland, Michael, and Richard Yeo, eds. *Telling Lives in Science: Essays on Scientific Biography.* Cambridge University Press, 1996.

Smeaton, William. "The Early History of Laboratory Instruction in Chemistry at the École Polytechnique, Paris, and Elsewhere." *Annals of Science* 10 (1954): 224–233.

Smith, Crosbie, and Jon Agar, eds. *Making Space for Science: Territorial Themes in the Shaping of Knowledge.* Macmillan, 1998.

Smith, Robert J. *The École Normale Supérieure and the Third Republic.* Albany: SUNY Press, 1982.

Söderqvist, Thomas. "Existential Projects and Existential Choice in Science: Science Biography as an Edifying Genre." In *Telling Lives in Science,* ed. M. Shortland and R. Yeo. Cambridge University Press, 1996.

Sordes, R. *Histoire de l'enseignement de la chimie en France.* Paris: Chimie et Industrie, 1928.

Strohl, Henri. *Le protestantisme en Alsace.* Strasbourg: Éditions Oberlin, 1950.

Turner, R. Steven. "The Growth of Professorial Research in Prussia, 1818–1848—Causes and Contexts." *Historical Studies in the Physical Sciences* 3 (1971): 137–182.

Turner, R. Steven. "Justus Liebig versus Prussian Chemistry: Reflections on Early Institute-Building in Germany." *Historical Studies in the Physical Sciences* 13 (1982): 129–162.

Turner, R. Steven, E. Kerwin, and D. Woolwine. "Careers and Creativity in Nineteenth-Century Physiology: Zloczower Redux." *Isis* 75 (1984): 523–539.

Van Zanten, David. *Building Paris: Architectural Institutions and the Transformation of the French Capital, 1830–1870.* Cambridge University Press, 1994.

Velluz, Léon. *Vie de Berthelot.* Paris: Plon, 1964.

Virtanen, Reino. *Marcelin Berthelot: A Study of a Scientist's Public Role.* Lincoln: University of Nebraska Studies, no. 31, 1965.

Volhard, Jacob. *Justus von Liebig,* two volumes. Leipzig: Barth, 1909.

Weiss, John. *The Making of Technological Man: The Social Origins of French Engineering Education.* MIT Press, 1982.

Weisz, George. The Academic Elite and the Movement to Reform French Higher Education. Ph.D. dissertation, SUNY Stony Brook, 1976.

Weisz, George. "Le corps professoral de l'enseignement supérieur et l'idéologie de la réforme universitaire en France, 1860–1885." *Revue française de sociologie* 18 (1977): 201–232.

Weisz, George. "Reform and Conflict in French Medical Education, 1870–1914." In *The Organization of Science and Technology in France, 1808–1914,* ed. R. Fox and G. Weisz. Cambridge University Press, 1980.

Weisz, George. *The Emergence of Modern Universities in France, 1863–1914.* Princeton University Press, 1983.

Willink, Bastiaan. "On the Structure of a Scientific Golden Age: Social Change, University Investments and Germany's Discontinuous Rise to 19th Century Scientific Hegemony." *Berichte zur Wissenschaftsgeschichte* 19 (1996): 35–49.

Zloczower, Avraham. *Career Opportunities and the Growth of Scientific Discovery in Nineteenth-Century Germany.* Arno, 1981.

Zwerling, Craig. *The Emergence of the École Normale Supérieure as a Center of Scientific Education in Nineteenth-Century France.* Garland, 1990.

Zwerling, Craig. "The Emergence of the École Normale Supérieur as a Centre of Scientific Education in the Nineteenth Century." In *The Organization of Science and Technology in France, 1808–1914,* ed. R. Fox and G. Weisz. Cambridge University Press, 1980

Index